Environmental Footprints and Eco-design of Products and Processes

Series Editor

Subramanian Senthilkannan Muthu, Head of Sustainability - SgT Group and API, Hong Kong, Kowloon, Hong Kong

Indexed by Scopus

This series aims to broadly cover all the aspects related to environmental assessment of products, development of environmental and ecological indicators and eco-design of various products and processes. Below are the areas fall under the aims and scope of this series, but not limited to: Environmental Life Cycle Assessment; Social Life Cycle Assessment; Organizational and Product Carbon Footprints; Ecological, Energy and Water Footprints; Life cycle costing; Environmental and sustainable indicators; Environmental impact assessment methods and tools; Eco-design (sustainable design) aspects and tools; Biodegradation studies; Recycling; Solid waste management; Environmental and social audits; Green Purchasing and tools; Product environmental footprints; Environmental management standards and regulations; Eco-labels; Green Claims and green washing; Assessment of sustainability aspects.

More information about this series at http://www.springer.com/series/13340

Subramanian Senthilkannan Muthu
Editor

Sustainable Packaging

Editor
Subramanian Senthilkannan Muthu
Head of Sustainability
SgT and API
Kowloon, Hong Kong

ISSN 2345-7651 ISSN 2345-766X (electronic)
Environmental Footprints and Eco-design of Products and Processes
ISBN 978-981-16-4611-9 ISBN 978-981-16-4609-6 (eBook)
https://doi.org/10.1007/978-981-16-4609-6

This Springer imprint is published by the registered company Springer Nature Singapore Pte Ltd.
The registered company address is: 152 Beach Road, #21-01/04 Gateway East, Singapore 189721, Singapore

Contents

About the Editor

Dr. Subramanian Senthilkannan Muthu currently works for SgT Group as Head of Sustainability, and is based out of Hong Kong. He earned his Ph.D. from The Hong Kong Polytechnic University, and is a renowned expert in the areas of Environmental Sustainability in Textiles & Clothing Supply Chain, Product Life Cycle Assessment (LCA) and Product Carbon Footprint Assessment (PCF) in various industrial sectors. He has five years of industrial experience in textile manufacturing, research and development and textile testing and over a decade's of experience in Life Cycle Assessment (LCA), carbon and ecological footprints assessment of various consumer products. He has published more than 100 research publications, written numerous book chapters and authored/edited over 100 books in the areas of Carbon Footprint, Recycling, Environmental Assessment and Environmental Sustainability.

The Environmental Performance of Glass and PET Mineral Water Bottles in Italy

Annarita Paiano, Teodoro Gallucci, Andrea Pontrandolfo, Tiziana Crovella, and Giovanni Lagioia

Abstract Worldwide the environmental weight of the packaging has overtaken the threshold, both due to the waste and the emissions generated. This issue stimulated the European Union (EU) to provide for a stringent regulation to tackle this burden. Particularly, the consumption of mineral water packed is very significant, as regards the use of plastic bottles, especially in the small size, which stresses the need for a boosted management of packaging by the governments, industries and consumers (Botto et al. in Environ Sci Policy 14:388–395, 2011). Over the years, the EU has shown an increasing consumption of mineral water packed, and Italy, with 222 L per capita is the first European consumer country and the third worldwide. This chapter investigated the glass and Polyethylene Terephthalate (PET) packaging to analyse their environmental impact and undertake a comparison among them (Vellini and Savioli in Energy 34:2137–2143, 2009). Particularly the research provides a twofold analysis. Firstly, it assesses the impacts of 1 kg of hollow glass through the Life Cycle Assessment methodology (Schmitz et al. in Energy Policy 39:142–155, 2011;Vinci et al. in Trends in beverage packaging 16:105–133, 2019;) and makes a comparison with a 1 kg of PET (Marathe KV, Chavan K, Nakhate P (2017) Lifecycle Assessment (LCA) of Polyethylene Terephthalate (PET) Bottles—Indian Perspective. http://www.in-beverage.org/lca-pet/ICT%20Final%20Report%20on%20LCA%20of%20PET%20Bottles_for%20PACE_01_01_2018.pdf. Accessed 2 March 2021). Secondly, the Greenhouse gas emissions of still water bottled based

A. Paiano (✉) · T. Gallucci · A. Pontrandolfo · T. Crovella · G. Lagioia
Department of Economics, Management and Business Law, University of Bari Aldo Moro, Largo Abbazia Santa Scolastica, 53-70124 Bari, Italy
e-mail: annarita.paiano@uniba.it

T. Gallucci
e-mail: teodoro.gallucci@uniba.it

A. Pontrandolfo
e-mail: andrea.pontrandolfo@uniba.it

T. Crovella
e-mail: tiziana.crovella@uniba.it

G. Lagioia
e-mail: giovanni.lagioia@uniba.it

S. S. Muthu (eds.), *Sustainable Packaging*, Environmental Footprints and Eco-design of Products and Processes, https://doi.org/10.1007/978-981-16-4609-6_1

on the current Italian consumption is evaluated using the Carbon Footprint methodology, to highlight which among the glass and PET mineral water bottles have the better environmental performance (Kouloumpis et al. in Sci Total Environ 727, 2020). Finally, according to the European 2030–2050 climate and energy framework, an improved eco-friendly performance scenario based on post-consumption options for both materials, was investigated regarding the Italian mineral water bottles consumption.

Keywords Life cycle assessment · Carbon footprint · Packaging · PET · R-PET · Glass · Bottled water

1 Introduction

Nowadays, bottles play a fundamental role in protecting the integrity of a product and guarantee the quality and safety of drinking water. All containers commonly used in the water industry such as glass, PET and aluminium are recyclable. Furthermore, these materials are safe and comply with the food contact regulations.

PET is the plastic material most used in the food industry, especially for water and soft drinks [86]. In recent years, the PET bottle has been increasingly used as food packaging, because it provides an excellent barrier, preserves the characteristics of the liquid contained, is hygienic, safe, light, impact-resistant, transparent, and economic. Furthermore, from an environmental perspective, the PET bottle is easily recyclable, through a mechanical process.

Mostly, packaging preserves food safety and quality during transportation, distribution and storage along the supply chain.

PET is one of the most diffused thermoplastic polymers available on the plastic market [23] mostly used for the consumption of mineral water.

Bottles, like other rigid plastic packaging in PET, continue to record growth in demand despite the strong environmental pressure in terms of emissions during production and post-consumer waste.

According to [73], by 2025 the global demand for PET will reach 22.65 million tons, with an annual growth rate post pandemic Covid Sars-19 of 3.7% and consumption up to 27.13 million tons. The impact of Covid-19 on PET is not yet clear, probably the overall value could fall by up to 17% compared to 2019.

Nowadays, the global market of PET producers is still fragmented, despite the consolidation of the sector in recent years, in which some of them are acquiring material recycling capabilities to achieve sustainability goals.

The global PET packaging market in 2019 was dominated by bottled water (35%) and carbonated soft drinks (27%), followed by no-food (12%), all other drink (11%), food (8%), thermoforming (7%) as [73] highlighted. Among the most consumer PET, Asia–Pacific (40%) represents the first area, North America (21%) and Western Europe (18%) followed [73].

The analysis of the European market conducted by Nisticò [62] has highlighted that PET packaging covers almost the 16% of European plastic consumption, which, in 2019, was 50.7 million tonnes, of which plastic packaging represents roughly 40%. About 42% of the collected plastic packaging waste, was recycled whereas 58% was sent to energy recovery or landfill [66].

Constantly growth in living standards, urbanization, growth of retail infrastructure (as large-scale-retail) and the replacement of traditional packaging will drive the growth of PET consumption for the next few years, especially in developing countries.

Regarding glass containers for food and beverage consumption, their market is undergoing significant changes to being more customer focused and to become more customer friendly [26].

Over time, the weight of glass containers has reduced significantly, guarantying decreased logistics costs. However, the success of this material is due to its unique characteristics such as chemical durability, optical properties, transparency and low cost.

In terms of retail and consumer choices, it has to be pointed out that PET is preferred by the commercial consumers, glass remains a product for the luxury segment for e.g. for important brands of restaurants or lounge bars. For example, in 2019, Ardagh Group introduced a sparkling wine bottle for Allure Winery for the luxury edition. Additionally, glass is heavily influenced by the alcoholic beverage industry as most manufacturers sell alcoholic beverages in glass packaging [26].

The global market value of glass packaging in 2019 was 51 billion Euro and for 2025 will reach 65 billion Euro [29].

The leader market of glass packaging is Asia–Pacific, followed by Europe which increases the consumption of alcoholic drinks [54].

The rate of collection for the recycling of glass packaging at the global level varies from 2 to 100%. On the contrary, in the EU it is 74%, ranging from 25% in Romania to 93% in Sweden [70]. At the European level, in 2016, 12 million tonnes of glass containers were recycled, of which 90% glass bottles [24]. Glass packaging contributes to a circular model implementation because glass does not lose its properties either in closed-loop or open-loop reuse/recycling models [70].

In recent years, most beverage producers have identified packaging as a fundamental means of innovation in the sector. Hence, in order to adapt and satisfy jointly the changing needs of consumers in terms of size, design, materials and the role of packaging as a communication tool, many companies have increased their investments in packaging research. This issue has consequently become a place of comparison and competition between the numerous players in the market in terms of sustainability too, directing the research and use of packaging by the lowest environmental impact.

The present chapter provides an evaluation of the sustainability of the most used packaging for mineral still water. Particularly it investigated the glass and Polyethylene Terephthalate (PET) materials to analyse their environmental impact and undertake a comparison among them.

In detail, this chapter is organized into six sections.

Section 2 identifies the methodology adopted, focusing on the Life Cycle Assessment (LCA) and Carbon Footprint (CF) tools;

Section 3 analyses the literary review, split into two subsections regarding the LCA and CF analysis of packaging respectively;

Section 4 displays both the assessment and comparison of PET and glass packaging, based on the LCA methodology;

Section 5 is split into five subsections, the first and the second of which overviewed the European and Italian water bottles sector, whereas the third analyses the CF of water bottles and makes a comparison between glass and PET materials. The last subsections introduce the post-consumption options for glass and PET water bottles, assessing the CF for both materials, and making a comparison of reuse and recycling alternatives. The Italian assessment of water PET bottles is undertaken according to the target of the EU [12]/904. The perspective scenarios are built to underline the reduction of CF by increasing the rate of recycled PET (R-PET).

Finally, Sect. 6 highlights the main findings of the chapter and some recommendations to the stakeholders involved in the packaging sector.

2 Methodology

This study is based on a literature review and a comparative analysis of the most packaging materials of still water (Fig. 1). The literature review investigated the

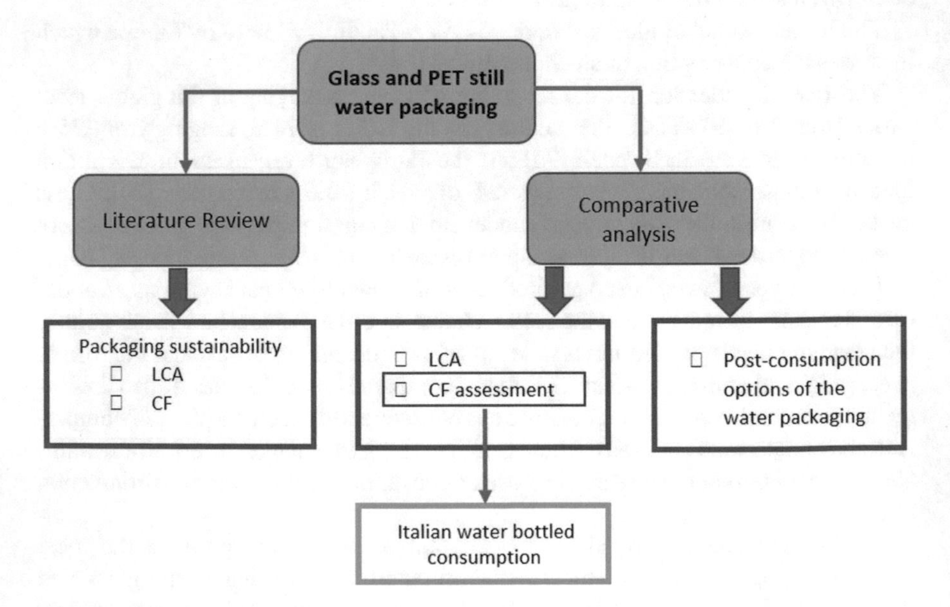

Fig. 1 Overview of methodology

research topic of the sustainability of the packaging sector, focusing on the LCA and CF methods of assessments and analysis.

The comparative analysis between the PET and glass packaging has been split into twofold. Firstly, through the LCA was assessed the environmental impacts of water packaging, comparing mass (1kg) and volume (single bottle) units. Then, the evaluation of the CF of the still water bottles, based on the current Italian consumption, has been undertaken. Further analysis of the improvements according to the end-of-life systems of the water packaging has been carried out with reference to reuse for glass and recycling for PET. The goal is to highlights which among the glass and PET still water bottles have the better environmental performance.

2.1 Environmental Impact Assessment Methodologies. An Overview of Life Cycle Assessment and Carbon Footprint

The LCA is based on the international technical standards of ISO 14040:2006 (environmental management—life cycle assessment—principles and framework) [41] and ISO 14044:2006 (environmental management e life cycle assessment e requirements and guidelines) [42], which allow to evaluate and measure the environmental impacts of products along their entire life cycle, within the system boundary defined as well.

It is a "cradle to the grave" method, which has acquired an important role since representing the most efficient approach to provide the data and information required to implement eco-sustainable strategies.

The definition of the functional unit (FU) gain significance to set up comparative analyses within alternative scenarios. The FU is indented to indicate the reference object of the study to which all input and output data have to be normalized. As the Fig. 2. displays, the LCA procedure involves four phases: (1) Goal and Scope Definition, which identifies the goal, scope, functional unit and system boundary, inter alia; (2) Life Cycle Inventory (LCI) involves the inventory analysis, containing the main issues of data collection and validation, and the allocation too; (3) Life Cycle Impact Assessment (LCIA), which defines the category, characterization and

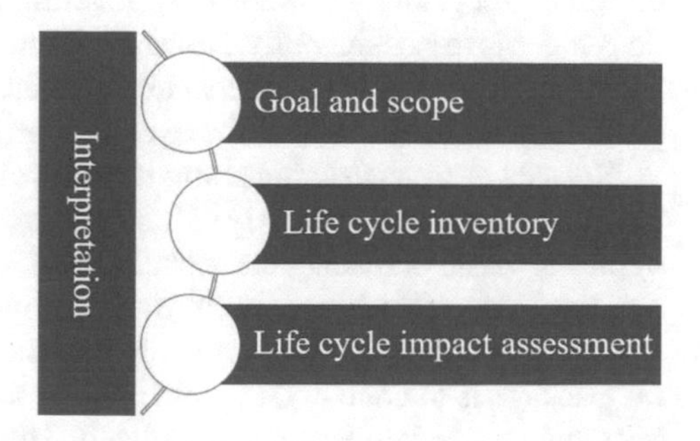

Fig. 2 Phases of LCA according to the ISO standards

weighting of the environmental impacts; (4) Life Cycle Interpretation identifies the main environmental issues and evaluates the results.

As regards the CF, it is a subset of LCA and quantifies the Greenhouse Gas (GHG) emissions, expressed in the unit of CO_2eq, generated by a product, service, activity or organization. LCA encompasses multiple "impact categories" along the life cycle, such as e.g. acid rain, summer smog, cancer effects and land use, if the end-point approach is considered, whereas CF, a more flexible and suitable tool for benchmarking different products or services, is a mono-criterion analysis focused on the emissions which affected the climate change.

To assess the CF there are some voluntary reference standards, such as the Public Available Specification 2050 based on BSI (British Standards Institution guideline for the assessment of the life cycle GHG emissions of goods and services), or on the GWP protocol corporate standard ISO 14067:2018 "Greenhouse gases—Carbon footprint of products—Requirements and guidelines for quantification", based on LCA methodology. The methodology commonly used in calculating CF, follows the ISO standards (14067: 2018), which is based on the LCA methodology and provides more specific guidelines for calculating the value of this indicator. This standard provides the guidance to the quantification and reporting of the CF of products along the supply chain [92].

In our case study the CF has been calculated according to this standard ISO.

3 Literature Review

In the last years, scientific literature focused on the analysis of waste minimization in the food and drink industry and of innovation in packaging design. The main goal has been the reduction in food losses and, at the same time, in the environmental impact of the food-packaging system.

In the early'90, certain sustainability criteria embedding in the packaging production began to be highlighted. Due to the first EU directive on packaging and packaging waste in 1994, the EU member countries have adjusted the national legislation to regulate the production, recovery and recycling of packaging waste, providing for policies and measures too for the allocation of the packaging costs. Today, the product design for recycling is promoted by governments as one of the most important practices to achieve sustainability. However, it must be stressed that the success of design for the environment largely depends on financial incentives for companies to design products with more content of recycled or recyclable material.

Nowadays, the packaging is ubiquitous in our daily lives and plastic continues to be the most used material, mainly in food packaging. As consequence, a steady increase in plastic waste outpacing the global growth rates and waste-collection systems has occurred [65]. This is especially true for multi-material packaging, which poses a significant challenge in the recycling process. Traditionally, the primary function of the package is to contain and protect the product, but its role has evolved over time becoming more complex and articulated. The growing issues of climate change and the issue of plastic waste disposal are pushing industries and researchers in studying solution to redesigning packaging, proposing new materials (as biodegradable) or substituting it with glass or paper. The challenge is to reverse the trend of plastic packaging and find materials by lower environmental impacts. The main question is whether glass or unconventional materials are better than plastic from an environmental point of view. To address this issue, we have analysed the scientific literature with the goal to benchmark the main environmental impacts of packaging from the extraction of raw materials, manufacturing phase until post-consumption disposal.

Particularly, to decrease the environmental impact of food packaging, the Life Cycle Assessment and Carbon Footprint are the most used analytical tools. These are based on the international standards ISO 14040-14044-14067 [41–43] and provide guidance for performing transparent and robust environmental calculations.

For this reason, the following literature review has been organized into two subsections related to (a) LCA analysis of packaging and (b) CF analysis of packaging.

3.1 LCA Analysis of Packaging

The Life Cycle Assessment, which is a methodology for calculating the environmental impacts of a product over its life cycle and based on ISO 14040 and ISO 14044, has been the most used for these comparative analyses.

Lee and Xu [51] reviewed different sustainable product packaging through LCA emphasizing the importance of eco-design and light weighting in minimizing environmental performance. Furthermore, they underlined how the environmental burdens are affected both by packaging materials and the impact of legislation on the disposal of used packaging.

Also, Dominic et al. [13] recommended eco-design in practice as a key factor to leverage sustainability gains. Differently, Del Monte et al. [11] studied the environmental performances of alternative packaging systems for retail sales of coffee showing how the replacement of plastic components can reduce the packaging weight enhancing the energy efficiency of production lines. The paper of [3], comparing the environmental burdens associated with drink packaging, underlined how "packaging is the main hotspot for most environmental impacts, contributing between 59 and 77%". Another study focused on the evaluation of the environmental impacts of packaging systems for milk and dairy products always through LCA analysis [31]. The findings show which the most impacting phase of the milk life cycle in the packaging, underlying the importance of the selection of the post-consumption materials.

Del Borghi et al. [10], have benchmarked the differences in the environmental performance between glass bottles and steel tin cans, emphasising that the impact of packaging production accounting for over 70% of the total environmental impact indicators. The LCA assessment has been used also in an exploratory study to correlate the environmental impact of food across the whole supply chain [89]. In this study, the authors benchmarked food packaging, pointing out that the use of appropriate packaging can reduce the percentage of food waste and ensure better preservation of food quality.

Manfredi and Vignali [57] studied the life cycle of tomato puree production including packaging and transport to suggest potential improvements. Also, in this study the analysis shows that lightweight glass packaging could reduce the environmental impact. Guiso et al. [33] assessed the impacts of extra virgin olive oil packaging concluding that the largest environmental benefit for environmental sustainability is represented by the glass while tin-plated steel cans are preferable for long-distance transport. Accorsi et al. [1] compared, through LCA methodology, glass versus plastic packaging associated with the bottled extra-virgin olive oil (EVOO) and underlined the potential of PET packaging in reducing the environmental impact of EVOO supply chains.

Garfí et al. [28] compared the environmental impacts between tap water consumption through different treatment scenarios (conventional drinking water treatment and domestic reverse osmosis treatment) and bottled mineral water in Spain. The results showed how the packaging of bottled water had the worst results mainly due "to the high consumption of raw materials and energy for bottle manufacturing, and to the higher weight of glass bottles per volume of water". Similar findings have been obtained by [49] emphasizing that bottled water has higher environmental impacts than tap water because it requires much higher material inputs than tap water and generates more waste.

Horowitz et al. [34] analysed the environmental impacts of bottled water with different bottle materials, such as Polylactic acid (PLA), corn based, R-PET, and regular (petroleum based) PET showing that the recycled PET is the more environmentally friendly for bottled water production, whereas the regular PET and PLA bottle are less environmentally favorable.

In terms of sustainability in the drinking water sector, [78] highlighted that PET bottles if, properly recycled, can assure an environmental benefit due to virgin material usage causing lower burdening on natural resources depletion.

Other papers investigated the environmental advantages of the recycling process of packaging.

Toniolo et al. [80] underlined the environmental benefits of assessing the convenience of an innovative recyclable package compared to an alternative package that is not recyclable. Furthermore, the study of [72], allowed to highlight the main environmental loads of a multilayer polymer bag for food packaging showing that the most impacting phase is the production of the polymer granules and can be reduced by thinning the thickness of the polymer.

Landi et al. [50] investigated the reuse of glass bottles to quantify the potential environmental performance highlighting how the glass recycling phase can reduce the environmental load whereas.

It is important to underline that literature focused too on studies concerning influence of green packaging on the consumer choices. For example, a recent study has highlighted how consumer before their purchases, gather information on the internet about the sustainability of packaging, changing their preferences to pro-environmental ones [32]. Other researchers have analysed the increasing influence on consumer decisions of sustainable product solutions [91]. Despite the slowness towards sustainable business practices, in conjunction with the environmental issue for the growing rate of packaging waste, nowadays, consumer's material needs-oriented towards more natural, high-quality materials and presents a growing demand for food packaging that does not do increase pollution. Globally, consumer demands driving the research and development of new materials in order to find alternatives to conventional materials made from fossil resources [81].

According to [5, 60], given the impact on different ecosystems, there is a potential role of the packaging within the Circular Economy with the aim of designing a sustainable pattern, extracting as much value as possible from products, components and long-term materials.

Conversely, it is important to underline that consumers are suspicious about the health risk of recycled or unknown materials composing the packaging [6, 8].

3.2 Carbon Footprint of Packaging

The CF indicator, spread in recent decades, is presented as an intuitive and easily understandable indicator even for non-expert users. However, [25] has supported that CF is not a new concept, existing for decades but differently named, e.g. as the result of the GWP impact indicator, LCA based. In the same line, [45] suggested that "The carbon footprint is quantified using indicators such as the Global Warming Potential (GWP)", which defined by the Intergovernmental Panel on Climate Change (IPCC), as "GWP is an indicator that reflects the relative effect of a greenhouse gas in terms of climate change considering a fixed time period, such as 100 years (GWP100)". Moreover, [45] explained as CF is one of the main impact categories in the LCA methodology, which typically uses IPCC characterization factors for CO_2 equivalents. Particularly, the CF comprehends a life cycle assessment limited to the emissions affecting climate change.

Matthews et al. [59] underlined that CF is one of the most important conceptual extensions of the Ecological Footprint, although a univocal definition was still missing.

However, until more than ten years ago, there is no unanimous consensus on the CF methodology. The spectrum of definitions varied from direct CO_2 emissions to GHG over the entire life cycle and the units of measurement were not even clear [88].

Among the most recent definitions of CF, [67] have defined the CF as an indicator of the total Greenhouse Gas emissions of a product expressed as CO_2equivalents.

The methodology commonly used in calculating CF follows the ISO standard (14067: 2018) based on the LCA methodology, as mentioned above, and provides more specific guidelines for calculating the value of this indicator.

According to Navarro et al. [61], CF represents today one of the most used environmental and sustainability indicators at the company or product level.

Within the plastic packaging industry, the CF indicator has been applied to various food-grade plastic packaging products [14, 55], using a cradle-to-cradle LCA methodology, compared PET containers for strawberries with PLA and Polystyrene (PS) ones and argued that "PET showed the highest overall values for all the impact categories, mainly due to the higher weight of the containers". Furthermore, to minimize the ecological footprint of a packaging system, the authors stressed that the production and procurement of inputs have to take place locally, reducing the high impacts of the transport.

In the same year, [35] compared the glass and the polypropylene (PP) plastic used to produce the baby food cups. The authors highlighted that the end-of-life impact of plastic packaging was lower than glass. Moreover, plastic packaging having a lighter mass than glass, is responsible for the reduction of CF.

A few years earlier, [46] examined the CF of PP yogurt cups and highlighted that, thought an LCA analysis, a 32 oz (0.9 L) instead of one by 6 oz (0.17 L) container can save energy, waste and reduce the impacts. Instead, [64] quantified the CF of mineral water in PET and glass bottles of various sizes: they recommended to recycling all beverage packaging materials more as they have less impact on the environment. Also, larger packs should be preferred which always have a lower environmental impact than smaller packs. However, when choosing the packaging material for a less impactful beverage, transport and secondary materials must be considered. Maga et al. [56] benchmarked environmental burdens for different meat packaging. The result pointed out the importance of recycling materials to reducing the negative effects on the environment highlighting how the use of "recycled PET instead of virgin PET allows reducing the carbon footprint by approximately 40%". Another study carried out by [48] analysed the substitution of Polyethylene Terephthalate (PET) with glass bottling liquids in the domestic sector. The substitution of PET with glass can reduce the Global Warming Potential by 18.9%.

Vinci et al. [85] suggested, for glass packaging, that the product innovations allowed to reduce the thickness, weight of packaging, decreasing the CO_2 emissions by about 4–5%.

Recently, Wong et al. [90] emphasized that the increase in the production and consumption of beverages is receiving considerable attention in terms of environmental sustainability especially from the point of view of carbon emissions.

Among the few published studies on this item, [9], have highlighted that replacing glass bottles with PET ones, in the case of beer production, did not lead to a significant reduction in CF emissions, unless refillable PETs or Glass are used. According to [39], virgin glass is the best option for tomato puree.

In some studies, it has been suggested that the result of carbon emissions mapping could be indicated within a carbon label to allow consumers to make more environmentally friendly choices [79]. A few years earlier, [47] already claimed that a CF label is expected to make consumers attend to how their product choices affect GHG emissions and help them to identify low-carbon alternatives.

According to Roibás et al. [67] among the alternatives to reduce the CF of packaging, the company should be encouraged the use of renewable energy sources or recycled packaging materials.

Botto et al. [7] conducted a comparison between two types of Italian drinking water, focused on tap water and PET bottled natural mineral water, applying the carbon footprint methodology and quantifying the emissions in terms of CO_2 equivalent. The results showed that 1.5 L of tap water saves 0.34 kg CO_2eq in comparison with the PET water-bottle consumption. Consequently, tap water consumption (for the 2 L per day recommended) could prevent 163.50 kg CO_2eq of greenhouse gas emissions per year. Thus, the substitution of PET water bottles with tap water avoids a significant amount of GHG emissions associated above all with the production of PET bottles, equal to 59% of the total impact.

Paiano et al. [63] stated that the CF methodology allows to assess some of the impacts of the water and beverage packaging.

4 Assessment and Comparison Through the LCA Methodology of PET and Glass Packaging

In order to make the comparison between glass and PET packaging, the results based on a previous LCA analysis undertaken by the authors on the hollow glass production of a company in Southern Italy [27], have been benchmarked with data and results of an LCA study related to the production of PET [58]. Most data are comparable between them. Both analyses have been carried out in accordance with ISO 14040: 2006 and 14040: 2006 and 14044: 2018 as well as the software used was GaBi thinkstep (a specific LCA software to calculate the environmental impacts). The environmental impact categories used are the same: GWP (Global Warming Potential), AP (Acidification Potential), EP (Eutrophication Potential) POCP (Photochemical Oxidant Creation) ADP (Abiotic Resource Depletion) and ADPel (Abiotic Resource Depletion Elements); Abiotic Depletion Potential–Fossil Fuels (ADP-fossil fuels); Water Depletion (WD). For both studies, a midpoint approach was adopted. The system boundary adopted was "cradle to grave", using the CML 2001 impact categories (January 2016 version) without considering the label and cap of the bottles. Furthermore, the inventory phase was split into the parts of same stages: the upstream phase which considered the raw material acquisition and transport to the factories; the core phase which included the consumption of energy (electricity, natural gas and diesel), the depletion of substances for maintenance and treatment in the manufacturing stage as well as the waste produced to the landfill; the downstream phase which encompassed the transport of both glass and PET post consumption, waste collection, recycling and disposal.

The study of hollow glass was carried out considering the use of 66% of recycled material, while the LCA analysis of PET was carried out considering a recycling rate of 70% which is compensated in the lifecycle of the next related product.

Both functional units were 1 kg of finished hollow glass and 1 kg of PET resin and the assessment of environmental metrics has been scaled down to calculate the impact corresponding to the various packaging sizes. It has to be underlined that in our analysis, a cut-off of 1% in relation to the material incoming and outgoing flows (raw materials, primary packaging, waste products and substances for maintenance and treatment) was adopted, while no allocations or parameterizations were implemented. Differently, in their analysis [58] considered a cut-off under the tolerance limit of 5% for most data, whereas the authors carried out the credit allocation methods for the PET recycling system.

The assessment of the environmental impacts of the different packaging materials for bottled water is first assessed and discussed with regards to both the mass and volume units of reference material.

PET is a polymer of the polyester family and is a thermoplastic resin produced by the polymerization of ethylene glycol and terephthalic acid. This material was created in 1941, but only in 1973, following the patent of the chemist Nathaniel Wyeth, the bottle production began. Hollow glass, instead, a very ancient packaging material, is produced by mixing different raw materials (cullet, yellow sand, soda, calcium carbonate, dolomite), in a melting furnace.

To make an exhaustive comparative analysis of these materials, the environmental impact was firstly analysed comparing the production of 1 kg of glass and 1 kg of PET and then was carried out a comparison between the production of a single bottle made by glass and PET. A twofold analysis is significant because the quantification of environmental impacts based only on the production of 1 kg of PET and 1 kg of hollow glass can be misleading: as from 1 kg of hollow glass, it is possible to obtain approximately only 2 glass bottles, with an average weight of 0.46 kg each whereas from 1 kg of PET roughly 52 bottles, with an average weight of 0.019 kg each, can be produced. Hence, the comparison based on a single bottle produced has been also provided.

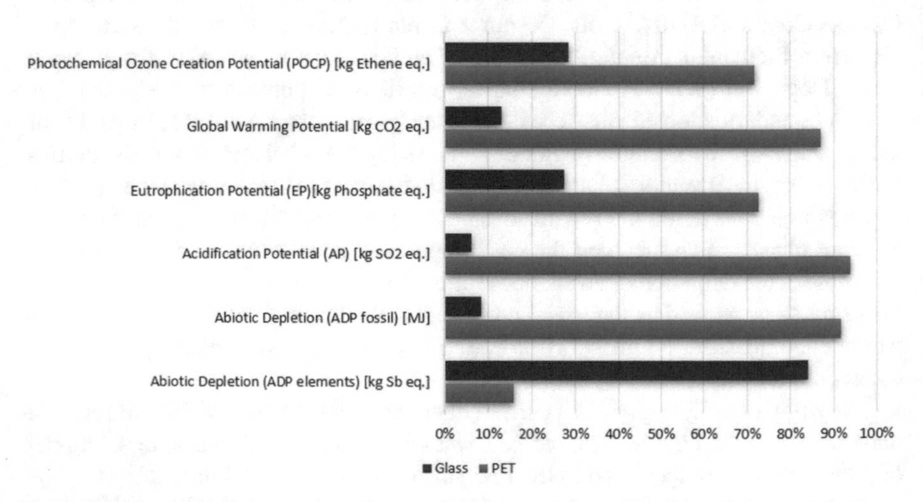

Fig. 3 LCIA per 1 kg glass and 1 kg PET. *Source* Personal elaboration by Gallucci et al. [27] and Marathe et al. [58]

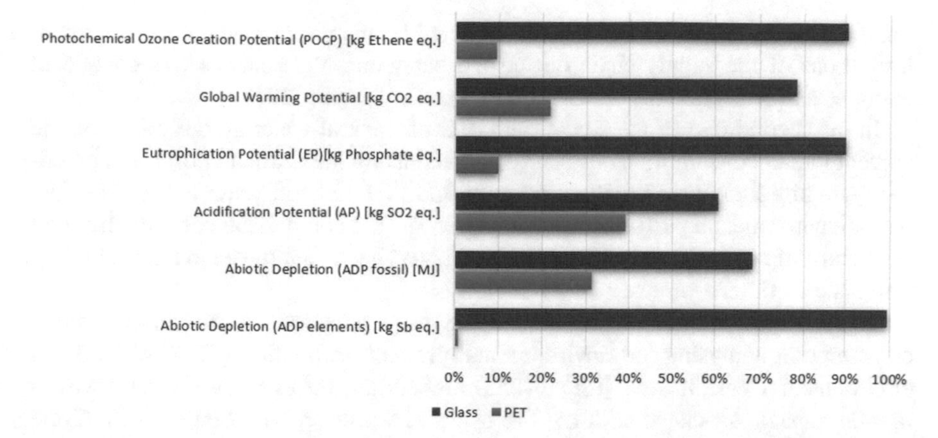

Fig. 4 LCIA per 1 L empty bottle. *Source* Personal elaboration by Gallucci et al. [27] and Marathe et al. [58]

Figure 3 shows the comparison between environmental impact categories in hollow glass and PET production.

As expected, the production of 1 kg of hollow glass has a lower environmental impact than the production of 1 kg of PET.

In particular, the benchmarking analysis shows that all the indicators, except ADPel, have less environmental impacts in the glass production. Specifically, comparing the GWP indicator it occurs that 1 kg of glass is equal to 0.656 kg CO_2eq, mainly due to natural gas consumption and raw materials like calcium carbonate, while 1 kg of PET is equal to 4.38 kg CO_2eq, due to the polymerisation process of PET production which is very energy intensive and strongly affected the result.

Differently, comparing the results of a single empty bottle, emerges a converse result: for all the impact indicators, the glass bottle has a significantly higher impact (Fig. 4). Particularly, the EP is 89% higher and the GWP (0.3 kg CO_2eq) roughly 73% higher than the PET bottle (0.08 kgCO_2eq) too. This is due to the previous consideration of the large number of bottles produced from 1 kg of PET and this evidence significantly lowers the environmental scores.

5 Carbon Footprint of the Water Packaging. The Italian Assessment

This section provides the CF of the Italian water packaging based on the total still water bottled consumption.

In the first part of this analysis, the current state of the bottled water sector was investigated at a European and Italian levels. Data relating to production, consumption, exports, sales channels, consumer preferences and sustainability of the entire sector were gathered to quantify future prospects, based on the Italian and European regulations. To obtain robust data, not affected by the production and consumption limitations of the supply chain during the pandemic, the analysis was conducted using water production and consumption data for the year 2019.

In the second part of the study, the GHG of mineral water bottles based on the current Italian consumption has been evaluated using the Carbon Footprint methodology, to highlight which among the glass and PET mineral water bottles have the better environmental performance. Moreover, further comparison between different post-consumption systems of these packaging has been undertaken to highlight their influence on the CF values.

The GHG of mineral water bottles has been assessed on the current Italian consumption analysing the Environmental product declaration (EPD) which is an environmental certification (ISO 14025) embedding information on the environmental impact associated with the life cycle of a product based on the Life Cycle Assessment of the product.

In order to understand the current situation of the sector in Italy, five EPDs and a Carbon footprint report of several important brands operating in the water bottling sector were used for the comparison.

To carry out the analysis, only the still water bottles consumed in 2019 were used, which amounted to approximately 9,300,000,000 L.

All selected EPDs were made with a cradle to grave approach according to the reference Product Category Rules (PCR) "2010: 11 Bottled waters, not sweetened or flavoured" and the reference Central Product Classification (CPC) code 24410. The functional unit (FU) was 1 L of water bottled in different sizes.

5.1 European Bottled Water Sector

In Europe, the packaging sector is the main manufacturing sector for plastics, indeed it represents about 40% of the entire demand, equal to over 20 Mt [66].

In recent decades, due to the growing need for the conservation and protection of goods, the plastics sector has become one of the main fields responsible for environmental and health impacts [53]. Considering that, according to the latest EU statistics, in 2018 on average, only 41.5% of plastic packaging waste was recycled [22], significant improvements are required.

In the global debate on the use of plastic, one of the main problems is linked to the use of single-use formats, e.g. it is estimated that globally, in 2021 over 583 billion PET bottles will be produced [77], this will represent an important challenge in terms of recovery and reuse of marketed bottles due to the current limited recycling rates.

In 2019, according to our estimates, over 51 billion litres of bottled water were consumed in the European Union [76] and, taking into account that, 87.4% of the water placed on the market is packaged in PET bottles, whereas 12.4% in glass, 0.1% in cardboard and 0.1% aluminium [87], the amount of PET consumed annually in this sector is significant.

Notwithstanding the high-quality standards of tap water in European countries, many people prefer to drink bottled water, in many cases due to the negative perception of tap water [30]. At the same time, this preference is certainly a more expensive and less environmentally friendly choice considering the huge amount of plastic produced to bottle water every year.

Figure 5 shows the per capita consumption of bottled water in Europe for the different countries; in 2019 the average consumption of bottled water in Europe was 118 L per capita [75]. In particular, the main European consumer country is Italy with average water consumption of 222 L per capita [2].

This figure not only ranks Italy in first place in Europe, but even in second place in the world, preceded only by Mexico, where annually about 244 L per capita are consumed [52]. Differently, Finland and Sweden are the European countries that consume fewer litres of bottled water, their consumption represents only 8.5% and 5% of Italian consumption, respectively.

Fig. 5 European per capita consumption of bottled water (L). *Source* Personal elaboration by Acquitalia [2] and Statista [75]

5.2 *Italian Bottled Water Sector*

The Italian production and consumption of bottled water are represented in Fig. 6. In the last decade, the trend shows an increase in the consumption of over 17% and in the exports by 77%. With over 15 billion litres of water bottled, exports of 1,6 billion litres, and 160 plants across the country, the Italian bottled water annual revenue, was amounted to roughly 3 billion in 2019 [2]. In the same year, based on data from large retailers, it emerged that water represented 77% of the volume and 30% of the turnover of the entire beverage sector except for wines (ISMEA, 2020).

The Italian leadership in bottled water consumption is probably due to the awareness that bottled water is qualitatively better and more controlled than that of our tap. To confirm this thesis, in a recent survey it emerged that 29% of Italian families do not trust drinking tap water [44]. Even the significantly lower cost of tap water has not limited the use of bottled water: in this regard, it has been estimated that the cost per litre of water is about 250 times lower than the average price of water bottle currently sold in a supermarket [52].

According to the habits and preferences of consumers, the consumption of bottled water in Italy is mainly represented by still water, which represents 69% of total consumption, whereas the consumption of sparkling water is equal to 31%.

Since the 1980s, plastic has gradually monopolized the market, replacing glass, and becoming the first material used for bottling water. However, the ever increasing attention to the environmental sustainability of packaging and awareness of the possible risks associated with the use of plastic are forcing companies to find new solutions.

In Italy, 82% of the water is sold in PET bottles (Table 1), among which 86% is linked to the use of 1.5 and 2 L sizes, generally used for domestic and family

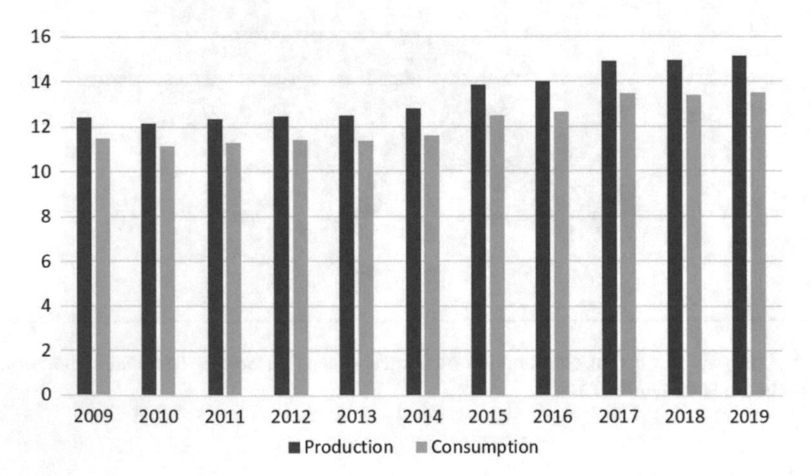

Fig. 6 Italian production and consumption of bottled water (billion litres). *Source* Personal elaboration by Acquitalia [2]

Table 1 Packaging material for still water bottled consumption. *Source* Personal elaboration by Acquitalia [2]

Material	Bottle size (L)	Water 10^9 L	Percentage
PET		7.6	82
	0.5	0.61	8
	1.5	5.34	70
	2	1.22	16
	Others	0.46	6
Glass	1	1.5	16
Paper, aluminium cans	Various	0.2	2

consumption, whereas, the 0.5 L bottles, mainly used for outdoor consumption, represent about 8%. With a much lower diffusion than PET and characterized by consumption mainly into the HORECA (hotels, restaurants and catering) sector, glass is currently used to 16%. It has to be noted that, due to a very limited quantity of 0.75 L glass bottles consumed, the glass category includes both 0.75 L and 1 L size. Finally, with a very low share, but great growth in recent years, 2% of the water is sold in aluminium cans or cardboard packaging.

5.3 *Carbon Footprint Assessment*

As cited above, the Carbon Footprint is a measure that expresses the total greenhouse gas emissions generated by a product, over its entire life cycle. These impacts are expressed in units of CO_2 equivalent and the functional unit chosen is 1L water. Figure 7 shows the average values of CO_2eq per 1 L of water bottled in different

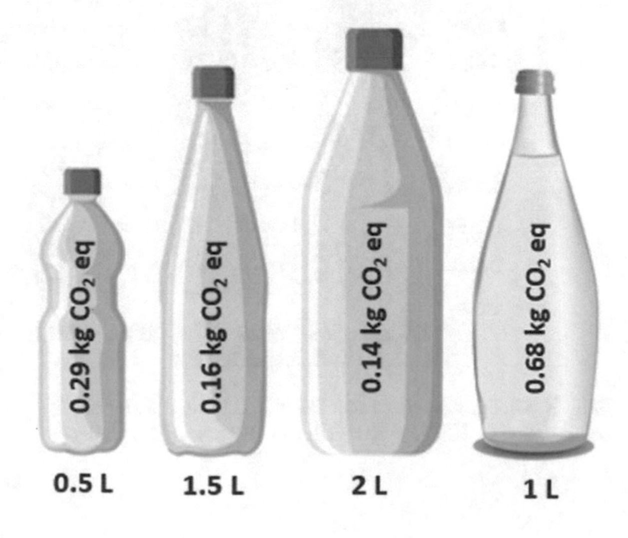

Fig. 7 Carbon footprint per material and size on average. *Source* Personal elaboration by EPD Vera [17], EPD Lete and Sorgesana [18], EPD Cerelia [19], EPD Levico [20] and EPD Frasassi [21]

sizes, encompassing caps (aluminium, steel, PP or HDPE), labels (paper or PP) and glue.

Considering the sizes of PET bottles, the 0.5 L one is the most impactful: however, if compared with the glass bottle, it emerges that its impact is significantly lower. By an impact of 0.14 kg CO_2eq, the most convenient size among those analysed is the 2 L one.

Then, considering the life cycle of the two materials, it is better to use PET for bottling water if the volume is chosen over the mass.

5.4 A Comparison as Regards to the Post-Consumption Options of Water Bottles

It must be stressed that the post-consumption options (reuse, recycling or disposal) of the packaging can determine a different impacting result.

PET bottles are traditionally considered single-use, and this facilitates the principle of use and throw-away, which has generated several environmental problems. Conversely, glass bottles can be reused multiple times especially in the household and this aspect must be considered in the whole environmental analysis. In this regard, a further comparison was made between bottled water in PET and glass bottles. The functional unit for both analyses was one litre of still bottled water again.

In order to identify the environmental impact of glass post-consumption systems, the authors amounted the CF of 1L water bottled (Fig. 8).

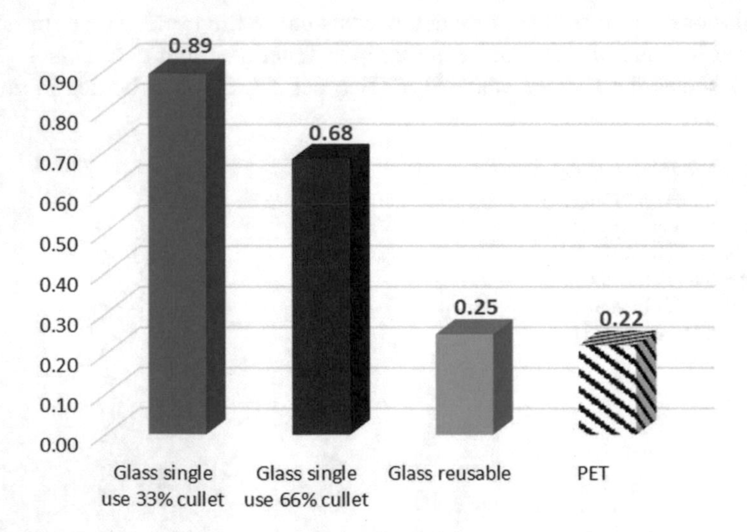

Fig. 8 Global warming potential of 1 L bottled water per 1 L size (kg CO_2eq). *Source* Personal elaboration by EPD Levico [20], EPD Frasassi [21], Gallucci [27] and San Benedetto [69]

Specifically, two glass bottles containing different percentages of glass cullet (33 and 66%) were compared to evaluate the reduction in CF linked with the increase in the amount of recycled material of the single bottle. Furthermore, it was useful to evaluate the performance of the reuse system of a glass bottle to make a comparison with recycling. Finally, the CF of a 1L PET water bottled amounted.

Notwithstanding the use of cullet is a key factor in reducing the CF for single-use bottles, reuse is still the best choice for glass bottles, assuming reuse by three times.

Figure 8 indeed displays the significantly lower impact of the reusable glass bottle, equal to about 64% and 72% in comparison to the single-use glass bottles with 66% and 33% of cullet respectively. Furthermore, results shown in Fig. 8 pointed out that the life cycle of the single-use glass bottles, is considerably more impactful than the PET one. Furthermore, despite the analysis emerged that the CF of a PET bottle is less impactful than a reusable glass bottle, it must be stressed that results suggest a very slight difference between the two bottles confirming the goodness of the reuse system.

The benchmarking among environmental impact indicators shows the importance of interpreting data and results based on LCA metrics to supporting environmental choices and finding a solution to the encountered critical issues. As Fig. 8 shows the improvement of the reuse process of the glass bottle can compensate for the higher CF value of its single use.

The LCA analysis has encouraged the analysis of the reusable system of glass bottles comparing this system with the Italian traditional packaging separate collection system.

Although in Italy the practice of returning glass bottles to the manufacturer in the past was widespread, the introduction of plastic as packaging has changed the behaviours. The lightness, resistance, the simplicity of processing compared to other materials, jointly with the removal of the costs of transport, handling and storage of the empties bottle have made plastic bottles more advantageous for producers. At the same time, consumers were attracted by the deposit cost to zero and the responsibility of returning the empty bottles. The disposable model has increasingly consolidated in the market, although it is converse to the waste hierarchy provided by the EU legislation.

A reusable bottle (it can be a glass bottle, but also a PET plastic bottle), can be returned to the supplier, so that it can be reused. In Italy, in 2017, as it happens in other countries, the Italian institutions organized the management of returnable bottles without being successful. In a market where the introduction of single-use bottles is constantly increasing, the theoretical reusing the same bottle up to 30 times [74], could represent a simple, but effective choice for the protection of the environment.

Currently, the glass bottle reuse system in Italy is in disuse, indeed, only 10% of the glass bottles for water consumed in the country are returned to the same distributor to be reused after the washing and sterilization and filling processes. This figure is in contrast with the northern European countries, where the use of this system is widespread. Based on data in Fig. 8, particularly the CF of the glass 33% recycled and the glass reused, we estimated that if all water glass bottled in the Italian consumption, were returned to the distributor for the reuse, saving of approximately 852,000 t CO_2eq, could be achieved compared to the current situation. This amount could be comparable with the yearly emissions of roughly 570,000 cars [82].

Furthermore, the process mainly used in Italy for glass recycling involves numerous steps that contribute to increasing the impact of the product. Once the bottles have been delivered to the special container for glass collection, a vehicle collects and takes them to the collection centre where the material is stored, cleaned, divided by colour, and crushed. Then, the cullet will be transported to the glass factory for melting and manufacturing of the new product. The new bottle returns to the water producer for the bottling phase. Therefore, this system requires many steps that contribute to a high impact, mainly generated by the transport of the glass according to its weight, but by the high temperatures in the melting phase too which will contribute to very high energy consumption.

As we can note, as distribution plays a key role in impact assessment, it is necessary to take into account the distance between the bottling plant and the distributor. In the reuse process, it is necessary to consider the impacts generated both by the transport for recovery of the glass bottle and by the industries which provide for the preparing and cleaning of the bottles [83]. According to Amienyo et al. [3], the best performance occurring in the range of 1–5 times of the glass reuse, because of the operational phases affecting the result, whereas [74] identify to 8 times the glass reuse, suitable period before scuffing of bottles.

In Landi et al. [50] made an environmental comparison for glass wine bottles, between two end-of-life scenarios: recycling and reuse. From the analysis emerged significant environmental savings in all the impact categories for the reuse scenario, mainly due to the avoided impact generated by glass melting and bottle forming which represents very energy-intensive phases.

A further enhancement of the glass bottle reuse system consists in the reduction of the new packaging production and waste post consumption as well.

5.5 PET and R-PET in the Beverage Industry

In recent years, consumer awareness about the environmental issues associated with PET single-use bottles has led the entire sector to support new recycling solutions to reduce waste and maximize the reuse of materials.

The industry, also thanks to the recently stringent EU regulations on the reuse of recycled materials, is making efforts to reduce the amount of plastic needed for bottle manufacturing, and, at the same time, increasing the R-PET content in order to minimize the environmental impact.

To achieve these results, water bottling companies are constantly investing in new recycling technologies to use greater quantities of recycled materials.

Plastic recycling is one of the most important aspects in the field of waste disposal. Through the separate collection, it is possible to have both economic advantages and benefits for the environment, creating employment and creating new products from waste materials ready to be re-marketed. In this context, above all recycling PET plastic bottles could represent an important step forward in reducing packaging waste and avoid the dispersion of the bottles in the environment, particularly into the sea.

Although a lot of thermoplastics can be recycled, recycling PET bottles is much more convenient than many others. This is due to the plastic bottles for drinks and water bottles are made almost exclusively of PET, which makes them more easily identifiable in a recycling stream.

Economically, the entire petrochemical industry and, as a consequence also the PET market, are heavily affected by oil price trends, particularly if global threats occur. Since February 2020, due to the Covid-19 pandemic, there was a global collapse in the oil price and consequently in the PET prices.

Compared to R-PET, virgin PET has become much cheaper [37]. For this reason, the effects of this sharp reduction in the PET prices have temporarily slowed down the market development and the R-PET use.

The R-PET generation process involves the recycling of plastic post-consumption. After the collection phase, PET is sorted, cleaned, and transformed into tiny flakes.

It is important to underline that R-PET requires less energy for production than virgin PET bottles [4] and also guarantees minimization of the use of natural resources.

In recent years, the second life of PET plastic bottles regarded a wide range of uses, from the furniture sector to clothing. It can be used to produce new plastic beverage bottles, but also in the production of tubes and containers or used in the textile industry [68] for fleeces, knits and carpets manufacturing. In Europe, in 2018, it has been estimated that the largest use of R-PET (30%) occurred for sheet production, whereas 24% for fibre. Instead, due to a higher quality of R-PET required in food and beverage packaging, only 18% of the total was sold for food bottle production [15].

Over the years, the efforts of the bottled beverage industry have made it possible to rationalize the amount of material used for the primary packaging. Indeed, the weight of both PET and glass water bottles has been significantly reduced. Compared to twenty years ago, a 0.5 L PET bottle is currently 51% lighter [36]; whereas glass bottles are about 30% lighter today. This has allowed producers to obtain important improvements in terms of efficiency in the use of material resources and reduction of energy consumption in the manufacturing process and greenhouse gas emissions as well.

According to the EU [12]/904—on the reduction of the impact of certain plastic products on the environment, the collection targets for plastic bottles have significantly increased in comparison with current rates: 77% of plastic bottles should be collected since 2025 and 90% since 2029.

Currently, as we can see from Fig. 9, in the EU in 2019 about 64% of the PET bottles put on the market were collected. Despite this, collection rates vary significantly across regions: some countries such as Germany and Finland exceed 93% of collected PET bottles, while other countries such as Bulgaria reach a maximum of 20%.

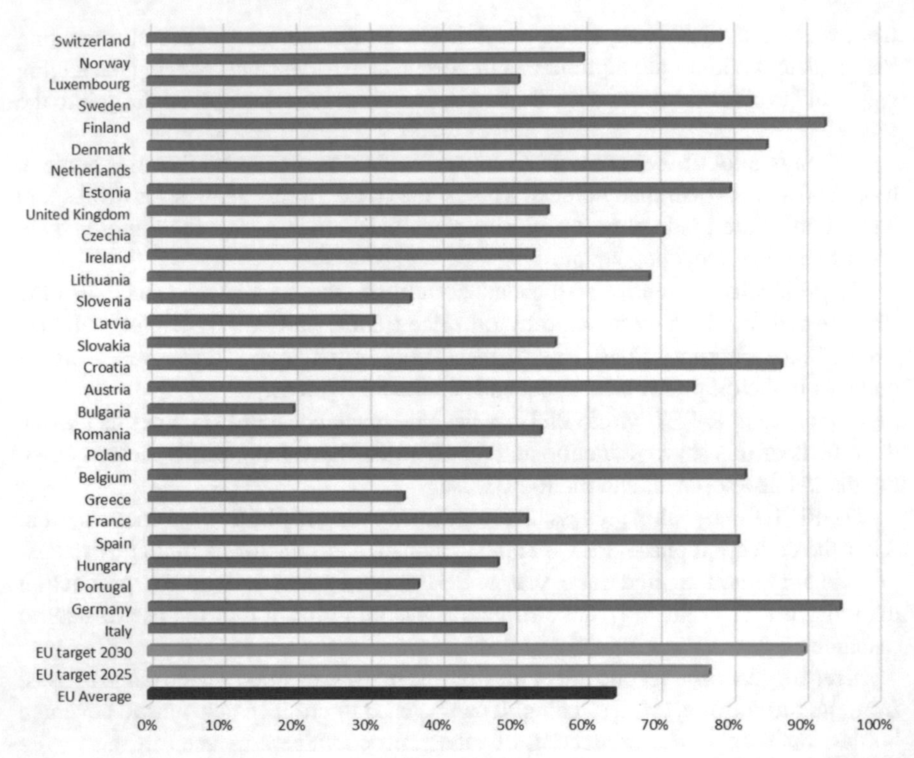

Fig. 9 PET bottle collection rate per country in 2019. *Source* Personal elaboration by ICIS [38]

Furthermore, according to the EU Directive above, since 2025, in the EU Member States, all PET bottles placed on the market must contain at least 25% of recycled plastic with a goal of 30% by 2030. In 2017, the average amount of R-PET contained in the new water bottles was between 11 and 14.5% [16, 38].

Within this context, in order to evaluate and quantify the effects and benefits generated by European legislation on the Italian market, the current CO_2eq emission of the entire sector was estimated and compared with the objectives set by the EU for 2025 and 2030 in terms of R-PET rate of the new PET bottles. In the analysis, the main bottled water sizes consumed in Italy were 0.5, 1.5 and 2 L, which were used as a reference. Taking into account that the EPDs available in Italy displayed an average use of R-PET of 11% in the production of water bottles, we measured the CO_2eq emissions associated with each format referring to the current R-PET rate.

The analysis showed that, despite the high impact generated by the single 0.5 L PET bottle in comparison with the other sizes, in the Italian market, due to the very high percentage of 1.5 L bottles sold, the total emissions of the entire market are mainly generated by the latter format. We estimated that the entire consumption of PET water bottled (7.6 billion L, Table 1) generates annually about 1,201,857.6 t CO_2eq. Particularly, the 1.5 L size represents 71.1% of the total amount of CO_2eq

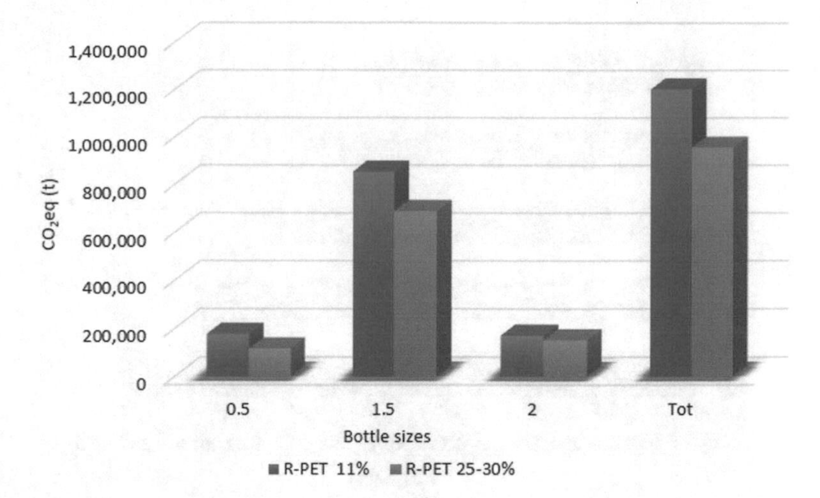

Fig. 10 CO_2eq emissions of the Italian still water bottled

emissions whereas the 0.5 L and 2 L sizes represent 14.7% and 14.2% of the total respectively.

Furthermore, considering the EU targets about the R-PET content, mentioned above, the corresponding average emissions have been estimated. Figure 10 shows the current and perspective CO_2eq emissions. Based on these results, a comparison highlights that the reduction of CO_2eq emissions is significant. In regard to the value per category, the increase in R-PET would allow the greatest rate reduction, equal to 34.5%, in the smaller size (0.5 L) whereas a reduction in CO_2eq emissions of 11.4% occurring in the larger size (2 L). In absolute terms, the 1.5 L size, which is the most marketable, allows saving approximately 165,484.2 t CO_2eq per year. Overall, compared with the current scenario, a reduction in total emissions by about 20.5%, equivalent to saving by approximately 246,014 t CO_2eq emissions, has been estimated.

In Italy, a further step forward for the use of recycled PET was made in October 2020. Indeed, in 2020, law n.126 was presented aimed at guarantee producers the opportunity to manufacture bottles in 100% recycled PET previously limited to 50%.

Therefore, even in Italy, as in the rest of Europe, it will be possible to produce fully recycled bottles, reducing harmful emissions and thus respecting the environment. Inevitably, this will also provide a great boost to Italian companies engaged in the Circular Economy.

Based on this initiative, an analysis to evaluate the potential improvement of the environmental performance of the entire sector was carried out. The analysis provides indications on the use of increasing rates of R-PET in the production of new bottles.

This analysis, unlike the previous ones, aims to evaluate exclusively the use of PET and R-PET in the production of bottles, without evaluating the impact associated with the cap, label and water. This is necessary to quantify more accurately the CO_2eq emissions avoided due to the R-PET use.

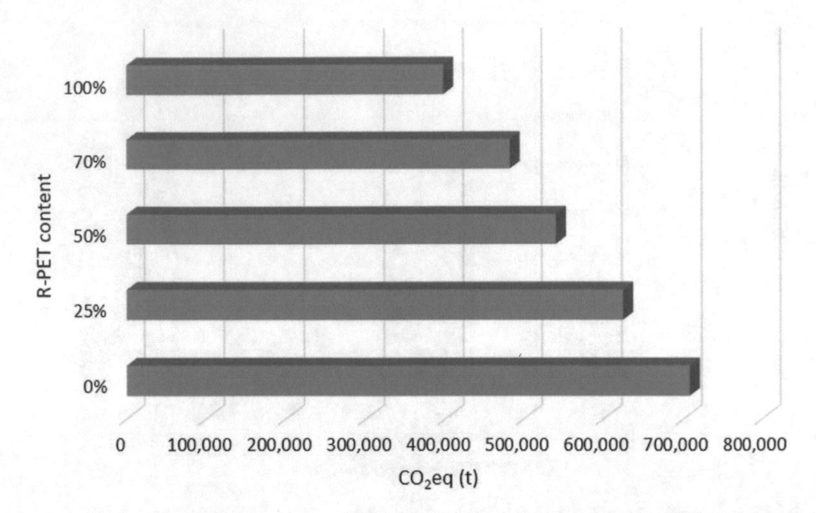

Fig. 11 Decreasing of CO_2eq emissions due to R-PET use for the entire Italian PET requirement in the still water bottle production. *Source* Personal elaboration by Acquitalia [2] and Marathe et al. [58]

Then, we estimated the amount of PET needed annually to satisfy the entire market. About 5.5 billion still water PET bottles are consumed annually in Italy, which is equivalent to about 110 million tons of PET. Taking into account the CO_2eq emissions of 1 kg PET according to [58], we assessed the CF of the Italian PET water bottles. As the Fig. 11 displays, the increase of R-PET content boosted to a significant reduction in CO_2eq emissions, achieving a minimum level of roughly 400,000 t CO_2eq if the R-PET is increased to 100%.

Considering that, until 2020, many Italian companies had been limited by the previous legislation in terms of percentages of R-PET usable in the production of the bottles, it is desirable that better results than those required by European legislation can be achieved, overcoming the minimum use of 30% of R-PET by 2030.

6 Conclusions

The packaging sector spun to a very critical debate about its sustainability. The challenges to be addressed encompass the more stringent targets by the EU legislation, the environment and economics as well.

This chapter provided an evaluation of the sustainability of the most used packaging for mineral still water. Particularly it investigated the glass and PET to analyse their environmental impact and undertake a comparison among them in regard to their manufacturing and post-consumption options too. LCA methodology has been used to assess the impact indicators of the different packaging and the CF to evaluate one of the mineral water bottles based on the current Italian consumption.

The results stressed the better performance of the 1 kg glass than 1 kg PET, conversely, the resulting performance from 1 L of water bottled.

Furthermore, post-consumption methods have been investigated to understand their influence on environmental impact for each material. Particularly the reuse system positively affects the glass Carbon Footprint, indeed, compared to a glass bottle made of 33% of recycled material, the reusable glass bottle has a 72% lower impact than the first one. Notwithstanding 1 L PET bottle is lower impactful than the glass one, results suggest a very slight difference between the two bottle materials, if the glass reuse is applied. Particularly, the last option could save roughly 852,000 t CO_2eq concerning the current Italian consumption of water glass bottled.

Further analysis provides suggestions on the use of increasing rates of R-PET in the production of new bottles. Results highlighted the decrease of the Carbon Footprint of PET bottles by the increasing R-PET content, with reference to the Italian level. Particularly, if the 5.5 billion still water PET bottles yearly consumed in Italy were produced increasing the R-PET content to 100%, over 300,000 t CO_2eq could be saved.

The efforts to counteract the GHG emissions provided for the stringent EU legislation, need a deep analysis about both the sustainability of the packaging materials but also their end-of-life options.

This framework provided useful guidelines to better manage the sustainability of the water packaging and suggest to researchers and policymakers the issues and challenges being faced in order to move towards the path of the green new deal.

CRediT Authorship Contribution Statement

Annarita Paiano (corresponding author): conceptualization, methodology, writing—original draft preparation, writing—review and editing and supervision.

Teodoro Gallucci: scientific literature and reviewing.

Andrea Pontrandolfo: investigation, data curation and validation, writing - original draft preparation, writing—review and editing.

Tiziana Crovella: resources and visualisation.

Giovanni Lagioia: reviewing and editing.

References

1. Accorsi R, Versari L, Manzini R (2015) Glass vs plastic: life cycle assessment of extra virgin olive oil bottles across global supply chains. Sustainability 7:2818–2840. https://doi.org/10.3390/su7032818
2. Acquitalia (2020) Natural mineral water industry 2020–2021. https://www.beverfood.com/downloads/acquitalia-annuario-acque-minerali/. Accessed 1 Mar 2021
3. Amienyo D, Guiba H, Stichnothe H, Azapagic A (2013) Life cycle environmental impacts of carbonated soft drink. Int J Life Cycle Assess 18:77–92. https://doi.org/10.1007/s11367-012-0459-y
4. APR—The Association of Plastic Recyclers (2018) Life cycle impacts for postconsumer recycled resins: PET, HDPE, and PP. https://plasticsrecycling.org/images/library/2018-APR-LCI-report.pdf. Accessed 10 Mar 2021

5. Batista L, Bourlakis M, Smart P, Maull R (2018) In search of a circular supply chain archetypeea content-analysis-based literature review. Prod Plan Control 29:438–451. https://doi.org/10.1080/09537287.2017.1343502
6. Baxter W, Aurisicchio M, Childs P (2017) Contaminated interaction: another barrier to circular material flows. J Ind Ecol 21:507–516. https://doi.org/10.1111/jiec.12612
7. Botto S, Niccolucci V, Rugani B, Nicolardi V, Bastianoni S, Gaggi C (2011) Towards lower carbon footprint patterns of consumption: the case of drinking water in Italy. Environ Sci Policy 14:388–395. https://doi.org/10.1016/j.envsci.2011.01.004
8. Chen YS, Chang CH (2013) Greenwash and green trust: the mediation effects of green consumer confusion and green perceived risk. J Bus Ethics 114:489–500. https://doi.org/10.1007/s10551-012-1360-0
9. Cimini A, Moresi M (2018) Mitigation measures to minimize the cradle-to-grave beer carbon footprint as related to the brewery size and primary packaging materials. J Food Eng 236:1–9. https://doi.org/10.1016/j.jfoodeng.2018.05.001
10. Del Borghi A, Strazza C, Magrassi F, Taramasso AC, Gallo M (2018) Life cycle assessment for eco-design of product-package systems in the food industry- The case of legumes. Sustain Prod Consum 13:24–36. https://doi.org/10.1016/j.spc.2017.11.001
11. Del Monte M, Padoano E, Pozzetto D (2005) Alternative coffee packaging: an analysis from a life cycle point of view. J Food Eng 66:405–411. https://doi.org/10.1016/j.jfoodeng.2004.04.006
12. Directive 2019/904/EU of the European Parliament and of the Council of 5 June 2019 on the reduction of the impact of certain plastic products on the environment (Text with EEA Relevance). Off J Eur Union. Brussels. Belgium. https://eur-lex.europa.eu/eli/dir/2019/904/oj. Accessed 3 Mar 2021
13. Dominic CAS, Östlund S, Buffington J, Masoud MM (2015) Towards a conceptual sustainable packaging development model: a corrugated box case study. Packag Technol Sci 28:397–413. https://doi.org/10.1002/pts.2113
14. Dormer A, Finn DP, Ward P, Cullen J (2013) Carbon footprint analysis in plastics manufacturing. J Clean Prod 51:133–141. https://doi.org/10.1016/j.jclepro.2013.01.014
15. EFBW, Petcore Europe, Plastics Recyclers Europe (2020) PET market in Europe state of play, production, collection and recycling data. https://743c8380-22c6-4457-9895-11872f2a708a.filesusr.com/ugd/dda42a_c4c772a57d6b4fcaa3ab7b7850cb536c.pdf. Accessed 20 Mar 2021
16. EPBP—European PET Bottle Platform (2020) How to keep a sustainable PET recycling industry in Europe. https://www.epbp.org/. Accessed 10 Mar 2021
17. EPD Vera (2015) Acqua minerale Nestle' Vera (PET) Fonte in bosco. Registration number No: S-P-00745. https://portal.environdec.com/api/api/v1/EPDLibrary/Files/24f41de9-01b9-477a-a926-0b7f7782fee8/Data. Accessed 20 Mar 2021
18. EPD Lete and Sorgesana (2018) Environmental product declaration—Lete and Sorgesana acqua minerale effervescente naturale Lete (PET 0,5 litri - 1,5 litri) e naturale Sorgesana (PET 2,0 litri). Registration number No: S-P-00394. https://portal.environdec.com/api/api/v1/EPDLibrary/Files/d43824c1-c766-4778-a0eb-aede8df9b177/Data. Accessed 28 Mar 2021
19. EPD Cerelia (2019) Environmental product declaration—Cerelia mineral water, dichiarazione ambientale di prodotto dell'acqua minerale naturale Cerelia imbottigliata in pet 0,5L, pet 1,5L, vetro a perdere 1L). Registration number No: S-P-00123. https://portal.environdec.com/api/api/v1/EPDLibrary/Files/c963e647-7409-47f4-beed-e88cc46e9374/Data. Accessed 10 Mar 2021
20. EPD Levico (2019) Environmental product declaration - Levico mineral water, bottled mineral water, still, sparking, slight sparkling, in glass bottle format of 1L, 0.75L, 0.5L, 0.25L. Registration number No: S-P-01712. https://portal.environdec.com/api/api/v1/EPDLibrary/Files/4d626f77-0095-4caa-ba3c-8e853cef0c1f/Data. Accessed 5 Mar 2021
21. EPD Frasassi (2021) Environmental product declaration—Frasassi acqua minerale, Mineral water: still, lightly sparkling and sparkling in PET bottles in formats: 0.5–1–1.5–2 liters. Registration number No: S-P-02682. https://portal.environdec.com/api/api/v1/EPDLibrary/Files/21f4eeed-8e47-4b99-b8c7-08d8d1855218/Data. Accessed 5 Mar 2021

22. Eurostat (2021) More than 40% of EU plastic packaging waste recycled. https://ec.europa.eu/eurostat/web/products-eurostat-news/-/ddn-20210113-1. Accessed 2 Mar 2021
23. Fakirov S (2017) Fundamentals of polymer science for engineers. Wiley-VCH, Weinheim
24. FEVE—The European Container Glass Federation (2020) Glass Packaging Industry's Position on the new Circular Economy Action Plan. https://feve.org/wp-content/uploads/2020/06/FEVE-position-Circular-Economy-Action-Plan-May-2020.pdf. Accessed 28 Mar 2021
25. Finkbeiner M (2009) Carbon footprinting—opportunities and threats. Int J Life Cycle Assess 14:91–94. https://doi.org/10.1007/s11367-009-0064-x
26. FMI—Future Market Insight (2021) Glass Container Market. Analysis and review: glass container market by end-use industry—cosmetics & perfumes, pharmaceuticals, food packaging, and beverage packaging for 2021–2031. https://www.futuremarketinsights.com/reports/container-glass-market. Accessed 28 Mar 2021
27. Gallucci T, Lagioia G, Piccinno P, Lacalamita A, Pontrandolfo A, Paiano A (2020) Environmental performance scenarios in the production of hollow glass containers for food packaging: an LCA approach. Int J Life Cycle Assess 26:785–798. https://doi.org/10.1007/s11367-020-01797-7
28. Garfí M, Cadena E, Sanchez-Ramos D, Ferrer I (2016) Life cycle assessment of drinking water: Comparing conventional water treatment, reverse osmosis and mineral water in glass and plastic bottles. J Clean Prod 137:997–1003. https://doi.org/10.1016/j.jclepro.2016.07.218
29. Garside M (2020) Glass containers and bottles global market value 2019 & 2025 in Statista. https://www.statista.com/statistics/700364/glass-bottles-and-containers-market-value-worldwide/. Accessed 20 Mar 2021
30. Geerts R, Vandermoere F, Van Winckel T, Halet D, Joos P, Van Den Steen K, Van Meenen E, Blust R, Borregan-Ochando VSE (2020) Bottle or tap? Toward an integrated approach to water type consumption. Water Res 173:115578. https://doi.org/10.1016/j.watres.2020.115578
31. Chaouki G (2012) Life cycle assessment of packaging materials for milk and dairy products. Int J Thermal Environ Eng 4:117–128. https://doi.org/10.5383/ijtee.04.02.002
32. Grebitus C, Roscoe RD, Van Loo EJ, Kula I (2020) Sustainable bottled water: how nudging and internet search affect consumers' choices. J Clean Prod 267:121930. https://doi.org/10.1016/j.jclepro.2020.121930
33. Guiso A, Parenti A, Masella P, Guerrini L, Baldi F, Spugnoli P (2016) Environmental impact assessment of three packages for high-quality extra-virgin olive oil. J Agric Eng 47:191–196. https://doi.org/10.4081/jae.2016.515
34. Horowitz N, Frago J, Mu D (2018) Life cycle assessment of bottled water: a case study of Green2O products waste management 76:734–743. https://doi.org/10.1016/j.wasman.2018.02.043
35. Humbert S, Rossi V, Margni M, Jolliet O, Loerincik Y (2009) Life cycle assessment of two baby food packaging alternatives: glass jars vs. plastic pots. Int J Life Cycle Assess 14:95–106. https://doi.org/10.1007/s11367-008-0052-6
36. IBWA—International Bottled Water Association (2021) Recycling. https://bottledwater.org/recycling/. Accessed 10 Mar 2021
37. ICIS—Independent Commodity Intelligence Services (2020a) INSIGHT: Different views on pricing, supply and demand split Europe R-PET market. https://www.icis.com/explore/resources/news/2020/05/08/10505492/insight-different-views-on-pricing-supply-and-demand-split-europe-r-pet-market. Accessed 8 Mar 2021
38. ICIS—Independent Commodity Intelligence Services (2020b) INSIGHT: European plastic bottle recycling held back by structural shortage of feedstocks. https://www.icis.com/explore/resources/news/2020/12/30/10590184/insight-european-plastic-bottle-recycling-held-back-by-structural-shortage-of-feedstocks. Accessed 10 Mar 2021
39. Ingrao C, Faccilongo N, Valenti F, De Pascale G, Di Gioia L, Messineo A, Arcidiacono C (2019) Tomato puree in the Mediterranean region: an environmental life cycle assessment, based upon data surveyed at the supply chain level. J Clean Prod 233:292–313. https://doi.org/10.1016/j.jclepro.2019.06.056

40. ISMEA—Istituto di Servizi per il Mercato Agricolo Alimentare (2020) Consumi alimentari, i consumi domestici delle famiglie italiane. http://www.ismeamercati.it/flex/cm/pages/ServeAttachment.php/L/IT/D/c%252Ff%252Fe%252FD.75eab906e872db688dd8/P/BLOB%3AID%3D10214/E/pdf. Accessed 8 Mar 2021
41. ISO (International Organization for Standardization) (2006a) 14040 Environmental management—life cycle assessment—principles and framework
42. ISO (International Organization for Standardization) (2006b) 14044 Environmental management—life cycle assessment—requirements and guidelines
43. ISO (International Organization for Standardization) (2018) 14067 Greenhouse gases - Carbon footprint of products—requirements and guidelines for quantification
44. ISTAT—Istituto nazionale di statistica (2020) Rapporto SDGs 2020. Informazioni statistiche per l'Agenda 2030 in Italia. Garantire a tutti la disponibilità e la gestione sostenibile dell'acqua e dei servizi igienico sanitari. https://www.istat.it/storage/rapporti-tematici/sdgs/2020/goal6.pdf. Accessed 10 Mar 2021
45. JRC-EU (2007) EPLCA—Carbon footprint—What it is and how to measure it. European Platform on Life Cycle Assessment, European Commission. http://www.to-be.it/wp-content/uploads/2015/07/Carbon-footprint.pdf. Accessed 3 Mar 2021
46. Keoleian GA, Phipps AW, Dritz T, Brachfeld D (2004) Life cycle environmental performance and improvement of a yogurt product delivery system. Packag Technol Sci 17:85–103. https://doi.org/10.1002/pts.644
47. Kimura A, Wada Y, Kamada A, Masuda T, Okamoto M, Goto SI, Tsuzuki D, Cai D, Oka T, Dan I (2010) Interactive effects of carbon footprint information and its accessibility on value and subjective qualities of food products. Appetite 55:271–278. https://doi.org/10.1016/j.appet.2010.06.013
48. Kouloumpis V, Pell RS, Correa-Cano ME, Yan X (2020) Potential trade-offs be- tween eliminating plastics and mitigating climate change: an LCA perspective on Polyethylene Terephthalate (PET) bottles in Cornwall. Sci Total Environ 727:138681. https://doi.org/10.1016/j.scitotenv.2020.138681
49. Lagioia G, Amicarelli V, Calabrò G (2012) Empirical study of the environmental management of Italy's drinking water supply. Resour Conserv Recycl 60:119–130. https://doi.org/10.1016/j.resconrec.2011.12.001
50. Landi D, Germani M, Marconi M (2019) Analysing the environmental sustainability of glass bottles reuse in an italian wine consortium. Proc CIRP 80:399–404. https://doi.org/10.1016/j.procir.2019.01.054
51. Lee SG, Xu X (2005) Design for the environment: life cycle assessment and sustainable packaging issues. Int J Environ Technol Manag 5:14–41. https://doi.org/10.1504/IJETM.2005.006505
52. Legambiente and Altreconomia (2018) Acque in bottiglia 2018, un'anomalia tutta italiana. https://www.legambiente.it/wp-content/uploads/dossier-acque_in_bottiglia_2018.pdf. Accessed 10 Mar 2021
53. Lombardi M, Rana R, Fellner J (2021) Material flow analysis and sustainability of the Italian plastic packaging management. J Clean Prod 287. https://doi.org/10.1016/j.jclepro.2020.125573
54. M&M—Market and Market (2021) Glass packaging market by applications (Alcoholic beverage packaging, non-alcoholic beverage packaging, food packaging, pharmaceutical and personal care packaging) and by Geography (North America, Europe, Asia-Pacific and Rest of the World)—Global Trends & Forecast to 2019. https://www.marketsandmarkets.com/Market-Reports/glass-packaging-market-149119613.html. Accessed 28 Mar 2021
55. Madival S, Auras R, Singh SP, Narayan R (2009) Assessment of the environmental profile of PLA, PET and PS clamshell containers using LCA methodology. J Clean Prod 17:1183–1194. https://doi.org/10.1016/j.jclepro.2009.03.015
56. Maga D, Hiebel M, Aryan V (2019) A comparative life cycle assessment of meat trays made of various packaging materials. Sustainability 11:5324. https://doi.org/10.3390/su11195324

57. Manfredi M, Vignali G (2014) Life cycle assessment of a packaged tomato puree: a comparison of environmental impacts produced by different life cycle phases. J Clean Prod 73:275–284. https://doi.org/10.1016/j.jclepro.2013.10.010
58. Marathe KV, Chavan K, Nakhate P (2017) Lifecycle assessment (LCA) of polyethylene terephthalate (PET) Bottles—Indian perspective. http://www.in-beverage.org/lca-pet/ICT%20Final%20Report%20on%20LCA%20of%20PET%20Bottles_for%20PACE_01_01_2018.pdf. Accessed 2 Mar 2021
59. Matthews HS, Hendrickson CT, Weber CL (2008) The importance of carbon footprint estimation boundaries. Environ Sci Technol 42:5839–5842. https://doi.org/10.1021/es703112w
60. Meherishi L, Narayana SA, Ranjani KS (2019) Sustainable packaging for supply chain management in the circular economy: a review. J Clean Prod 237. https://doi.org/10.1016/j.jclepro.2019.07.057
61. Navarro A, Puig R, Fullana-i-Palmer P (2017) Product vs corporate carbon footprint: some methodological issues. A case study and review on the wine sector. Sci Total Environ 581–582:722–733. https://doi.org/10.1016/j.scitotenv.2016.12.190
62. Nisticò R (2020) Polyethylene terephthalate (PET) in the packaging industry. Polym Test 90. https://doi.org/10.1016/j.polymertesting.2020.106707
63. Paiano A, Crovella T, Lagioia G (2020) Managing sustainable practices in cruise tourism: the assessment of carbon footprint and waste of water and beverage packaging. Tour Manage 77. https://doi.org/10.1016/j.tourman.2019.104016
64. Pasqualino J, Meneses M, Castells F (2011) The carbon footprint and energy consumption of beverage packaging selection and disposal. J Food Eng 103:357–365. https://doi.org/10.1016/j.jfoodeng.2010.11.005
65. PlasticsEurope (2019) Plastics—The facts 2019. https://www.plasticseurope.org/it/resources/publications/1804-plastics-facts-2019. Accessed 16 Mar 2021
66. PlasticsEurope (2020) Plastics—The facts 2020 An analysis of European plastics production, demand and waste data. https://www.plasticseurope.org/download_file/force/4261/181. Accessed 10 Mar 2021
67. Roibás L, Rodríguez-García S, Valdramidis VP, Hospido A (2018) The relevance of supply chain characteristics in GHG emissions: the carbon footprint of Maltese juices. Food Res Int 107:747–754. https://doi.org/10.1016/j.foodres.2018.02.067
68. Sadeghi B, Marfavi Y, AliAkbari R, Kowsar E, Ajdari FB, Ramakrishna S (2021) Recent studies on recycled PET fibers: production and applications: a review. Mater Circ Econ 3:4. https://doi.org/10.1007/s42824-020-00014-y
69. San Benedetto (2018) Linea Acqua Minerale San Benedetto Ecogreen Italia, 2018 Summary Report. https://www.sanbenedetto.it/wpcontent/uploads/2019/10/Summary_report_2018_ECOGREEN.pdf. Accessed 15 Mar 2021
70. Sazdovski I, Bala A, Fullana-i-Palmer P (2021) Linking LCA literature with circular economy value creation: a review on beverage packaging. Sci Total Environ 771. https://doi.org/10.1016/j.scitotenv.2021.145322
71. Schmitz A, Kamiński J, Scalet BM, Soria A (2011) Energy consumption and CO_2 emissions of the European glass industry. Energy Policy 39:142–155. https://doi.org/10.1016/j.enpol.2010.09.022
72. Siracusa V, Ingrao C, Lo Giudice A, Mbohwa C, Dalla Rosa M (2014) Environmental assessment of a multilayer polymer bag for food packaging and preservation: an LCA approach. Food Res Int 62:151–161. https://doi.org/10.1016/j.foodres.2014.02.010
73. Smithers (2021) The Future of PET Packaging to 2025. https://www.smithers.com/resources/2020/sept/global-pet-packaging-demand-to-reach-$44-1-billion. Accessed 27 Mar 2021
74. Stefanini R, Borghesi G, Ronzano A, Vignali G (2021) Plastic or glass: a new environmental assessment with a marine litter indicator for the comparison of pasteurized milk bottles. Int J Life Cycle Assess 26:767–784. https://doi.org/10.1007/s11367-020-01804-x
75. Statista (2019) Per capita consumption of bottled water in Europe in 2019, by country. https://www.statista.com/statistics/455422/bottled-water-consumption-in-europe-per-capita/. Accessed 5 Mar 2021

76. Statista (2020) Per capita consumption of packaged water in the European Union (EU) from 2010 to 2019. https://www.statista.com/statistics/620191/packaged-water-consumption-in-the-european-union-per-capita/. Accessed 2 Mar 2021
77. Statista (2021) Production of polyethylene terephthalate bottles worldwide from 2004 to 2021. https://www.statista.com/statistics/723191/production-of-polyethylene-terephthalate-bottles-worldwide/. Accessed 2 Mar 2021
78. Tamburini E, Costa S, Summa D, Battistella L, Fano EA, Castaldelli G (2021) Plastic (PET) vs bioplastic (PLA) or refillable aluminum bottles—what is the most sustainable choice for drinking water? A life-cycle (LCA) analysis. Environ Res 196. https://doi.org/10.1016/j.envres.2021.110974
79. Thøgersen J, Nielsen KS (2016) A better carbon footprint label. J Clean Prod 125:86–94. https://doi.org/10.1016/j.jclepro.2016.03.098
80. Toniolo S, Mazzi A, Niero M, Zuliani F, Scipioni A (2013) Comparative LCA to evaluate how much recycling is environmentally favourable for food packaging. Resour Conserv Recycl 77:61–68. https://doi.org/10.1016/j.resconrec.2013.06.003
81. Trajkovska Petkoska A, Daniloski D, D'Cunha NM, Naumovski N, Broach AT (2021) Edible packaging: Sustainable solutions and novel trends in food packaging. Food Res Int 140. https://doi.org/10.1016/j.foodres.2020.109981
82. Transport & Environment (2018) CO_2 Emissions from cars: the facts. https://www.transportenvironment.org/sites/te/files/publications/2018_04_CO2_emissions_cars_The_facts_report_final_0_0.pdf. Accessed 20 Apr 2020
83. Tua C, Grosso M, Rigamonti L (2020) Reusing glass bottles in Italy: a life cycle assessment evaluation. Proc CIRP 90:192–197. https://doi.org/10.1016/j.procir.2020.01.094
84. Vellini M, Savioli M (2009) Energy and environmental analysis of glass container production and recycling. Energy 34:2137–2143. https://doi.org/10.1016/j.energy.2008.09.017
85. Vinci G, D'Ascenzo F, Esposito A, Musarra M (2019) Glass beverage packaging: innovation by sustainable production. Trends in beverage packaging 16:105–133. https://doi.org/10.1016/B978-0-12-816683-3.00005-0
86. Welle F (2011) Twenty years of PET bottle to bottle recycling—an overview. Resour Conserv Recycl 55:865–875. https://doi.org/10.1016/j.resconrec.2011.04.009
87. Welle F (2018) The facts about PET. https://www.petcore-europe.org/images/news/pdf/factsheet_the_facts_about_pet_dr_frank_welle_2018.pdf. Accessed 8 Mar 2021
88. Wiedmann T, Minx J (2008) A definition of 'carbon footprint'. In: Pertsova CC (ed) Ecological economics research trends. Nova Science Publishers, Hauppauge NY, USA
89. Wohner B, Pauer E, Heinrich V, Tacker M (2019) Packaging-related food losses and waste: an overview of drivers and issues. Sustainability 11:264. https://doi.org/10.3390/su11010264
90. Wong EIC, Chan FFY, So S (2020) Consumer perceptions on product carbon footprints and carbon labels of beverage merchandise in Hong Kong. J Clean Prod 242. https://doi.org/10.1016/j.jclepro.2019.118404
91. Zeng T, Durif F (2019) The influence of consumers' perceived risks towards eco-design packaging upon the purchasing decision process: an exploratory study. Sustainability 11:10–13. https://doi.org/10.3390/su11216131
92. Zhong S, Chen R, Song F, Xu Y (2019) Knowledge mapping of carbon footprint research in a LCA perspective: a visual analysis using citespace. Processes 7:818. https://doi.org/10.3390/pr7110818

Health and Eco-Innovations in Food Packaging

Antonella Cammarelle, Francesco Bimbo, Mariarosaria Lombardi, and Rosaria Viscecchia

Abstract Packaging plays a pivotal role in preserving food quality, integrity and safety along the whole food supply chain. Its importance is also linked to the possible reduction of food loss and waste aimed at promoting more sustainable production and consumption patterns. Actually, at the end of food product use, a large amount of packaging is wasted and often it escapes formal collection and recycling systems and eventually it end-ups polluting our environment. Hence, there is the need to contribute to packaging innovations able to minimize food loss and waste by optimizing the use of the materials such as active, intelligent and sustainable packaging. In this context, there is a large room for innovation in the packaging sector in the attempt to enhance food safety and to maintain the quality of products. Also, innovative packaging may have higher chances to satisfy the social needs in increasing the sustainability of individual choices, reaching the Sustainable Development Goals indicated by the 2030 UN Agenda. In the light of these premises, the aim of this chapter is twofold. First, it is to explore whether consumers are willing to purchase food products packaged with innovative solutions as well as define the determinants of their intentions. Second, it is to investigate if the food and drink manufacturers are willing to invest in such packaging innovations. In order to reach the aforementioned objectives, 260 Italian consumers were surveyed and 20 Italian micro and small-medium entrepreneurs were interviewed. Preliminary results show that most of the consumers are interested in buying food products packed with innovative solutions, with a particular emphasis on sustainable packaging such as biodegradable and compostable ones. Finally, most of

A. Cammarelle · F. Bimbo · R. Viscecchia
Department of Agriculture, Food, Natural Resource and Engineering (DAFNE), University of Foggia (Italy), Via Napoli, 25, 71122 Foggia, Italy
e-mail: antonella.cammarelle@unifg.it

F. Bimbo
e-mail: francesco.bimbo@unifg.it

R. Viscecchia
e-mail: rosaria.viscecchia@unifg.it

M. Lombardi (✉)
Department of Economics, University of Foggia (Italy), Via R. Caggese, 1, 71121 Foggia, Italy
e-mail: mariarosaria.lombardi@unifg.it

S. S. Muthu (eds.), *Sustainable Packaging*, Environmental Footprints and Eco-design of Products and Processes, https://doi.org/10.1007/978-981-16-4609-6_2

the interviewed manufacturers are willing to invest at least one packaging innovation, mainly preferring between the active packaging and the compostable one.

Keywords Active packaging · Intelligent packaging · Sustainable packaging · Food loss and waste · Manufacturer's willingness to invest · Consumers' willingness to purchase

1 Introduction

1.1 The Role of Packaging to Reduce Food Loss and Waste

Food packaging is an essential component of the Food Supply Chain (FSC) aiming to preserve quality and ensure a longer shelf-life of foods [16, 63, 79]. The main functions of primary packaging are to contain food and to protect it from the external environment, in order to preserve its nutritional properties for a longer period of time and of course to facilitate distribution, sale and consumption [18, 132, 137]. Providing complete food containment and protection could prevent dangerous leakage and mechanical damage during transport and storage. Then, appropriate containment and protection must be assured through the numerous handling stages that occur from the packaging line to the final consumer use, in order to avoid food losses and waste (FLW) [40]. Reference [51] defines FLW as the decrease in quantity or quality of foods along the whole FSC [51]. In detail, food loss (FL) takes place at production, postharvest and processing stages in the FSC. Food waste (FW), on the other hand, occurs at retail and final consumption [51, 125]. According to Ref. [52], every year almost one-third of the whole foods produced for human consumption is lost or wasted [52]. The associated global economic cost is estimated to be equal to USD 750 billion [50]. However, the social impact is far greater considering that worldwide some 850 million people live with chronic hunger [53]. Moreover, FLW has also an important environmental impact [60]. It is estimated that it is responsible for the generation of approximately 8% of the total greenhouse gas (GHG) emissions [49]. Then, reducing FLW is a global priority in order to avoid financial losses, enhance food security and reduce environmental risks. Furthermore, it provides a critical contribution to achieve the world's sustainable development goals (SGDs), specifically the target 12.3 calls for "helving per capita global FLW by 2030" [50, 51]. Reducing FLW also has the potential to contribute to reaching the SDG 2 (End Hunger) and the SDG 13 (Climate Change) [51, 141]. Among the various causes of the generation of FLW, deficient packaging material as well as the total absence of packaging is considered some of the main ones, as shown in Table 1 [50, 114]. Indeed, as already mentioned, food packaging provides protection against physical damage to preserve food from aesthetic defects that may be cause of rejection from retailers and consumers [40, 51]. Moreover, it provides also chemical and biological protection [105]. Chemical protection minimizes compositional changes caused by external environmental influences such as exposure to gases (e.g., oxygen), change

Table 1 Type of packaging-related food loss and waste from processing and packaging stage to the final consumption

Food supply chain stage	Type of packaging-related food loss and waste	References
Processing and packaging	– Problem in the filling process; – Packaging failures while sealing; – Packaging damage; – Packaging problems leading to food spoilage; – Irregular sized packaging; – Packaging changes due to marketing reasons	– [138] – [114] – [108] – [51] – [168] – [171]
Distribution and retail	– Packaging does not provide enough mechanical protection (inappropriate packaging material, poor stackability, no packaging at all); – Damage to barcodes on packaging; – Aesthetic issues or packaging defects;	– [114] – [51] – [171]
Final consumption	– Difficult to open packaging;¼ – Difficult to empty packaging; – Inappropriate packaging size (e.g., oversized portions); – Deficient packaging methods and materials that impact the longevity of foods	– [31] – [114] – [142] – [169] – [93] – [109] – [67] – [51] – [171]

Source our elaboration from [171]

in temperature, relative humidity or exposition to light [63, 105, 128]. For example, color change commonly happens when packaged foods are exposed to retail lighting. This could be considered by consumers as aging symptom and consequently food could be discarded in favor of an apparently newer alternative [40]. Instead, biological protection provides a barrier to microorganisms, insects, rodents and other animals that could determine serious product infestation with potential risks for human health [40, 105]. Finally, the shelf-life of foods is also influenced by intrinsic factors of food such as water activity, pH value, available oxygen, nutrients and preservative [63, 128]. Then, the shelf-life of packed foods is influenced by the interactions between extrinsic and intrinsic factors [63].

However, the main functions of packaging are not only to contain and to protect food, improving its shelf-life, but also to communicate information and to provide convenience features (Fig. 1). Indeed, packaging must provide details required by law such as ingredients list, nutritional information, weight/volume, manufacturer's or seller's, expiration date (best before/use by dates) and any relevant warning statements [40, 171]. Additionally, information on packaging could help consumers to minimize FW giving instructions on how to store, open or cook foods as well as encouraging consumers to freeze leftovers [171]. Finally, packaging plays a vital role in enabling convenient food, contributing to saving time and minimizing effort

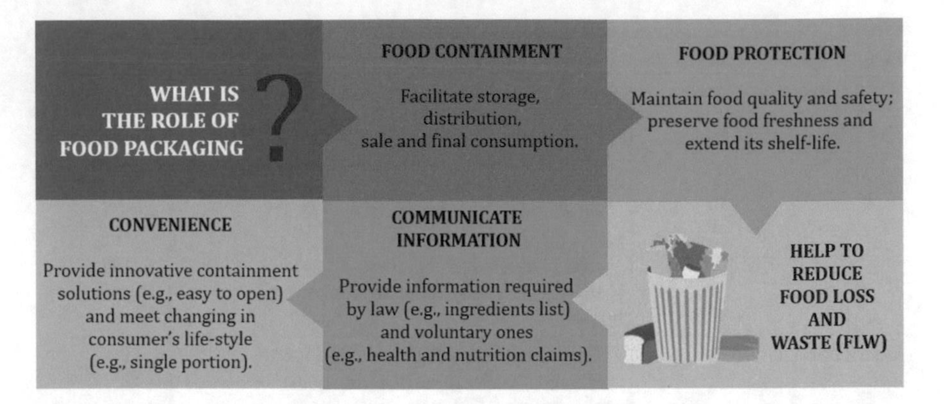

Fig. 1 Functions of food packaging. *Source* our elaboration from [171] and [56]

during food preparation [47, 105]. Change in socio-demographics characteristics, such as the emergence of single person households, as well as the change in lifestyle of consumers have led food companies to develop innovative packaging solutions, like single portion and microwavable packaging [47]. Convenience features like easy to handle, to open, to access and to reseal are important not only to meet consumer needs but also to reduce the generation of household FW. For example, consumers may spill food or drink if the packaging opening is difficult. This is particularly true for old people or for those with disabilities [171]. According to an explanatory study on Swedish households, yogurt and sour milk in liquid packaging boards contributed 75% of the "difficult to empty" waste [169].

In this context, improvement in packaging features as well as the adoption of health innovations such as active packaging, able to extend the shelf-life of foods, and the intelligent one, which provides information to the users about the remaining time to buy and consume foods, could help to prevent the generation of FLW as well as to reach the SDGs, adopted by the United Nation Member States in 2015 [47, 56, 163]. However, whenever the options for prevention are exhausted, FLW has the potential to be recovered into other production systems (e.g., bio-refineries) [111]. According to the "FW recovery hierarchy", introduced by the European Directive on waste (2018/851), FLW could not only be reused as animal feed or compost but also diverted to other industrial uses like, energy recovery via anaerobic digestion (Fig. 2) [97, 124].

Moreover, increasing efforts are currently being focused on the production of bio-products such as biodegradable and compostable polymers for eco-innovations in food packaging [64, 111, 136]. This last option could provide great benefits for the environment thanks to a reduction of methane gas emissions from landfills as well as to a preservation of non-renewable raw materials (e.g., fossil fuels) [27, 33, 43, 64]. Furthermore, the adoption of eco-innovations in food packaging can also help to reduce the so-called issue "plastic soup" [104]. Indeed, according to [53], plastic is the second most used material in the packaging market [53], the most recent data show a global production of 368 Mt in 2019, where China was the biggest

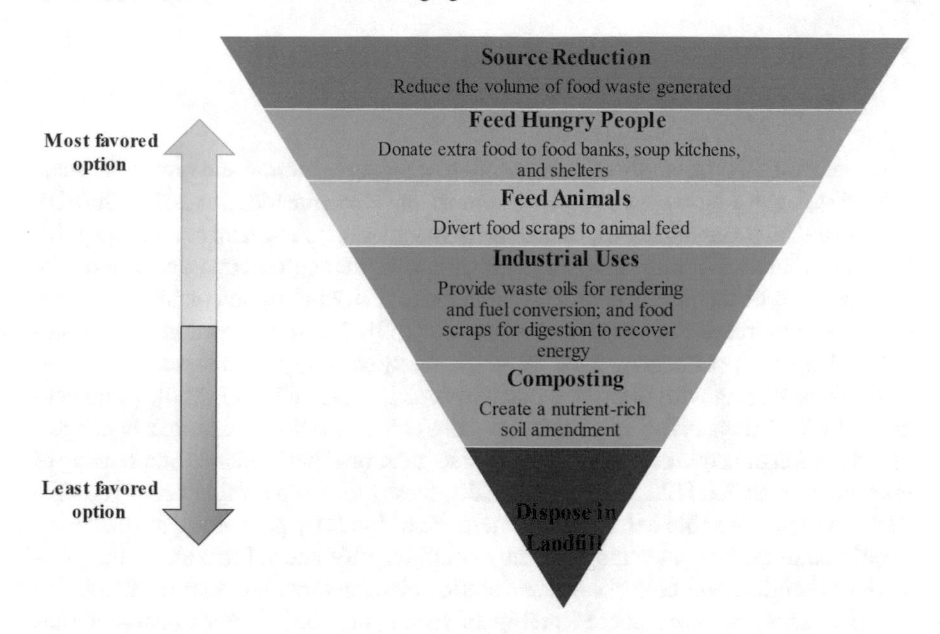

Fig. 2 Food Waste Recovery Hierarchy. *Source* our elaboration from Garcia-Garcia et al. [124]

producer (31%. Europe (EU-28 plus Norway and Switzerland reached, indeed, 57.9 Mt highlighting a reduction of 6% [129]). Generally, plastic packaging represents the most important industrial application (on average 30% of the total) in the world [38]. Specifically, polymers such as polystyrene (PS), polyethylene (PE), polypropylene (PP) and polyethylene terephthalate (PET) were widely used as packaging materials in the food and drink (F&D) industry because of their relatively low cost, easy availability and their mechanical characteristics able to provide good barriers to oxygen and carbon dioxide [26, 27, 102, 148]. However, a large amount of plastic packaging, at the end of its useful life, often escapes formal collection and recycling systems and eventually it leaks away polluting our environment [38,98]. In 2016, it is reported that this amount was equal to 41% of the global plastic packaging production [39]. Hence, there is the need to adopt eco-innovations in primary packaging able to reduce the environmental impact of packages, while maintaining food quality and safety. The chapter goes on to illustrate the principal trends in food and drink packaging with a specific focus on the role of active, intelligent and sustainable packaging. Finally, it ends with two empirical studies on the consumers' willingness to purchase foods packed by innovative solutions as well as on the manufacturers' willingness to invest in packaging innovations, respectively.

2 Recent Trends in Food and Drink Packaging: Nanotechnology

Nanotechnology is the study of manipulation of matter in atomic and molecular scale [23, 157]. According to the European Commission (Recommendation 2011/696/EU), "nanomaterial means a natural, incidental or manufactured material containing particles, in an unbound state or as an aggregate or as an agglomerate and where, for 50% or more of the particles in the number size distribution, one or more external dimensions is in the size range 1 nm–100 nm" [59]. The incorporation of nanoparticles along with packaging material gives the opportunity to generate new types of F&D packaging with improved barrier properties [19, 63, 147, 150]. Nanotechnology applications in food packaging are also able to provide two more advantages: incorporation of active compounds to provide functional performance and sensing of relevant information [122, 157]. Specifically, these types of packaging could contain substances that are able to extend the shelf-life of foods by preventing the causes of deterioration (active packaging) or they could identify and inform about the presence of chemical and biological deteriorating elements that are able to change the internal packaging atmosphere (intelligent packaging) [63]. Finally, nanomaterials are mainly used for plastics, considering that currently, the majority of packaging materials are petroleum-based [110]. However, the application of nanotechnology could open the new possibilities to improve the mechanical, thermal and gas barrier properties for bio-based packaging materials [89, 122, 151]. Figure 3 synthetizes the nano food packaging applications, stressing the main functions and features.

The main risk related to nanotechnology applications is their potential toxicity [157]. Indeed, recent studies found that nanoparticles can migrate from packaging to foodstuff [19]. For instance, Echegoyen and Nerin [36] showed that the migration of nanosilver from commercially existing food containers, which are claimed to be microwavable, was higher when time and temperature were increased. Although the observed amount of nanomaterial migration seems to be lower than the limitation of the European Union (EU) legislation [36, 58], the potential risks for the human health are still unclear [19]. The EU Regulation 450/2009/EC, which integrated the Regulation 1935/2004/EC, established that the single substances, or the combination of substances, used to make an active or intelligent component should be authorized after the European Food Safety and Authority (EFSA) has performed a risk assessment of substance migrations from food contact materials into food [26, 35, 137]. Once this safety assessment passes, the nanomaterial will be listed as an approved food contact material [19]. Moreover, the same regulation specifies that active and intelligent materials should be labeled, with the word "do not eat", to inform consumers about non-edible parts and then avoid accidental consumption [159]. Despite the concern over toxicity effects [147], new food packaging technologies based on nanomaterials are receiving increasing attention by the F&D industry. In 2014, the global nanotechnology-related F&D packaging market was equal to USD $7.3 billion, and active technology represents the largest share of the market with USD $4.35 billion sales [122]. According to a more recent market overview, the

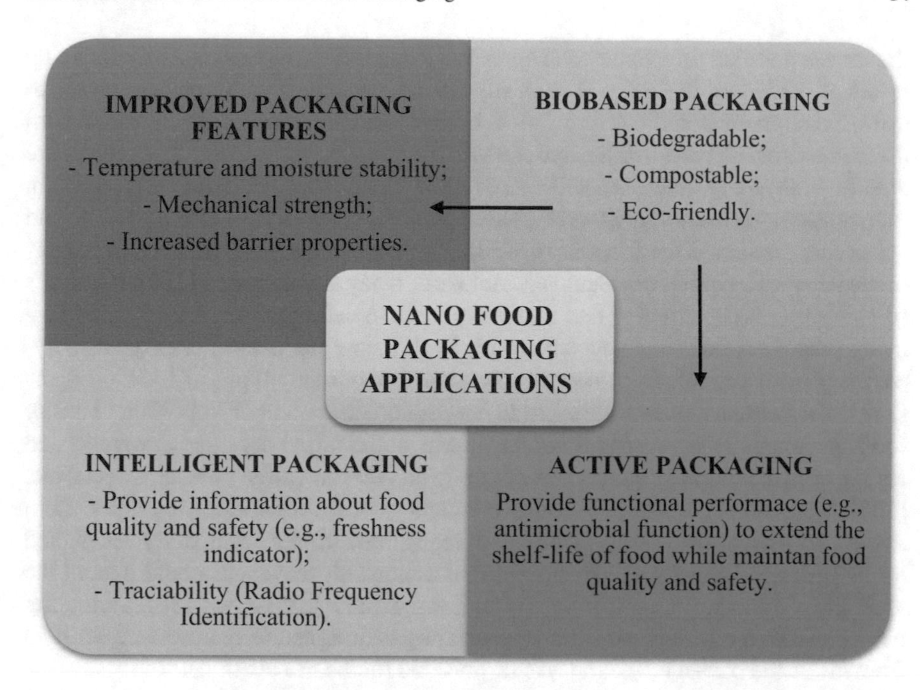

Fig. 3 Nano food packaging applications, functions and features. *Source* our elaboration from Kuswandi et al. [89]

active and intelligent packaging market was valued at USD $17.5 billion in 2020 and is expected to reach a value of USD $25.16 billion by 2026, registering a CAGR of 6.78% during the forecast period of 2021–2026 [115]. Thus, nanotechnology applications in the F&D industry will play an increasingly important role in the next future [47].

3 Health Innovations: Active and Intelligent Packaging

3.1 Active Packaging

The EU Regulation 450/2009/EC defines active materials as: "materials that are intended to extend the shelf-life of foods and to maintain or improve the condition of packaged food. They are designed to deliberately incorporate components that may release substances into the packaged food or the surrounding environment or absorb some substances from food or the environment" [13, 137]. Therefore, active packaging controls the quality and safety of the packaged product and it is able to change the environmental conditions inside the package whenever it is necessary [16]. Furthermore, this technology has the potential to reduce FW, giving consumers the largest possible time to buy and consume foods [163]. In detail, active packaging refers to the incorporation of several substances, directly into the packaging material or in separate container (e.g., sachet, label) inside package, which can absorb (scavenger) or release (emitter) gaseous matter [16, 63, 170]. Then, scavengers are able to remove any undesired substances found inside the packaging (e.g., oxygen, moisture, ethylene, carbon dioxide) guarantying stable conditions during storage and enhancing the shelf-life of foods [16, 170, 172]. For instance, ethylene is a phytohormone released by most fruits and vegetables after they are harvested [131]. It induces ripening, quickens softening and inevitably leads to deterioration of fresh and minimally processed fruits and vegetables [13, 131]. Therefore, the use of ethylene scavengers plays a key role in prolonging the shelf-life of many types of fresh products [158]. The presence of moisture inside the packaging is another important cause of food spoilage [13]. It causes microbial growth, softening of dry crispy products and caking of hygroscopic products like milk powder, instant coffee powder, sweets, etc. [164]. Moisture absorbent pads are also commonly used for controlling liquid from foods like fish, meat, poultry, fruits and vegetables. Moreover, a study conducted by PortoConte Ricerche [130] showed that a moisture scavenger could extend the shelf life of "spianate", typical Sardinian bread, from 6 to 28 days [130]. Today, the most popular and widely used active packaging is designed to remove oxygen (O_2) from inside the packet [13, 63]. O_2 is responsible for the growth of aerobic microbes, off flavor and odor development, color changes and nutritional losses [13, 18, 70]. Therefore, the incorporation of oxygen scavengers into food packages is important to decrease the oxidative effects of foods such as meats, sausages, milk powder or spices [13, 63]. O_2 can also be avoided through antioxidant agents added into packaging materials [131]. For instance, lipid oxidation is one of the main causes of deterioration of fish during its processing and storage. The use of film with antioxidant properties could help prevent and minimize the lipid oxidation in food products, while maintaining nutritional quality and extending their shelf life [5]. Antioxidants can also be used for nuts, butter, fresh meat, meat derivatives, bakery products, fruits and vegetables [131]. Moreover, anti-oxidant layers may be utilized in combination with plastic materials to reduce the oxygen content dissolved in the beverage and by also limiting oxygen ingress and increasing the shelf-life [6, 135]. At the same time, PET packaging with improved UV-light barrier properties helps to preserve the shelf life of light-sensitive products (e.g., dairy products, nuts, meat products and wine), avoiding color changes, flavors and aroma degradation [91]. Antimicrobial packaging is another example of active agents in thin polymeric films to counter the growth of pathogens and ensure the safety of consumers [16]. For instance, the addition of essential oils with antimicrobial actions into transparent plastic films can extend the shelf life of various perishable goods. Tests showed that the packaging prevented the growth of mold in bread for at least 3 weeks, expanded the sealing quality of fresh cherries by 40% and extended the shelf life of cheddar cheese by 50% [48]. Indeed, oregano and rosemary extracts showed antioxidant and antimicrobial effects and increased the display life of lamb meat without any color and flavor changes for 8–13 days compared with control samples [126]. The extension of the shelf-life and the reduction of possible microbiological risks for salads and fresh salmon filets were also, respectively, confirmed by Refs. [116]. Furthermore, active packaging can also release desired compounds in order to decrease the deterioration effects of the food inside the package [16, 135]. Indeed, carbon dioxide (CO_2) is commonly added to suppress the microbial growth in certain products such as fresh meat, poultry, fish, cheese and baked goods [13, 99] and to reduce the respiration rate of fresh produce [13, 99]. Finally, active packaging plays an important role in protecting the specific aroma of foods by removing unwanted odors and flavors such as the Cryovac Freshness Plus Odor Scavenging system that has been targeted at the processed cheese and meat market [68]. Many other studies can be found regarding the application of active packaging in the F&D industry. Table 2 summarizes the principal applications of this technology for five different mainstream products.

Table 2 Some applications of active packaging technologies

Active packaging technologies	Bakery goods	Dairy products	Fruit and vegetables	Meat and fish	Wine	References
Antimicrobial packaging	X	X	X	X		– [48] – [18] – [126] – [116]
Antioxidant packaging	X	X	X	X	X	– [13] – [18, 131] – [135] – [6, 63]
Carbon dioxide releasers	X	X	X	X		– [13] – [99] – [18] – [92]
Ethylene scavengers			X			– [13] – [18] – [131] – [158]
Improved UV-light barrier		X		X	X	– [91]
Moisture scavengers	X	X	X	X		– [13] – [63, 164] – [130]
Odors and flavor scavengers		X		X	X	– [18, 68] – [16]
Oxygen scavengers	X	X	X	X		– [13] – [131] – [18] – [63]

Source our elaboration, 2021

3.2 Intelligent Packaging

The EU Regulation 450/2009/EC defines intelligent materials as: "materials which monitor the condition of packaged food or the environment surrounding the food" [137]. Intelligent packaging is also defined as the: "system that provides the user with information on the conditions of the food and should not release its constituents into the food" [26]. In contrast with active packaging, intelligent packaging systems should, in no way, release chemicals into the packaged food [137]. Indeed, an intelligent system is attached outside of the package and it is separated from food by a functional barrier, which prevents the migration of substances into the food [26, 137]. Intelligent packaging is able to sense, monitor the conditions of packaged food and to give information about food quality and safety to manufacturer, retailer and consumer [63, 73, 79]. Therefore, this technology could help in minimizing waste, reducing the risk of throwing away foods that are still edible [62, 163]. For instance, factors that contribute to FW include the confusion over "use-by" and "best-before" dates. Unawareness of the correct meaning of these terms results in edible products being thrown away at retailers, food services and at consumers' levels [163]. Thus, food quality could be attained using freshness indicators able to sense microbial growth or chemical alteration inside the food product [16]. The underlying concept is that microbial growth is the cause of irreversible changes such as a variation in pH, which determines food spoilage [79]. Freshness indicators working based on colorimetric approaches detect an increase in pH by reacting with the volatile amines formed [16]. Another evidence of food spoilage is the production of CO_2 in meat products, fresh fruits and fresh cut vegetables [16, 79]. With an increase in CO_2 concentration, freshness indicators used in form of labels show a rose color change from yellow–green to orange [24]. The ripeSense® indicator is another intelligent tool, which can indicate the quality and freshness of foods [79]. It is the world's first intelligent label that changes color to indicate the ripeness rate of fruit. It works by reacting to the aromas released by the fruit as it ripens. The freshness indicator initially is red, graduates to orange, and finally yellow (www.ripesense.co.nz). Moreover, considering that time to storage and fluctuation in temperature are common reasons of food spoilage, several studies pointed out that time–temperature indicators (TTIs) significantly help in monitoring changes in food attributes [16, 90, 79]. TTI can give a solution for maintaining the cold chain and can be an important action point to reduce FW [112, 156]. Essentially, TTIs are labels that undergo an irreversible change in color when food or drinks are exposed to a temperature exceeding the critical one [76, 90]. For instance, the Keep-it® indicator constantly monitors temperature over time and shows the actual remaining shelf-life of a product. When the dark stripe on the indicator is equal or is less than zero, the food is no longer edible (http://keep-it.com). It is an economical device that can help to control the quality of perishable foods such as dairy products, seafood, frozen and chilled meat [79]. Other examples of TTIs available on the market are 3 M MonitorMarkTM, Check-Point®, and OnVu® [79, 156]. Moreover, temperature indicator (TI) is able to inform consumers when

Table 3 Some applications of intelligent packaging technologies

Intelligent packaging technologies	Bakery goods	Dairy products	Fruit and vegetables	Meat and fish	Wine	References
Freshness indicator		X	X	X		– [16] – [79] – [24]
Gas sensor		X	X	X	X	– [88] – [79]
Leak indicator				X		– [16] – [79]
Time–temperature indicator (TTI)	X	X	X	X		– [13] – [16] – [79] – [24] – [90]
Temperature indicator (TI)			X	X	X	– [79] – [83]

Source our elaboration, 2021

the product has the right temperature to be drunk or eaten. For instance, the thermographic label shows when the Matua wine is ready to be drunk (www.matua.co.nz). Other examples of TI are the Coca-Cola Turkey and Coors Light aluminum cans, where the thermo-chromic ink design will be visible only when the cans are chilled [83]. The BlindSpotz™ cold chain sensor is another example of TI for meat products and fresh lettuce (www.ctiinks.com.). The leakage indicator, instead, monitors the integrity of the package and informs on whether the package has been delivered without being damaged through the packaging, storage, distribution, and retailing stages of the FSC [16, 79]. Indeed, defects in packaging can affect the quality of foods and can also result in variations in optimal concentration of gases like O_2 and CO_2 that could be attained using gas sensors like Ageless Eye® [16]. Furthermore, gas sensors were also developed for the detection of beverage contaminants, such as allergens, and adulterants [88]. Finally, several studies can be found regarding the application of intelligent packaging in the F&D Industry. Table 3 summarizes the principal applications of this technology for five different mainstream products.

4 Eco-innovations: Compostable Packaging

According to the EN 13,432 standard, "Packaging—Requirements for packaging recoverable through composting and biodegradation—Test scheme and evaluation criteria for the final acceptance of packaging", the compostability is the characteristic of a material to turn into compost within 3 months by the industrial composting process [42, 160]. Compostable packaging could be made from petrochemical materials [e.g., Poly (butylene adipate-co-terephthalate) (PBAT), Polycaprolactone (PCL)], partly bio-based (e.g., starch blends) or by completely bio-based materials [e.g., Polylactic acid (PLA), Polyhydroxyalkanoates (PHA)] [43, 160]. Bio-based products are defined in the European standard EN 16,575 as "derived from biomass". Despite the biological origin, bio-based products are not always intrinsically "renewable". According to Van den Oever [160], bio-based feedstock can be called renewable as long as new crop cultivation balances harvesting [160]. The renewable feedstocks could be derived from plants that are rich in carbohydrates (e.g., corn, sugar cane), plants that are not eligible for food or feed production, and from organic waste feedstocks [44]. Biopolymers' production is an option with a growing interest in the packaging sector [64]. In 2020, global production capacities of bioplastics amounted to about 2.11 Mt with almost 47% (0.99 Mt) of the volume destined for the packaging market [45]. Indeed, for almost every conventional plastic material and application, there is a bioplastic alternative available on the market that has the same properties and potentially offers additional advantages [44]. In this context, a broad spectrum of these products is available for the F&D industry. Due to this, companies are starting to introduce eco-innovation into the market. For example, Sarchio® pasta launched the new packaging in PLA (www.sarchio.com) as well as Nonno Nanni® made the stracchino's package completely compostable. This last is made from the use of agro-industrial residues (www.nonnonanni.it). Moreover, Almaverde Bio® frozen food packages are Ok Compost certified in accordance with the EN 13,432 (www.fruttagel.it) as well as the new packaging for meat launched by Fileni® (www.fileni.it). Finally, a British designer created the "Greenbottle" that is a paper wine bottle completely compostable [81]. Furthermore, compostable packaging could also be obtained from the use of organic waste coming from the production and/or transformation process of the F&D industry. In this regard, there is a particular emphasis on recovering, recycling and exploiting industrial waste. F&D manufacturers could reduce the high costs of the treatment and the final disposal of solid and liquid waste [28]. For instance, it well known that the dairy industries generate large volumes of liquid waste (milk or cheese whey) as a by-product during coagulation of the casein process [173]. Its disposal poses the most critical pollution problem in the dairy industry, considering that it represents 85–95% of the original milk volume and contains about 55% of the whole milk nutrients [113]. Furthermore, a vast amount of evidence points out the possibility to use the cheese whey for the production of biopolymers such as PHB (Polyhydroxybutyrate) [35, 46, 165] . The PHB material could be used to produce economic and competitive packages for dairy products uses [www.wheypack.eu]. Moreover, WHEYLAYER-based films were validated for storing various foodstuffs such as, sausage, cheese and fresh pasta [46]. Furthermore, considering the large amounts of solid organic waste generated by the wine sector (e.g., skins of grapes), this beverage industry seems to be most sensitive to the ability to reuse process waste to reduce their disposal costs [75]. In this context, Naturally Clicquot is the first-ever Champagne secondary packaging produced from the skins of the grapes (www.comunicareilvino.it). However, strides are yet to be made in the research and development of primary packaging for wine made from the solid wastes of the winery industry. Other evidence pointed out that compostable packaging can also be made from peels of tomatoes (www.matrec.com), green peas, red lentils [146], peals and pulses discarded during the production phase, and unsold artichokes (www.wearepackagingfans.com). Finally, chitosan is another byproduct of the food industries that, due to its low toxicity, biodegradability, stability and relatively low cost, is suitable for the production of compostable packaging. Chitosan films have good mechanical and antimicrobial properties and moderate permeability to gasses (CO_2 and O_2) [27, 87]. Furthermore, they have antioxidant and antibacterial properties able to prevent food spoilage and extend food shelf life [86, 95]. Table 4 highlights the references found as examples of compostable packaging' applications for five different mainstream products.

5 Food Packaging Innovation: Evidence from Consumers

The commercial success of food products packaged in innovative packaging solutions is determined by the consumers' acceptance, their willingness to purchase and to pay a premium price for such technologies [175]. Therefore, the aim of this chapter was to collect evidence from existing studies focused on the consumers with respect to food packaging innovations able to extend the shelf-life of foods and improve foods' safety such as active and intelligent packaging, as well as able to reduce the environmental impact of the packages (e.g., sustainable packaging). An empirical case study aimed to identify the determinants of the respondent's willingness to purchase active and intelligent packaging will end this section.

5.1 Consumer's Acceptance

Active and intelligent packaging—The aggregate analysis of the literature pointed out that the consumer's acceptance of food products packaged with active and intelligent packaging is influenced by the consumer's knowledge of these technologies [9, 123, 147, 153]. In this regard, despite such packaging innovations are receiving a growing attention by the F&D industry, many research studies highlight a general lack of consumers' knowledge about these technologies [9, 65, 153, 167]. Specifically, consumers seem to be more aware about intelligent packaging (17%) rather than the active one (4%) [9]. This is likely due to the fact that intelligent packaging, like temperature indicators, usually contains color-based indicator, to display the quality of food products, which makes this technology easier to be recognized by consumers [9]. According to Ref. [161], women and consumers with a lower level of education were less familiar with active and intelligent packaging, while no difference across age groups was found [161]. However, after receiving general information about these packaging solutions, respondents showed a generally positive attitude toward such innovations [101]. Providing specific examples helps consumers to see

Table 4 Some applications of compostable packaging

Compostable packaging	Bakery Goods	Dairy products	Fruit and vegetables	Meat and fish	Wine
Plant-based	www.sarchio.com	www.nonnonanni.it	www.fruttagel.it	www.fileni.it	[81]
FLW-based	www.wearepackagingfans.com	[46], [35]	[86]	[75], [95]	

Source our elaboration, 2021

the functional benefits of these technologies [121, 127] and this was especially true among those who showed a greater knowledge about food risks [161].

Sustainable packaging—Results pointed out that consumers are generally aware about the negative impact on the environment of the food packaging [7, 155, 162], however not all consumers are effectively able to identify sustainable packaging during the shopping trip [7]. Packaging features able to recognize eco-friendly packaging were the eco-labels and logos, the packaging material used as well as the packaging design such as the color (brown or green) [96, 103, 145]. However, in a study conducted in South Africa, 12% of the respondents declared to not know the difference between sustainable and conventional packaging [145]. This result was also supported by Ref. [74] who showed that only 30% of the Polish students have already heard about sustainable packaging [74]. Finally, the high level of uncertainty was about the correct meaning of bioplastics, considered by consumers as already biodegradable in the environment [66], as well as about the consumer's perception of the packaging sustainability. For instance, glass and paperboard were considered more sustainable whereas their environmental impact is commonly higher [69].

5.2 *Consumer's Willingness to Purchase*

Active and intelligent packaging—Perceived benefit is considered the most important driver of the consumer's willingness to purchase active and intelligent packaging [153]. Consumers perceived several benefits from the use of these technologies such as the improvement of food safety and quality by the reduction of food perishability [140]. Moreover, consumers considered the role of intelligent packaging important to provide additional information about the freshness of the food product [123, 140]. The shelf-life extension of foods, obtained from active packaging, was seen by consumers as an additional benefit increasing time to consume food [127]. The latter, in fact, allows consumers to buy more foods during the shopping trip reducing its frequencies [100, 123]. Finally, consumer's perception of naturalness, freshness and healthiness of their foods increased more with the use of these packages rather than the use of additives [65, 123, 149]. Furthermore, benefits perceived by consumers are not only functional but also emotional, like the increasing trust in the supply chain and in the safety of their foods [123, 127, 128, 152], as well as the perception of the social value provided by active and intelligent packaging, which is related to the improvement of the social well-being through the reduction of FW [9, 65]. Instead, perceived risk for human health is one of the principal barriers to using both active and intelligent packaging [123]. Indeed, consumers seem to be more willing to purchase intelligent packaging rather than the active one considering that this technology doesn't interfere with the food product [123]. This result was also supported by Ref. [159], who considered the introduction of active packaging into the market more difficult [159] since consumers are often concerned about the toxicity effects of active substances added to polymer films, as well as are afraid of accidentally ingest active sachets, or whether their content gets disintegrated [1]. However, consumers were also concerned about the performance uncertainty of these technologies which means that active and intelligent packaging often do not provide the expected benefits [134]. Finally, the economic risk such as the additional cost to buy food items with active and intelligent packaging is another important barrier for their successful implementation into the market [1, 65, 152].

Sustainable packaging—According to [84], perceived benefits and risks were also found to be significant predictors of the consumer's willingness to purchase sustainable packaging. The protection of the environment, the reduction of the packaging waste and the improvement of human well-being were the main altruistic benefits perceived by consumers [103, 145]. Moreover, private benefits included convenience features (less packaging volume, easy to dispose), personal health benefits and the decrease in price thanks to the reduction of the packaging material used [103]. Perceived risks were instead linked to the increase in price and the consumer's perception of health risks due to a decline in quality, hygiene and product protection [103]. Then, a study conducted by Ref. [94] showed that 20% of the respondents usually avoid bu ying products without sustainable packaging. Also, according to Ref. [3], the most important aspect that affected the consumers' purchase decisions was the disposal aspect of the packaging and the biodegradability was the attribute most preferred by consumers. This result was also supported by Ref. [85] who found that North American students mostly preferred bio-based packaging rather than conventional one. In contrast with these findings, according to [11, 10] packaging attributes more considered by Poland consumers during the purchase decision were the expiry date and the origin of the product. At the same time, van Birgelen et al. [14] found that price and taste were more important than sustainable packaging among German consumers.

5.3 Consumer's Willingness to Pay

Active and intelligent packaging—Contrasting findings were found about the consumer's willingness to pay (WTP) for active and intelligent packaging. For instance, Ref. [127] highlighted that customers were willing to purchase such technologies when the price of foods was approximately the same [127]. Moreover, Ref. [177] found that respondents were willing to purchase a bottle of canola oil, able to keep the content fresher for longer and to alert consumers if the product starts to deteriorate, only when a price discount was provided to them [177]. However, Ref. [144] identified a group of consumers WTP a premium price for such packaging thanks to the increased perceived quality and freshness of the foods [144]. Moreover, according to Ref. [123], the consumers' WTP a higher price increased with the level of consumers' knowledge about these technologies, as well as Ref. [41] pointed out that consumers who were more concerned about food's risks were WTP approximately twice [41]. Finally, consumer's WTP for active and intelligent packaging varies with the respondents' socio-demographic characteristics. Specifically, younger consumers and females were generally WTP more than older individuals and males. The level of education, instead, did not affect consumers' WTP [123].

Sustainable packaging—Results from the aggregate analysis of the literature showed that most of the consumers were WTP a premium price for sustainable packaging, including 86% of the respondents of a survey conducted in Sweden [96], as well as 81% and 67% of the participants in studies conducted in the USA and Germany, respectively [119, 14]. Moreover, Ref. [8] found that the consumer's awareness about the environmental issues is the most important predictor of the intention to pay a higher price for such packaging. Finally, studies analyzed do not reveal any role of socio-demographic variables on the consumer WTP, a premium price for food products packaged with sustainable packaging [119, 145].

5.4 Empirical Case Study

This chapter aims at analyzing whether consumers are willing to purchase food products packaged with innovative solutions (e.g., active, intelligent and sustainable packaging), as well as to define the determinants of their intentions. In order to reach this objective, a convenient sample of 260 consumers was interviewed by a web-based survey conducted in April 2020 in Italy. The survey was targeted to Italians over 18 years old and who were responsible for the food shopping in their household. Most of the respondents were female (69.6%) with an average age of 35.8 (SD = 11.7). The sample was highly educated since 32.3% of consumers had completed high school. Most of the participants were employed (53.1%) with a family monthly income of between EUR 1,001–3,000 (46.5%). Households usually consist of three members (M = 3.4; SD = 1.2) with an inconsistent number of children under 14 years old (M = 0.4; SD = 0.7) (Table 5).

The survey was composed of three sections: consumers' willingness to purchase packaging innovations, determinants of the consumers' willingness to purchase active and intelligent packaging and socio-demographics characteristics. Then, to measure the willingness to purchase active, intelligent and sustainable packaging, respondents were asked to indicate their intention to buy, with a 7-point Likert item scale ranging from "totally not willing" (1) to "totally willing" (7). Moreover, we split the answers in two categories: scores 0–4, indicating low willingness to purchase and 5–7, indicating high willingness to purchase. Then, results showed that consumers were more willing to purchase foods packaged by sustainable packaging (N = 250), following by intelligent packaging (N = 241) and then the active one (N = 223) (Fig. 4).

Once observed that consumers were willing to purchase food packaging innovations, the goal of this research work was to analyze the key drivers of their intentions. Specifically, the goal was to analyze if consumers were willing to purchase active and intelligent packaging to reduce their household FW. Previous studies largely used the theory of planned behavior (TPB) introduced by [2] to explain food consumption

Table 5 Socio-demographics characteristics of respondents (N = 260)

Categorical variables	Sample %
Gender	
Male	30.4
Female	69.6
Education	
Primary school	0.4
Middle school	0.8
High school	32.3
Bachelor's degree	18.5
Master's degree	23.1
Postgraduate (e.g., PhD, master)	25
Occupation	
Not employed/student/housewife	46.9
Retired	0.8
Blue-collars	1.9
White-collars	31.2
Managers	6.5
Self-employed	12.7
Family monthly income	
Up to EUR 1,000	10.4
EUR 1,001–3,000	46.5
EUR 3,001–5,000	18.8
EUR 5,001–7,000	5.8
EUR 7,001 and over	18.5
Continuous variables	*Mean/SD/Range [min.–max.]*
Age	35.8/11.7/[20 min.–81 max.]
Household size	3.4/1.2/[0 min.–8 max.]
Number of children (under 14 years old)	0.4/0.7/[0 min.–4 max.]
Number of employed in family (excluding interviewed)	1.3/0.9/[0 min.–5 max.]

Source [20]

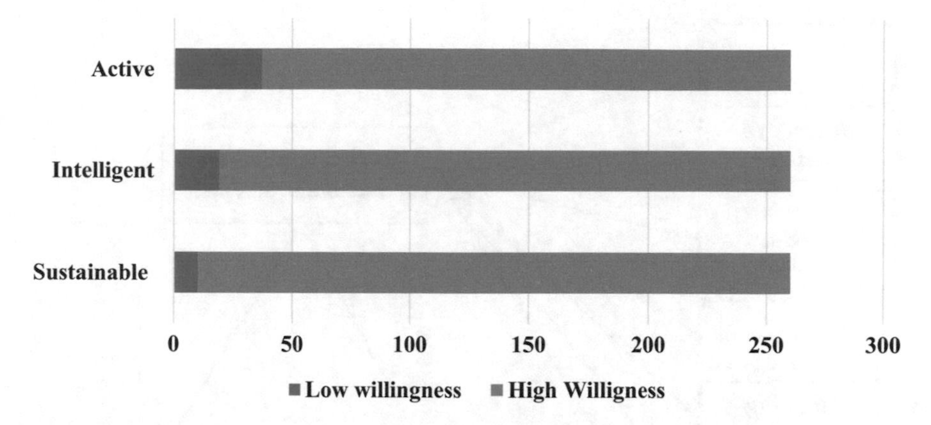

Fig. 4 Consumers' willingness to purchase foods packaged with innovative solutions. *Source* our elaboration, 2021

patterns as well as to analyze FW drivers. According to this theory, attitudes, subjective norms and perceived behavioral control are the only variables able to explain the intention to conduct a certain behavior. However, reviewed studies showed that individual awareness as well as household food management routines such as food shopping routines, planning routines and the ability to reuse food leftovers are also able to affect the individual willingness to reduce FW [133, 154]. These factors may also be related with the individual willingness to purchase active or intelligent packaging in attempting to further mitigate the household waste. Figure 5 shows the empirical link between these variables. Then, through the use of a questionnaire, respondents were asked to indicate their agreement or disagreement to some statements, listed in Table 6, generally scored on a seven-point Likert item scale ranging from "totally disagree" (1) to "totally agree" (7). Finally, the mean value was calculated for all the variables measured by using multiple items scale (Table 7).

The theoretical framework has been estimated using structural equation modeling (SEM), which allows to verify all the hypotheses done and to assess the magnitude and direction of the relations among the set of measured constructs [71, 82]. In this model, the correlation matrix between factors, reported in Table 7, was used as an input to estimate the structural coefficients between constructs and latent variables.

Results obtained by testing the conceptual models are shown in Table 8. The model converged well considering that the goodness-of-fit indicators were extremely close to the value 0.90 for the CFI and TLI and equal to 0.05 for RMSEA. Finally, the explained variance for the willingness to purchase active and intelligent packaging was 69.20% and 76.60%, respectively [20].

Results showed that for both the conceptual models, the intention to reduce household FW was a good predictor of the consumer's willingness to purchase active and intelligent packaging [20]. However, respondents were more willing to purchase intelligent packaging rather than the active one. This result can be deducted by the

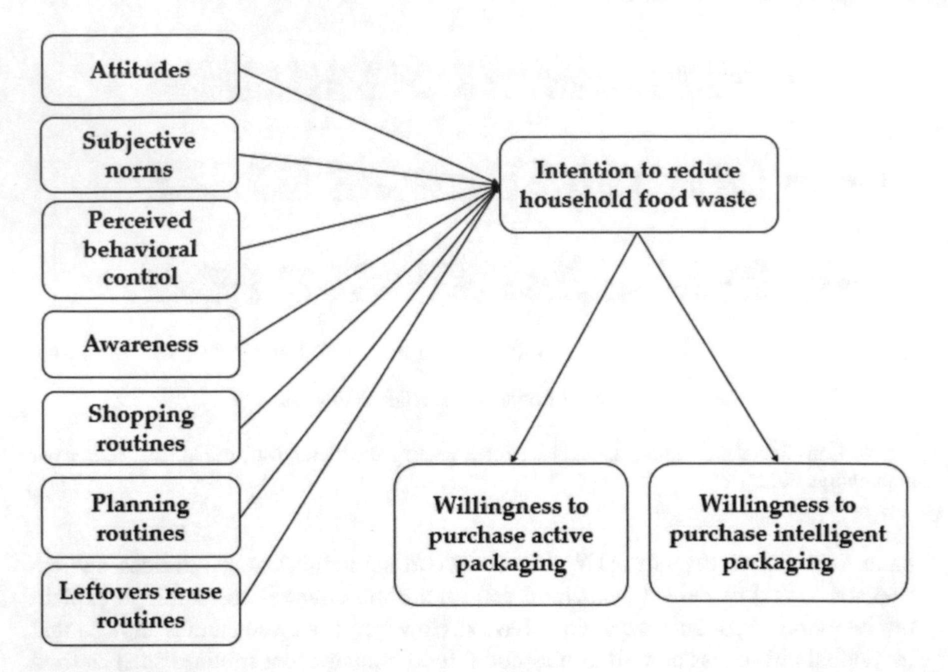

Fig. 5 Determinants of intention to reduce household food waste and willingness to purchase active and intelligent packaging. *Source* [20]

magnitude of the coefficient for the willingness to purchase active and intelligent packaging that is equal to 0.679 ($p < 0.001$) and to 0.812 ($p < 0.001$), respectively. Moreover, this result can also be confirmed by the correlation analysis where the mean value for the intention to buy intelligent packaging (6.29) is higher than that one for the intention to buy active packaging (5.81). With respect to the determinants of the intention to reduce household FW, almost all the variables considered in the conceptual models were significantly and positively related to the individual's intention [20]. For both the models, attitude toward FW was the best determinant of the intention to reduce wastes at home (0.400 and 0.384; $p < 0.001$). Perceived behavioral control was the second strongest driver with positive and significant coefficients equal to 0.295 ($p < 0.05$) and 0.327 ($p < 0.01$). The consumer's awareness about social and environmental issues was also an important predictor of the intention to reduce household FW, with magnitude of the coefficients equal to 0.218 ($p < 0.01$) and 0.239 ($p < 0.05$), respectively. Finally, with respect to the variables concerning the household food management, only planning routines made a significant contribution to the intention to lower household FW, with positive coefficients equal to 0.167 ($p < 0.001$) and 0.175 ($p < 0.05$). Then, subjective norms, shopping routines and leftovers reuse routines did not affect the intention to reduce household FW for both the conceptual models [20].

Table 6 Statements about consumers' intention to reduce household food waste and willingness to purchase active and intelligent packaging

Statements
Attitude
Throwing away food is an irresponsible behavior
Subjective norms
Most people important to me disapprove of me throwing out some food
Wasting food makes me feel guilty (e.g., for people who do not have enough food, for the environment, for the waste of money)
Perceived behavioral control
Household food waste is avoidable
Reducing household food waste help in solving waste issues
Awareness
The amount of food waste generated at home is a significant waste of money
The amount of food waste generated at home is a very important social and environmental problem
Shopping routines
I often buy unintended food products when shopping
I often buy food in packages that are too big for my household's needs
I usually buy higher amounts of food when there are special offers
Planning routines
The shopping trips are usually planned in advance (e.g., shopping lists are made)
The home meals are usually planned before going to the grocery store
Leftovers reuse routines
The leftovers are stored in appropriate conditions so they will consume later
Intention to reduce household food waste
I am not interested in reducing household food waste and I have not planned to reduce it in the next month
I am interested in reducing household food waste and I have planned to do so in the next month
I am interested in reducing household food waste and I have already started to do so in the last month
Willingness to purchase
Are you willing to purchase food products packed with active packaging?*

(continued)

Table 6 (continued)

Statements
Are you willing to purchase food products packed with intelligent packaging?*

* 7-point Likert item scale ranging from "totally not willing" (1) to "totally willing" (7)

6 Food Packaging Innovation: Evidence from Italian Manufacturers

The rising consumers' awareness about health and environmental issues has led F&D companies to increase their competitiveness by the introduction of packaging innovations into the market [27, 32]. However, in the EU-28, most of these innovations are developed by large enterprises that can rely on internal sources of knowledge [32]. Therefore, this chapter also aims at analyzing whether micro and small-medium entrepreneurs (SMEs) are willing to invest in F&D packaging innovations (active, intelligent and compostable packaging), helping in lower FLW. In detail, the study was focused on 20 micro and SMEs, located in the Apulia region. The Italian market was selected given the great importance of micro and SMEs in terms of employment and the number of companies [22]. Moreover, according to ISTAT (2010), most of these companies are located in the Apulia region (Italy) [72]. To the best of our knowledge, there are many studies focused on the technological aspect of packages [12, 55, 77], on the evaluation of the consumer's point of view, as seen in the previous section, as well as on their environmental sustainability assessment [29, 78, 176]. However, there are no studies on this topic with focus on the Italian market and specifically on the Apulia region. Therefore, this research work tries to show for the first time the F&D manufacturers' intentions to introduce packaging innovations into the market to reduce FLW.

6.1 Empirical Case Study

Qualitative research is considered a useful methodology during the early phases of the investigation of a phenomenon [37, 174]. In this case, we used a multiple case study methodology able to highlight similarities and differences between the preferences of the 20 F&D Apulian manufacturers related to the packaging innovations. Participants, selected through the icribs.com website, were involved in the production of five different mainstream products which are named in Fig. 6 with an alphabetical letter: fruit and vegetables (F), meat and fish (M), bakery foods (B), dairy products (D), wine (W).

Respondents were first contacted by phone and a questionnaire was sent to them, by email, after their approval. The survey was conducted from November 2019 to

Table 7 Descriptive statistics and correlations (N = 260)

Variables	*Mean*	*SD*	1	2	3	4	5	6	7	8	9	10
1. Intention to reduce household FW	4.6	0.87	1									
2. Attitudes	6.72	0.7	0.26**	1								
3. Subjective norms	6.27	0.95	0.18**	0.43**	1							
4. Perceived behavioral control	6.23	0.94	0.30**	0.47**	0.35**	1						
5. Awareness	6.37	0.86	0.23**	0.48**	0.34**	0.50**	1					
6. Shopping routines	3.57	1.34	−0.06	−0.00	−0.02	−0.08	−0.08	1				
7. Planning routines	5.2	1.45	0.14*	0.20**	0.14*	0.20**	0.19**	−0.4	1			
8. Leftovers reuse routines	5.63	1.65	0.15**	0.28**	0.12*	0.13*	0.14*	−0.13*	0.02	1		
9. Willing to purchase active packaging	5.81	1.38	0.11*	0.13*	0.00	0.08	0.11*	0.15*	0.09	−0.04	1	
10. Willing to purchase intelligent packaging	6.29	1.02	0.15**	0.24**	0.13*	0.34**	0.20**	0.04	0.10	0.00	0.47**	1

Note: * and ** indicate 5% and 1% significant levels, respectively.
Source our elaboration, 2021

Table 8 The structural model of intention to reduce household food waste and willingness to purchase active and intelligent packaging

Parameters	Willingness to purchase active packaging	Willingness to purchase intelligent packaging
	Coefficient	*Coefficient*
Intention to reduce household food waste	0.679***	0.812***
	Intention to reduce household food waste	Intention to reduce household food waste
Attitudes	0.400***	0.384***
Subjective norms	0.018	0.021
Perceived behavioral control	0.295**	0.327*
Awareness	0.218*	0.239**
Shopping routines	0.118	0.127
Planning routines	0.167***	0.175**
Leftovers reuse routines	0.094	0.109
Indexes of goodness-of-fit		
R^2	69.20%	76.60%
Likelihood ratio $\chi 2$ (6)	29.7 p-value < 0.001	34.7 p-value < 0.001
RMSEA	0.05	0.05
CFI	0.91	0.96
TLI	0.90	0.95

Note *, ** and *** indicate 10%, 5% and 1% significance levels, respectively; likelihood ratio test p-value equal to 0.10. *Source* [20].

April 2020 and allowed to collect general information about the companies interviewed (Table 9) as well as about their willingness to invest in packaging innovations (active, intelligent and compostable packaging).

Finally, the theoretical model proposed by [118] was used to verify whether manufacturer's stated willingness to invest for a specific packaging solution (active, intelligent and compostable packaging) could be considered as a potential, latent or a real demand [118]. According to this theory, the demand for innovation can be identified by a combination of need and technology, as shown in Table 10.

To complete the theoretical framework, the demand for innovation can be determined also by evaluating the capacity of the company to concretely translate the specific need into a packaging innovation. Then, the three typologies of demand for innovation are described in Table 11.

According to the adaptation of the taxonomy proposed by [118] to our case study, despite the company having a real demand for innovation, its capacity to translate the need into packaging innovation could be not only autonomous but also dependent. This is because micro and SMEs often cannot rely on internal sources of knowledge [4, 117]. Low investments in Research and Developments (R&D) as well as a general

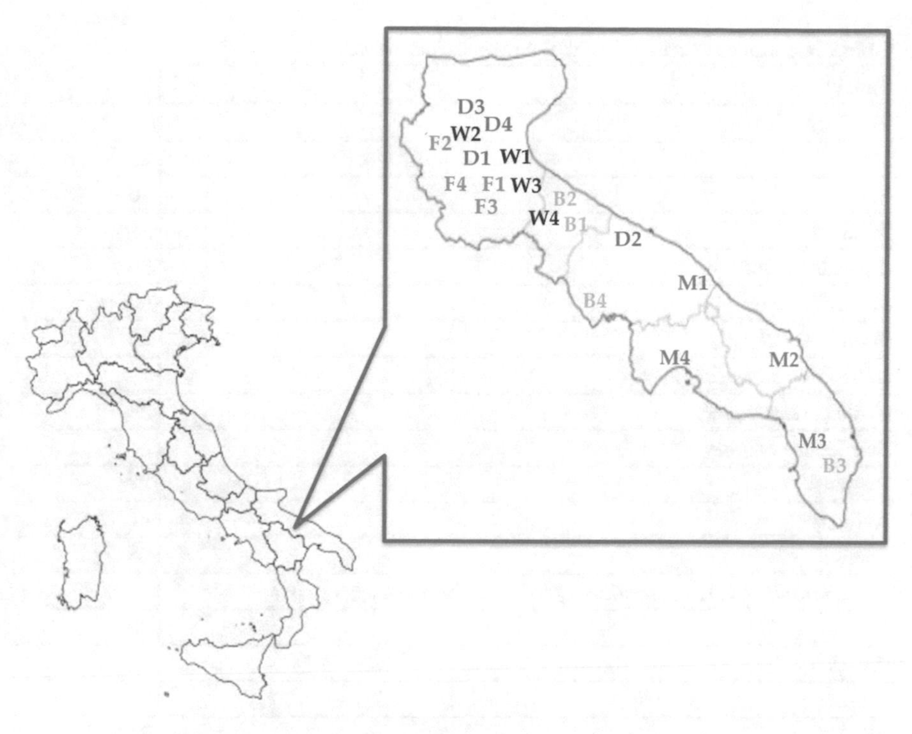

Fig. 6 Location of the 20 micro and SMEs Apulian F&D manufacturers. *Source* [21]

lack of high-skilled staff are common reasons for the necessity of an external source of knowledge. Therefore, companies must be "open" to collaborate with universities, research institutes, agencies, as well as suppliers and related industries (including chemicals and packaging sectors) to develop packaging innovations [4, 21, 117]. The analysis of the firms' awareness about their business innovation needs related to the packaging showed that 45% of the companies highlighted the need to "reduce the environmental impact of the packaging"; 20% to "extend the shelf-life of packaged foods or drinks", while the 15% showed the necessity to "provide information about food freshness and safety". The remaining 20% of the companies, instead, didn't express the need to continue improving the packaging features. With respect to the firms' knowledge about the existing technologies to address their needs, results showed that all the 20 F&D manufacturers were aware about compostable packaging; active packaging was known by 75% of the interviewed and the intelligent one by the 60% of them. Finally, after receiving general information about these packaging innovations most of the companies (75%) asserted to be willing to invest in at least one of these technological solutions, as shown in Fig. 7.

Then, 12 companies were willing to invest in active packaging. In detail, most of these manufacturers (67%) preferred the incorporation of the active components into the packaging materials. Instead, the remaining part (33%) was in favor of the incorporation of the active substances into sachets/labels to absorb or release

Table 9 General characteristics of the companies interviewed

Variable	%
Firm' size	
Micro	25
Small	40
Medium	*35*
Agricultural firms	30
Family Business	60
Multilocalized	20
External management	10
Graduate employees	22
Distribution of Sales	
Regional	*33*
National	41
Foreign	26
Investments in R&D as percentage of the sales (during lasts 3 years)	
0%	30
0.1–1%	50
1.1–2%	5
2.1–5%	5
5.1–10%	10
Product innovations (during lasts 3 years)	85
Point of weaknesses	
Administration and finance	*15*
Logistics	*25*
Personnel management	*10*
Production	*10*
Research and Development	*25*
Sales and marketing	*15*

Source our elaboration, 2021

Table 10 Firm's innovation needs and relative proposed packaging innovations

Business innovation need	Existing technology
Extend the shelf-life of packaged foods or drinks	Active packaging
Provide information about food's freshness and safety	Intelligent packaging
Reduce the environmental impact of the packaging	Compostable packaging

Source [21]

Table 11 Firm's demand for innovation

Type of demand	Description	Firm's awareness about own packaging need	Firm's awareness about the required packaging technology	Firm's capacity to translate the need into a packaging innovation
Real	Firm has a specific need and it is aware of the technological packaging solution to address its need	Aware	Aware	Autonomous/dependent
Latent	Firm has a specific need but it is not aware of the technological packaging solution to address its need	Aware	Not aware	Dependent
Potential	The firm doesn't express a specific need	Not aware	Not aware/Aware	Unconscious

Source [21]

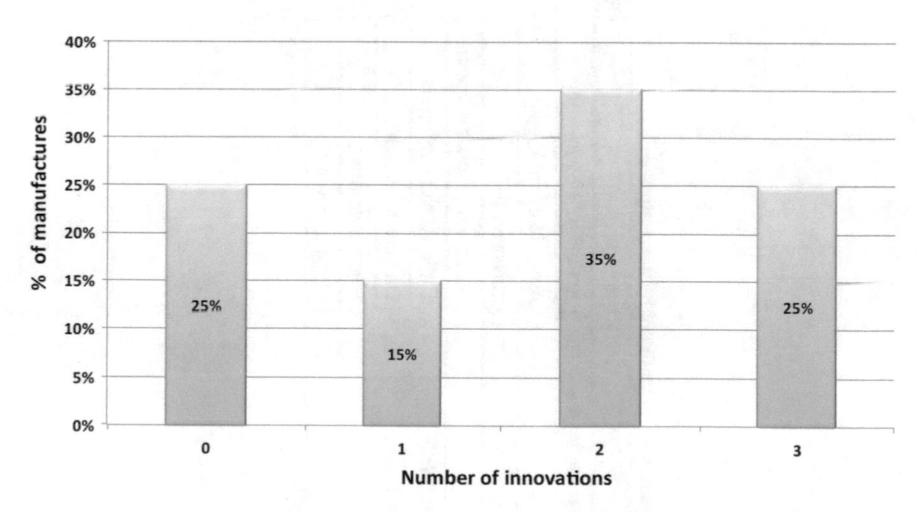

Fig. 7 Number of manufactures prompts to invest in one or more innovations. *Source* [21]

gaseous matters. Moreover, antimicrobial packaging was the most preferred technology; selected by almost 30% of the companies. Antioxidant packaging, improved UV-light barrier packages as well as ethylene scavengers and O_2 scavengers were instead the least chosen technologies, as shown in Table 12.

Table 12 Active packaging technologies preferred by the companies

Active technologies/cases		B1	B2	B3	D3	F2	F3	F4	M1	M2	M4	W1	W4
Active components into the packaging materials	Antimicrobial packaging		X	X		X				X			
	Antioxidant packaging												X
	Improved UV-light barrier											X	
	Odors and flavors scavengers				X						X		
Active substances into sachets/labels	Carbon dioxide releasers												
	Ethylene scavengers							X					
	Moisture scavengers	X					X						
	Oxygen scavengers								X				

Source our elaboration, 2021

Finally, by testing the theoretical model proposed by the authors, only two companies (Cases M1, M4) showed to have a real demand for active packaging and could be considered autonomous to translate their needs in a packaging innovation (Table 13).

Moreover, 10 companies stated to be willing to invest in intelligent packaging. The freshness indicator was the technological solution most preferred by the companies, followed by time–temperature indicators (TTIs). Instead, temperature indicator and gas sensors were the less selected technologies, as shown in Table 14.

However, only one company (Case D2) showed to have a real demand for intelligent packaging and could be considered autonomous to introduce this innovation into the market (Table 15).

Finally, 12 firms declared to be willing to invest for compostable packaging. Moreover, considering that compostable packaging could be made from a wide range

Table 13 Demand for active packaging

Cases	Business innovation need: "extend foods or drinks' shelf-life"	Awareness about active packaging	Firm's capacity to translate the need into a packaging innovation	Demand for active packaging
B1	Yes	No	Dependent	Latent
B2	No	No	Unconscious	Potential
B3	No	Yes	Unconscious	Potential
D3	No	Yes	Unconscious	Potential
F2	No	Yes	Unconscious	Potential
F3	No	No	Unconscious	Potential
F4	No	Yes	Unconscious	Potential
M1	Yes	Yes	Autonomous	Real
M2	No	Yes	Unconscious	Potential
M4	Yes	Yes	Autonomous	Real
W1	No	Yes	Unconscious	Potential
W4	No	Yes	Unconscious	Potential

Source [21]

Table 14 Intelligent packaging technologies preferred by the companies

Intelligent technologies/cases	B1	B2	B3	D2	D3	M1	M2	M3	M4	W1	W4
Freshness indicator	X	X	X		X	X					
Gas sensor											X
Leak indicator											
Time–temperature indicators (TTIs)				X			X	X	X		
Temperature indicator (TI)										X	

Source our elaboration, 2021

Table 15 Demand for intelligent packaging

Cases	Business innovation need: "provide information about food's freshness and safety"	Awareness about intelligent packaging	Firm's capacity to translate the need into a packaging innovation	Demand for intelligent packaging
B1	No	No	Unconscious	Potential
B2	No	No	Unconscious	Potential
B3	No	Yes	Unconscious	Potential
D2	Yes	Yes	Autonomous	Real
D3	No	No	Unconscious	Potential
M1	No	Yes	Unconscious	Potential
M2	Yes	No	Dependent	Latent
M4	No	Yes	Unconscious	Potential
W1	No	Yes	Unconscious	Potential
W4	No	No	Unconscious	Potential

Source [21]

of bio-based raw materials, interviewers were asked to indicate their preference. The results show that only two manufacturers (Cases B2 and M1) were interested in plant-based compostable packaging, for instance made from corn or sugar cane. The other 10 companies (Cases B1, B3, B4, D3, F1, F2, F3, F4, M2, and W1) preferred and the use of FLW for the production of compostable packaging. However, results showed that only 6 companies (Cases B2, B3, B4, D3, F1, and F3) showed to have a real demand for compostable packaging (Table 16).

7 Conclusions

This chapter provides useful information about those food packaging innovations predicted to play an increasingly important role in the upcoming years. The results from the aggregate analysis of the literature as well as the empirical case studies may have important policy and marketing implications. This is the first attempt to summarize and collect the evidence from the literature from both consumers' and manufacturers' point of view, contributing to the increase of the current knowledge about this topic. Then, the present study provides a first overview of the results about the consumer's acceptance, willingness to purchase and to pay for active, intelligent and sustainable packaging. Results suggest that consumers are willing to buy active and intelligent packaging to reduce food safety risks, improve food quality and lower household food waste. At the same time, consumers are willing to purchase sustainable packaging to protect the environment, reduce packaging waste and improve human well-being. However, consumer's acceptance is hindered by the

Table 16 Demand for compostable packaging

Cases	Business innovation need: "reduce the environmental impact of the packaging"	Awareness about compostable packaging	Firm's capacity to translate the need into a packaging innovation	Demand for compostable packaging
B1	No	Yes	Unconscious	Potential
B2	Yes	Yes	Autonomous	Real
B3	Yes	Yes	Dependent	Real
B4	Yes	Yes	Dependent	Real
D3	Yes	Yes	Dependent	Real
F1	Yes	Yes	Dependent	Real
F2	No	Yes	Unconscious	Potential
F3	Yes	Yes	Dependent	Real
F4	No	Yes	Unconscious	Potential
M1	No	Yes	Unconscious	Potential
M2	No	Yes	Unconscious	Potential
W1	No	Yes	Unconscious	Potential

Source [21]

lack of knowledge about such technologies. Moreover, despite many companies not having a clear understanding of their needs, results show that most of the interviewed manufacturers are willing to invest in at least one of the packaging innovations, mainly preferring between the active packaging and the sustainable one. However, most of these companies seem to be dependent on an external source of knowledge to develop the packaging solution. Then, policymakers and companies may develop informational campaigns to increase the level of consumers' knowledge about these technologies to encourage their adoption among consumers. Furthermore, policies such as informational and education campaigns should also try to increase the level of consumer's awareness about the negative impact of FW and packaging waste on the environment, and thus for human health, being an important driver of the intention to purchase such technological solutions. Finally, to facilitate the emergence of a real demand for packaging innovations among SMEs, there is the need to increase the collaboration between research institutes and industries as well as between all the actors of the FSC in order to implement circular economy strategies.

References

1. Aday MS, Yener U. Assessing consumers' adoption of active and intelligent packaging. Br Food J (2015) 117:157–177. http://doi.org/10.1108/BFJ-07-2013-0191
2. Ajzen I (1991) The theory of planned behavior. Organ Behav Hum Decis Process 50(2):179–211

3. Arboretti R, Bordignon P (2016) Consumer preferences in food packaging: CUB models and conjoint analysis. Br Food J 118(3):527–540. http://doi.org/10.1108/BFJ-04-2015-0146
4. Avermaete T, Viaene J, Morgan EJ, Pitts E, Crawford N, Mahon D (2004) Determinants of product and process innovation in small food manufacturing firms. Trends Food Sci Technol 15(10):474–483
5. BTSA (2020) BTSA collaborates in the development of an antioxidant packaging. Available online: https://www.btsa.com/en/btsa-colabora-desarrollo-packaging-antioxidante/ (accessed on 14 December 2020)
6. Bacigalupi C, Lemaistre MH, Boutroy N, Bunel C, Peyron S, Guillard V, Chalier P (2013) Changes in nutritional and sensory properties of orange juice packed in PET bottles: An experimental and modelling approach. Food Chem 141(4):3827–3836
7. Banterle A, Cavaliere A, Ricci EC (2012) Food labelled information: an empirical analysis of consumer preferences. Int J Food Sys Dyn 3:156–170
8. Barber N (2010) "Green" wine packaging: Targeting environmental consumers. Int J Wine Bus Res 22(4):423–444. http://doi.org/10.1108/17511061011092447
9. Barska A, & Wyrwa J (2016) Consumer perception of active intelligent food packaging. Prob Agric Econ 4(349):138–159
10. Baruk AI, Iwanicka A (2016) The effect of age, gender and level of education on the consumer's expectations towards dairy product packaging. Br Food J 118(1):100–118
11. Baruk AI, Iwanicka A (2015) Polish final purchasers' expectations towards the features of dairy product packaging in the context of buying decisions. Br Food J 117(1):178–194. http://doi.org/10.1108/DFJ-06-2014-0188
12. Bhargava N, Sharanagat VS, Mor RS, & Kumar K (2020) Active and intelligent biodegradable packaging films using food and food waste-derived bioactive compounds: A review. Trends Food Sci & Technol 105, Nov 2020, 385–401
13. Biji KB, Ravishankar CN, Mohan CO, Gopal TS (2015) Smart packaging systems for food applications: a review. J Food Sci Technol 52(10):6125–6135
14. Van Birgelen M, Semeijn J, Keicher M (2008) Packaging and proenvironmental consumption behavior. Environ Behav (2008) 41(1):125–146. http://doi.org/10.1177/0013916507311140
15. BlindSpotz^{TM} cold chain sensors. CTI's BlindSpotz^{TM} Chill Alert. Available online: https://www.ctiinks.com/blindspotz-food-cold-chain-alert (accessed on 11 September 2020)
16. Boarca B, Lungu I, & Holban AM (2019) Bioactive packaging for modern beverage industry. In Trends in Beverage Packaging, Grumezescu AN, Holban AM (Ed.), Duxford, United Kingdom: Woodhead Publishing, 51–71
17. Brody AL (2005) Commercial uses of active food packaging and modified atmosphere packaging systems. In Innovations in food packaging. Hann JH (ed.), Amsterdam, The Netherlands: Elsevier Academic Press, 457–474
18. Brody AL, Bugusu B, Han JH, Sand CK, Mchugh TH (2008) Innovative food packaging solutions. J Food Sci 73(8):107–116
19. Bumbudsanpharoke N, Ko S (2015) Nano-food packaging: an overview of market, migration research, and safety regulations. J Food Sci 80(5):R910–R923
20. Cammarelle A, Viscecchia R, Bimbo F (2021a) Intention to purchase active and intelligent packaging to reduce household food waste: Evidence from Italian Consumers. Sustain 13(8):4486
21. Cammarelle A, Lombardi M, & Viscecchia R (2021b) Packaging innovations to reduce food loss and waste: Are Italian Manufacturers Willing to Invest?. Sustain 13(4):1963
22. Caroli M, Brunetta F, & Valentino A (2019) L'industria alimentare in Italia. Sfide, traiettorie strategiche e politiche di sviluppo.Luiss Business School. Roma, Italy, pp. 1–15. Available online: http://www.federalimentare.it/documenti/IndustriaAlimentare_CuoreDelMadeInItaly/Abstract_8_5_19.pdf (accessed on 12 May 2021)
23. Chellaram C, Murugaboopathi G, John AA, Sivakumar R, Ganesan S, Krithika S, Priya G (2014) Significance of nanotechnology in food industry. APCBEE Proc 8:109–113
24. Chen HZ, Zhang M, Bhandari B, Guo Z (2018) Applicability of a colorimetric indicator label for monitoring freshness of fresh-cut green bell pepper. Postharvest Biol Technol 140:85–92

25. Comunicare il vino di Pamela Guerra (2020) Naturally Clicquot, il packaging fatto con la buccia. Available online: http://comunicareilvino.it/index.php/2015/11/naturally-clicquot-il-packaging-fatto-con-la-buccia/ (accessed on 15 December 2020)
26. Cruz RM, Alves V, Khmelinskii I, & Vieira MC (2017) New food packaging systems. In food packaging and preservation. Vol 9 1st edition, Grumezescu A, Holban AM (ed.), Elsevier Academic Press: Cambridge, MA, USA, 63–85
27. Da Rocha M, de Souza MM, Prentice C (2017) Biodegradable films: An alternative food packaging. In food packaging and preservation. Vol 9, 1st edition. Grumezescu A, Holban AM (ed.), Academic Press: Cambridge, MA, USA, 307–342
28. Di Pierro, P, Mariniello L, Giosafatto VL, Esposito M, Sabbah M, & Porta R (2017) Dairy whey protein-based edible films and coatings for food preservation. In Food Packaging and Preservation. Volume 9, 1st edition, Grumezescu A, Holban AM, Elsevier Academic Press: Cambridge, MA, USA, 439–456
29. Dilkes-Hoffman LS, Lane JL, Grant T, Pratt S, Lant PA, Laycock B. Environmental impact of biodegradable food packaging when considering food waste. J Clean Prod (2018) 180:325–334
30. Directive (EU) 2018/851 of the European Parliament and of the Council of 30 May 2018 amending Directive 2008/98/EC on waste (Text with EEA relevance). Official Journal L 312, 22.11.2008, p. 3–30. Available online: https://eur-lex.europa.eu/eli/dir/2018/851/oj (accessed on 8 March 2021)
31. Duizer LM, Robertson T, Han J (2009) Requirements for packaging from an ageing consumer's perspective. Packag Technol Sci: An Inter J 22(4):187–197
32. ECSIP Consortium (2016) The competitive position of the European Food and Drink Industry, Final Report, 2016, Publications Office of the European Union: Luxembourg, pp. 1–168. Available online: https://ec.europa.eu/growth/content/study-competitive-position-european-food-and-drink-industry-0_en (accessed on 15 September 2020)
33. EEA (2011) Waste opportunities. Past and future climate benefits from better municipal waste management in Europe, 2011. European Environment Agency: Copenhagen, Denmark, pp. 1–12. Available online: https://www.eea.europa.eu/publications/wasteopportunities-84-past-and (accessed on 5 April 2020)
34. EFSA (2009) CEF Panel. Guidelines of the panel of food contact materials, enzymes, flavourings and processing aids (CEF) on the submission of a dossier for the safety evaluation by the EFSA of active or intelligent substances present in active and intelligent materials and articles intended to come into contact with food. 2009. EFSA J 1208:1–11. https://doi.org/10.2903/j.efsa.2009.1208
35. ENEA (2018) Ambiente: dagli scarti caseari, arriva il packaging 100% biodegradabile e compostabile. Available online: http://www.enea.it/it/Stampa/news/ambiente-dagli-scarti-caseari-arriva-il-packaging-100-biodegradabile-e-compostabile (accessed on 3 February 2020).
36. Echegoyen Y, Nerín C (2013) Nanoparticle release from nano-silver antimicrobial food containers. Food Chem Toxicol 62:16–22
37. Eisenhardt KM (1989) Building theories from case study research. Acad Manag Rev 14(4):532–550
38. Ellen Macarthur Foundation. The new plastics economy: rethinking the future of plastics & catalysing action. 2017, Ellen MacArthur Foundation copyright, 1-68, Cowes, UK, Available online: https://www.ellenmacarthurfoundation.org/assets/downloads/publications/NPEC-Hybrid_English_22-11-17_Digital.pdf (accessed on 8 March 2021)
39. Ellen Macarthur Foundation (2020) 10 circular investment opportunities for a low-carbon and prosperous recovery. Ellen MacArthur Foundation copyright, Cowes, UK, Available online: https://www.ellenmacarthurfoundation.org/assets/downloads/Plastic-Packaging.pdf (accessed on 8 April 2021)
40. Emblem A (2012) Packaging functions. In Packaging technology. Emblem A. (Ed.), Woodhead Publishing, Sawston, UK, 24–49

41. Erdem S (2015) Consumers' preferences for nanotechnology in food packaging: a discrete choice experiment. J Agric Econ 66(2):259–279
42. European Bioplastics (2015) EN 13432 Certified bioplastics performance in industrial composting. European Bioplastics. Marienstraße eV, Berlin, Germany. Available online: https://docs.european-bioplastics.org/publications/bp/EUBP_BP_En_13432.pdf (accessed om 13 May 2021)
43. European Bioplastics (2016a) Accountability is key – Environmental communications guide for bioplastics. 2016a. European Bioplastics Marienstraße eV, Berlin, Germany. Available online: http://docs.european-bioplastics.org/2016/publications/EUBP_environmental_communications_guide.pdf (accessed on 6 October 2019)
44. European Bioplastics (2016b) Biobased plastics – fostering a resource efficient circular economy. 2016b. European Bioplastics Marienstraße eV, Berlin, Germany, Available online at https://docs.european-bioplastics.org/2016/publications/fs/EUBP_fs_renewable_resources.pdf (accessed on 24 October 2020)
45. European Bioplastics (2020) Nova-Institute. Bioplastics market development update 2020. 2020, European Bioplastics Marienstraße eV, Berlin, Germany. Available online: https://docs.european-bioplastics.org/conference/Report_Bioplastics_Market_Data_2020_short_version.pdf (accessed on 20 April 2021)
46. European Commission (2020) Final Report Summary – WHEYLAYER (whey protein-coated plastics films to replace expensive polymers and increase the recyclability). 2013. European Commission, Brussels, Belgium. Available online: https://cordis.europa.eu/project/id/218340/reporting (accessed on 3 February 2020)
47. European Commission (2016) The competitive position of the European food and drink industry. Final Report. 2016. Publications Office of the European Union, Luxembourg. 1-168. Available online: https://ec.europa.eu/growth/sectors/food/competitiveness/studies_en (accessed on 6 March 2021)
48. European Commission Extend (2019) ng food shelf-life with nanomaterials. 2019. European Commission, Brussels, Belgium. Available online: https://cordis.europa.eu/article/id/411736-extending-food-shelf-life-with-nanomaterials/it(accessed on 2 July 2020)
49. FAO. Food wastage footprint & Climate Change. 2015. FAO: Rome, Italy. Available online: http://www.fao.org/documents/card/en/c/7338e109-45e8-42da-92f3-ceb8d92002b0/ (accessed on 6 March 2021).
50. FAO. Transforming Food and Agriculture to Achieve the SDGs 20 Interconnected Actions to Guide Decisions-Makers, 2018, FAO: Rome, Italy, 1–79. Available online: http://www.fao.org/3/I9900EN/i9900en.pdf (accessed on 13 October 2020)
51. FAO. The State of Food and Agriculture 2019. Moving forward on Food Loss and Waste Reduction, 2019, FAO: Rome, Italy, 1–182. Available online: http://www.fao.org/3/ca6030en/ca6030en.pdf (accessed on 20 October 2020)
52. FAO (2018) Global food losses and food waste—extent, causes and prevention. 2011, FAO: Rome, Italy, 1–37. Available online: http://www.fao.org/3/a-i2697e.pdf (accessed on 16 June 2018)
53. FAO (2014) Building a common vision for sustainable Ffood and Agriculture. Principles and Approaches 2014 FAO: Rome, Italy, 1–56. Available online: http://www.fao.org/3/a-i3940e.pdf (accessed on 10 November 2020)
54. Fileni (2020) Fileni: un nuovo pack compostabile. 12 July 2020. Available online: https://www.fileni.it/blog/fileni-bio-nuovo-pack-compostabile/ (accessed on 29 December 2020)
55. Firouz MS, Mohi-Alden K, & Omid, MA (2021) critical review on intelligent and active packaging in the food industry: Research and development. Food Res Int 141:110113. http://doi.org/10.1016/j.foodres.2021.110113
56. Food Drink Europe (2018) Food Drink Europe's sustainable packaging roadmap. FoodDrinkEurope, Published October 2018, Brussels, Belgium. Available online: https://www.fooddrinkeurope.eu/wp-content/uploads/publications_documents/FoodDrinkEurope_Sustainable_Packaging_Roadmap.pdf (accessed on 7 March 2021)

57. Fortunati E, Peltzer M, Armentano I, Torre L, Jiménez A, Kenny JM (2012) Effects of modified cellulose nanocrystals on the barrier and migration properties of PLA nano-biocomposites. Carbohyd Polym (2012) 90(2):948–956
58. Fruttagel®. Fruttagel intera la gamma di verdure surgelate Almaverde Bio. 2 Maggio 2019. Available online: https://www.fruttagel.it/fruttagel-integra-la-gamma-di-verdure-surgelate-almaverde-bio/ (accessed on 14 September 2020).
59. Gallocchio F, Belluco S, Ricci A (2015) Nanotechnology and food: brief overview of the current scenario. Procedia food science 5:85–88
60. Garcia-Garcia G, Woolley E, Rahimifard S (2015) A framework for a more efficient approach to food waste management. Int J Food Eng (2015) 1(1):65–72
61. Garcia-Garcia G, Woolley E, Rahimifard S, Colwill J, White R, Needham L (2017) A methodology for sustainable management of food waste. Waste Biomass Valorization 8:2209–2227. https://doi.org/10.1007/s12649-016-9720-0
62. Ghaani M, Cozzolino CA, Castelli G, Farris S (2016) An overview of the intelligent packaging technologies in the food sector. Trends Food Sci Technol 51:1–11
63. Ghoshal G (2017) Recent trends in active, smart, and intelligent packaging for food products. In Food packaging and preservation. Vol. 9, 1st edition, Grumezescu A, Holban AM (Ed.), Elsevier Academic Press: Cambridge, MA, USA, 343–374
64. Girotto F, Alibardi L, Cossu R (2015) Food waste generation and industrial uses: a review. Waste Manage 45:32–41
65. Greehy G, McCarthy M, Henchion M, Dillon E, & McCarthy S (2011) An exploration of Irish consumer acceptance of nanotechnology applications in food. Proc Syst Dyn Innovation Food Networks 2011, Rickert U Gerhard Schiefer G (Ed.), 175–198
66. Guillard V, Gaucel S, Fornaciari C, Angellier-Coussy H, Buche P, Gontard N. The next generation of sustainable food packaging to preserve our environment in a circular economy context. Front Nutr (2018) 5:121
67. Gustavo JU Jr, Pereira GM, Bond AJ, Viegas CV, Borchardt M. Drivers (2018) opportunities and barriers for a retailer in the pursuit of more sustainable packaging redesign. J Clean Prod 187:18–28
68. Harrington R (2011) Sealed Air Cryovac touts benefits of odour-eating packaging. Available online: https://www.foodnavigator.com/Article/2011/04/22/Sealed-Air-Cryovac-touts-benefits-of-odour-eating-packaging (ac-cessed on 13 January 2020)
69. Herbes C, Beuthner C, Ramme I (2018) Consumer attitudes towards biobased packaging–A cross-cultural comparative study. J Clean Prod 194:203–218
70. Hogan SA, & Kerry JP (2008) Smart packaging of meat and poultry products. In Smart packaging technologies for fast moving consumer goods, 2008, John Wiley & Sons Ltd, The Atrium, Southern Gate, Chichester, West Sussex PO19 8SQ, England, 33–54
71. Iacobucci D (2010) Structural equations modeling: Fit indices, sample size, and advanced topics. J Consum Psychol 20(1):90–98
72. Intesa Sanpaolo (2016) Il settore agro-alimentare in Italia e in Puglia. 2016, Direzione Studi e Ricerche, Bari, Italy, 1–29. Available online: https://www.intesasanpaolo.com/content/dam/vetrina/landing/pdf/Il%20settore%20agroalimentare%20in%20Italia_Bari14062016.pdf (accessed on 20 August 2020)
73. Jang NY, Won K. New pressure-activated compartmented oxygen indicator for intelligent food packaging. Int J Food Sci Technol (2014) 49(2):650–654
74. Jerzyk E (2016) Design and communication of ecological content on sustainable packaging in young consumers opinions. J Food Prod Mark 22(6):707–716
75. Jin B, Kelly JM (2009) Wine industry residues. In Biotechnology for agro-industrial residues utilization. Singh nee' Nigam P, Pandey A (Eds.) Springer, Dordrecht, 293–311
76. De Jong AR, Boumans H, Slaghek T, Van Veen J, Rijk R, Van Zandvoort M (2005) Active and intelligent packaging for food: Is it the future? Food Addit Contam 22(10):975–979
77. Jõgi K, & Bhat R (2020) Valorization of food processing wastes and by-products for bioplastic production. Sustain Chem Pharm 18, December 2020, 100326

78. Kakadellis S, & Harris ZM (2020) Don't scrap the waste: the need for broader system boundaries in bioplastic food packaging life-cycle assessment–a critical review. J Cleaner Production 274, November 2020, 122831
79. Kalpana S, Priyadarshini SR, Leena MM, Moses JA, & Anandharamakrishnan C (2019) Intelligent packaging: Trends and applications in food systems. Trends in Food Science & Technology 93, November 2019, 145–157
80. Keep-it (2020) The expiration date stamp has EXPIRED. 2020. Available online: https://keep-it.com/ (accessed on 10 November 2020)
81. Kim S (2011) Biodegradable, compostable wine bottle made from paper. November 22, 2011. Available online: https://www.zdnet.com/article/biodegradable-compostable-wine-bottle-made-from-paper/ (accessed on 29 December 2020)
82. Kline R (2005) Principles and practice of structural equation modeling. Guilford Press: New York, NY, USA 1–534
83. Koe T (2018) Cool packaging: Coca—Cola Turkey launches new cans with thermochromic inks. Available online: https://www.foodnavigator-asia.com/Article/2018/07/04/Cool-packaging-Coca-Cola-Turkey-launches-new-cans-with-thermochromic-inks (accessed 5 July 2020)
84. Koenig-Lewis N, Palmer A, Dermody J, Urbye A (2014) Consumers' evaluations of ecological packaging–rational and emotional approaches. J Environ Psychol 37:94–105
85. Koutsimanis G, Getter K, Behe B, Harte J, Almenar E (2012) Influences of packaging attributes on consumer purchase decisions for fresh produce. Appetite 59:270-280. https://doi.org/10.1016/j.appet.2012.05.012
86. Kumbunleu J, Rattanapun A, Sapsrithong P, Tuampoemsab S, & Sritapunya T (2020) The study on chitosan coating on poly (Lactic Acid) film packaging to extend vegetable and fruit life. IOP Conf Ser: Mater Sci Eng 811:01202
87. Kurek M, Guinault A, Voilley A, Galić K, Debeaufort F (2014) Effect of relative humidity on carvacrol release and permeation properties of chitosan-based films and coatings. Food Chem 144:9–17
88. Kuswandi B (2017) Environmental friendly food nano-packaging. Environ Chem Lett (2017) 15(2):205–221
89. Kuswandi B (2016) Nanotechnology in food packaging. In Nanoscience in food and agriculture, Ranjan, Shivendu, Dasgupta, Nandita, Lichtfouse, Eric (Eds.) Springer, Cham, NY, USA 151–183
90. Kuswandi B, & Moradi M (2019) Sensor trends in beverages packaging. In Trends in Beverage Packaging, Grumezescu A, Holban AM (Eds), First Publisher: Elsevier Academic Press, 279–302
91. Kwon S, Orsuwan A, Bumbudsanpharoke N, Yoon C, Choi J, Ko S (2018) A short review of light barrier materials for food and beverage packaging. Korean J Packag Sci & Technol 24(3):141–148
92. Labuza TP, Breene WM (2008) Applications of active packaging for improvement of shelf-life and nutritional quality of fresh and extended shelf-life foods. J Food Proc Preserv 45:1–69
93. Lanfranchi M, Calabrò G, De Pascale A, Fazio A, & Giannetto C (2016) Household food waste and eating behavior: empirical survey. Br Food J 118(12):3059–3072
94. Lea E, Worsley A (2008) Australian consumers' food-related environmental beliefs and behaviours. Appetite 50(2-3):207–14
95. Lekjing S (2016) A chitosan-based coating with or without clove oil extends the shelf life of cooked pork sausages in refrigerated storage. Meat Sci 111:192–197
96. Lindh H, Olsson A, Williams H (2016) Consumer perceptions of food packaging: contributing to or counteracting environmentally sustainable development. Packag Technol Sci 29(1):3–23
97. Lombardi M, Costantino M (2021) A hierarchical pyramid for food waste based on a social innovation perspective. Sustain 13:4661. http://doi.org/10.3390/su13094661
98. Lombardi M, Rana R, Fellner J (2021) Material flow analysis and sustainability of the Italian plastic packaging management. J Cleaner Prod 287:125573. http://doi.org/10.1016/j.jclepro.2020.125573

99. Lopez-Rubio A, Almenar E, Hernandez-Muñoz P, Lagarón JM, Catalá R, Gavara R (2004) Overview of active polymer-based packaging technologies for food applications. Food Rev Intl 20(4):357–387
100. Loučanová E Kalamárová M, Parobek J, Dopico A. Simulation of intelligent and active packaging perceptions in slovakia. Acta Simulatio Int. Sci. J. Simulation (2016), 2:13–17
101. Macoubrie J. Nanotechnology: public concerns, reasoning and trust in government. Public Underst Sci (2006) 15(2):221–241
102. Madera-Santana TJ, Freile-Pelegrín Y, Azamar-Barrios JA (2014) Physicochemical and morphological properties of plasticized poly (vinyl alcohol)–agar biodegradable films. Int J Biol Macromol 69:176–184
103. Magnier L, Crié D (2015) Communicating packaging eco-friendliness: An exploration of consumers' perceptions of eco-designed packaging. Int J Retail & Distribs Manage 43(4/5):350–366
104. Magnier L, Mugge R, Schoormans J (2019) Turning ocean garbage into products–Consumers' evaluations of products made of recycled ocean plastic. J Clean Prod 215:84–98
105. Marsh K, Bugusu B. Food packaging—roles, materials, and environmental issues. J Food Sci (2007) 72(3):R39–R55
106. Matrec Srl (2019) Innovations ad reserach in the era of circular materials. Available online: https://www.matrec.com/en/trends-news/packaging-made-of-tomato-bio-plastic (accessed on 4 September 2020)
107. Matua® (2020) Chill check. Available online: https://www.matua.co.nz/chill-check (accessed on 10 September 2020)
108. Mena C, Adenso-Diaz B, Yurt O (2011) The causes of food waste in the supplier–retailer interface: Evidences from the UK and Spain. Resour Conserv Recycl 55:648–658
109. Meurer IR, Lange CC, Hungaro HM, Bell MJV, dos Anjos VDC, de Sá Silva CA, de Oliveira Pinto MA (2017) Quantification of whole ultra high temperature UHT milk waste as a function of packages type and design. J Clean Prod 153:483–490
110. Mihindukulasuriya SDF, Lim LT (2014) Nanotechnology development in food packaging: A review. Trends Food Sci Technol 40(2):149–167
111. Mirabella N, Castellani V, Sala S (2014) Current options for the valorization of food manufacturing waste: a review. J Clean Prod 65:28–41
112. Mohebi E, Marquez L (2015) Intelligent packaging in meat industry: An overview of existing solutions. J Food Sci Technol 52(7):3947–3964
113. Mollea, C., Marmo, L., & Bosco, F. Valorisation of cheese whey, a by-product from the dairy industry. In Food industry. 2013, Editors: Innocenzo Mazzalupo, Publisher: InTECH, Londn, UK, 549-588
114. Monier V, Mudgal S, Escalon V, O'Connor C, Gibon T, Anderson G, Montoux H, Reisinger H, Dolley P, Ogilvie S, & Morton G (2011) Preparatory study on food waste across EU 27. Technical Report - 2010 - 054. European Communities, Brussels, Belgium,1–213
115. Mordor Intelligence (2020) Active and intelligent packaging market – growth, trends, COVIS-19 impact, and forecast (2021–2026). Available online: https://www.mordorintelligence.com/industry-reports/active-and-intelligent-packaging-market-industry (aaccessed on 14 March 2020)
116. Muriel-Galet V, Cerisuelo JP, López-Carballo G, Aucejo S, Gavara R, Hernández-Muñoz P (2013) Evaluation of EVOH-coated PP films with oregano essential oil and citral to improve the shelf-life of packaged salad. Food Control 30(1):137–143
117. Muscio A, Nardone G, Stasi A (2017) How does the search for knowledge drive firms' eco-innovation? Evidence from the wine industry. Ind Innov 24(3):298–320
118. Muscio A, Nardone G, & Dottore A, Lerro D (2010) Understanding demand for innovation in the food industry. Measuring Bus Excellence 14(4):35–48
119. Neill LC, Williams R (2016) Consumer preference for alternative milk packaging. The case of an inferred environmental attribute. J Agri Appl Econ 48(3) 241–256
120. NonnoNanni® (2020) Piace anche alla natura. Available online: https://www.nonnonanni.it/compostabile/ (accessed on 14 September 2020)

121. Nosalova M, Loucanova E, & Parobek J (2018) Perception of packaging functions and the interest in intelligent and active packaging. Prob Agri Econ/Zagadnienia Ekonomiki Rolnej 4(357):141–152
122. Omanović-Mikličanin E, Maksimovic M, & Mulaomerović D (2017) Application of nanotechnology in food packaging. Front Microbiol December 12; 8:2517
123. O'Callaghan KA, Kerry JP (2016) Consumer attitudes towards the application of smart packaging technologies to cheese products. Food Packag Shelf Life 9:1–9
124. Papargyropoulou E, Lozano R, Steinberger JK, Wright N, bin Ujang, Z (2014) The food waste hierarchy as a framework for the management of food surplus and food waste. J Clean Prod 76:106–115
125. Parfitt J, Barthel M, Macnaughton S (2010) Food waste within food supply chains: Quantification and potential for change to 2050. Philos Trans R Soc B Biol Sci (2010) 365:3065–3081
126. Pateiro M, Domínguez R, Bermúdez R, Munekata PE, Zhang W, Gagaoua M, Lorenzo JM (2019) Antioxidant active packaging systems to extend the shelf life of sliced cooked ham. Curr Res Food Sci 1:24–30
127. Pennanen K, Focas C, Kumpusalo-Sanna V, Keskitalo-Vuokko K, Matullat I, Ellouze M, Pentikäinen S, Smolander M, Korhonen V, Ollila M (2015) European consumers' perceptions of time–temperature indicators in food packaging. Packag Technol Sci 28(4):303–323
128. Pereira de Abreu DA, Paseiro P, Maroto J and Cruz JM (2010) Evaluation of the effectiveness of a new active packaging film containing natural antioxidants (from barley husks) that retard lipid damage in frozen Atlantic salmon (Salmo salar L.). Food Res Int 43:1277–1282
129. PlasticEurope (2020) Plastics – the Facts 2020. An analysis of European plastics production, demand and waste data. Available online: https://www.plasticseurope.org/it/resources/publications/4312-plastics-facts-2020 (accessed on 8 March 2021)
130. PortoConte Ricerche (2020) Progetto Cluster – ACTIPACK. Relazione Divulgativa, 30 Settembre 2015. Available online: https://www.sardegnaricerche.it/documenti/13_398_20160511095347.pdf (accessed on 15 January 2020).
131. Prasad P, Kochhar A (2014) Active packaging in food industry: a review. J Environ Sci Toxicol Food Technol 8(5):1–7
132. Preda M, Popa MI, Mihai MM, Şerbănescu AA, & Holban AM (2019) Natural fibers in beverages packaging. In trends in beverage packaging. Grumezescu A, Holban AM (Eds) Elsevier Academic Press. pp. 409–424
133. Principato L, Mattia G, Di Leo A, Pratesi CA (2021) The household wasteful behaviour framework: A systematic review of consumer food waste. Ind Mark Manage (2021) 93:641–649
134. Ram S, Sheth JN (1989) Consumer resistance to innovations: the marketing problem and its solutions. J Consum Market 6:5–14. https://doi.org/10.1108/EUM0000000002542
135. Ramos M, Valdés A, Mellinas AC, Garrigós MC (2015) New trends in beverage packaging systems: A review. Beverages 1(4):248–272
136. Redlingshöfer B, Coudurier B, Georget M (2017) Quantifying food loss during primary production and processing in France. J Clean Prod (2017) 164:703–714
137. Restuccia D, Spizzirri UG, Parisi OI, Cirillo G, Curcio M, Iemma F, Puoci F, Vinci G, Picci, N (2010) New EU regulation aspects and global market of active and intelligent packaging for food industry applications. Food control (2010) 21(11):1425–1435
138. Ridgway JS, Henthorn KS, Hull JB (1999) Controlling of overfilling in food processing. J Mater Process Technol 92:360–367
139. RipeSense® (2020) The next revolution in fresh produce marketing. Available online: http://www.ripesense.co.nz/ (accessed on 15 De-cember 2020)
140. Robertson GL (2013) Food packaging: Principles and practice. 3rd edn. CRC Press, Taylor & Francis Group LLC., Boca Raton, FL
141. Roversi S, Laricchia C, Lombardi M (2020) Sustainable development goals and Agro-food system: The case study of the future food institute. In proceedings of the 3rd International Conference on economics and social sciences, innovative business models to restart the global economy. Sciendo publisher, Warsaw, Poland, 595–603

142. Rowson J, Yoxall A (2011) Hold, grasp, clutch or grab: Consumer grip choices during food container opening. Appl Ergon 42(5):627–633
143. Sarchio SpA (2021) Il nostro impegno per un mondo migliore. Available online: https://www.sarchio.com/sostenibilita (accessed on 05 April 2021)
144. Schnettler B, Crisóstomo G, Mora M, Lobos G, Miranda H, Grunert KG (2014) Acceptance of nanotechnology applications and satisfaction with food-related life in southern Chile. Food Science and Technology 34(1):157–163
145. Scott L, Vigar-Ellis D (2014) Consumer understanding, perceptions and behaviours with regard to environmentally friendly packaging in a developing nation. Int J Consum Stud (2014) 38(6):642–649
146. Selwood D (2016) Waitrose to launch pasta in packaging made in part from food waste. 2016. Available online: https://www.thegrocer.co.uk/sustainability-and-environment/waitrose-to-launch-pasta-in-packaging-made-in-part-from-food-waste/539935.article (accessed on 05 April 2020)
147. Siegrist M, Stampfli N, Kastenholz H, Keller C (2008) Perceived risks and perceived benefits of different nanotechnology foods and nanotechnology food packaging. Appetite 51(2):283–290
148. Siracusa V, Rocculi P, Romani S, Dalla Rosa M. Biodegradable polymers for food packaging: a review. Trends Food Sci Technol 19(12):634–643
149. Sodano V, Gorgitano MT, Verneau F, Vitale CD (2016) Consumer acceptance of food nanotechnology in Italy. Br Food J (2016) 118:714–733. https://doi.org/10.1108/BFJ-06-2015-0226
150. Sorrentino A, Gorrasi G, Vittoria V (2007) Potential perspectives of bio-nanocomposites for food packaging applications. Trends Food Sci Technol 18(2):84–95
151. Sozer N, Kokini JL. Nanotechnology and its applications in the food sector. Trends Biotechnol (2009) 27(2):82–89
152. Spence M, Stancu V, Elliott CT, Dean M (2018) Exploring consumer purchase intentions towards traceable minced beef and beef steak using the theory of planned behavior. Food Control (2018) 91:138–147
153. Stampfli N, Siegrist M, Kastenholz H (2010) Acceptance of nanotechnology in food and food packaging: a path model analysis. J Risk Res13(3):353–365
154. Stancu V, Haugaard P, Lähteenmäki L (2016) Determinants of consumer food waste behaviour: Two routes to food waste. Appetite 96:7–17
155. Steenis ND, van Herpen E, van der Lans IA, Ligthart TN, van Trijp HC (2017) Consumer response to packaging design: The role of packaging materials and graphics in sustainability perceptions and product evaluations. J Clean Prod 162:286–298
156. Stergiou F (2018) Effective management and control of the cold chain by application of time temperature indicators (TTIs) in food packaging. J Food Clin Nutr (2018) 1(1):12–15
157. Tarabella A, Masoni A, Trivelli L, Apicella A, Lombardi M, Rana R, Tricase C (2019) Innovation in the food industry: a comparison between new and traditional categories of foodstuffs. In Food Products Evolution: Innovation Drivers and Market Trends, Tarabella A (Ed.) Springer International Publishing, Cham, Switzerland, 23–39. http://doi.org/10.1007/978-3-319-23811-1_1
158. Terry LA, Ilkenhans T, Poulston S, Rowsell L, Smith AJ (2007) Development of new palladium-promoted ethylene scavenger. Postharvest Biol Technol 45:214–220
159. Tiekstra S, Dopico-Parada A, Koivula H, Lahti J, Buntinx M (2021) Holistic approach to a successful market implementation of active and intelligent food packaging. Foods 10(2):465
160. Van de Oever M, Molenveld K, Van der Zee M, Bos H (2017) Bio-based and biodegradable plastics – Facts and Figures. Focus on food packaging in Netherlands. 1722, Publisher: Wageningen Food & Biobased Research, Wageningen, The Netherlands, 1–67
161. Vandermoere F, Blanchemanche S, Bieberstein A, Marette S, Roosen J (2011) The public understanding of nanotechnology in the food domain: the hidden role of views on science, technology, and nature. Public Underst Sci 20(2):195–206
162. Venter K, Van der Merwe D, De Beer H, Kempen E, Bosman M (2011) Consumers' perceptions of food packaging: an exploratory investigation in Potchefstroom, South Africa. Int J Consum Stud 35(3):273–281

163. Verghese K, Lewis H, Lockrey S, Williams H (2013) The role of packaging in minimising food waste in the supply chain of the future. Final Report for CHEP, Publisher: RMIT University: Melbourne, Australia, 1–50
164. Vermeiren L, Devlieghere F, van Beest M, de Kruijf N, Debevere J (1999) Developments in the active packaging of foods. Trends Food Sci Technol 10(3):77–86
165. WHEYPACK Project (2020) PHB-based packaging from whey. 2020. Available online: http://www.wheypack.eu/documentos/EXECUTIVE_SUMMARY.pdf (accessed on 29 December 2020)
166. We are packaging fans (2020) Il packaging dagli scarti alimentari. Available online: https://wearepackagingfans.com/site/il-packaging-dagli-scarti-alimentari/ (accessed on 4 September 2020)
167. Van Wezemael L, Ueland Ø, Verbeke W (2011) European consumer response to packaging technologies for improved beef safety. Meat Sci 89(1):45–51
168. Whitehead P, Palmer M, Mena C, Williams, A, & Walsh C (2011) Resource maps for fresh meat across retail and wholesale supply chains. Etude WRAP NRSC009 Final report, Edited by: Justin FrenchBrooks, 1–105
169. Williams H, Wikström F, Otterbring T, Löfgren M, Gustafsson A (2012) Reasons for household food waste with special attention to packaging. J Clean Prod 24:141–148
170. Wilson CT, Harte J, Almenar E (2018) Effects of sachet presence on consumer product perception and active packaging acceptability-A study of fresh-cut cantaloupe. LWT 92:531–539
171. Wohner B, Pauer E, Heinrich V, Tacker M (2019) Packaging-related food losses and waste: an overview of drivers and issues. Sustain 11(1):264
172. Wyrwa J, Barska A (2017) Innovations in the food packaging market: active packaging. Eur Food Res Technol 243(10):1681–1692
173. Yadav JSS, Yan S, Pilli S, Kumar L, Tyagi RD, Surampalli RY (2015) Cheese whey: A potential resource to transform into bioprotein, functional/nutritional proteins and bioactive peptides. Biotechnol Adv 33(6):756–774
174. Yin RK (2003) Case study research: design and methods. Thousand Oaks, Calif.: Sage Publications, London
175. Young E, Mirosa M and Bremer P (2020) A systematic review of consumer perceptions of smart packaging technologies for food. Front Sustain Food Syst 4:63
176. Zhang BY, Tong Y, Singh S, Cai H, Huang JY (2019) Assessment of carbon footprint of nano-packaging considering potential food waste reduction due to shelf life extension. Resour Conserv Recycl 149:322–331
177. Zhou G, Hu W (2018) Public acceptance of and willingness-to-pay for nanofoods in the US. Food Control 89:219–226

Analyzing the Obstacles to Sustainable Packaging in the Context of Developing Economies: A DEMATEL Approach

Bhaskar B. Gardas, Vaibhav S. Narwane, and Nilesh P. Ghongade

Abstract The latest trend in the packing sector is to attain sustainability by accepting eco-friendly practices and harmonizing social aspects. In this article, the hindrances to the adoption of sustainable packaging practices have been identified and a decision-making methodology, namely, 'DEMATEL' has been applied to establish the causal and effect relations among them. Results revealed that a barrier, namely, 'Customer resistance to design changes' is the most significant one, whereas a dependent barrier, namely, 'Avoidance of sustainability ambition' is the least important one. This work would guide the managers, Governmental organizations, policy and strategy makers to formulate the policies for the packaging sector for achieving sustainability and to enhance the performance of the entire supply chain.

Keywords Sustainability · Packaging · Decision-making · Barriers · Green practices · Hindrances · MADM · Developing economies

1 Introduction

Recently, the environmental protection is considered as a major concern [1] and sustainable packing in this context plays a vital role in reducing pollution and waste [2]. It is also called as 'recyclable', 'green', 'eco-friendly' and 'eco-green' packaging [3]. The sustainable packing system uses materials that are safe for the environment and for mankind [4].

According to Grand View Research [5], sustainable packaging—(i) by application, food and beverages segment dominates the sustainable packaging market followed by personal care, healthcare and others and (ii) with China displacing the

B. B. Gardas (✉) · N. P. Ghongade
Department of Mechanical Engineering, University of Mumbai, M.H. Saboo Siddik College of Engineering, 8, Saboo Siddik Polytechnic Road, Mumbai, Maharashtra 400008, India

V. S. Narwane
Department of Mechanical Engineering, Somaiya Vidyavihar University, K.J. Somaiya College of Engineering, Ghatkopar East, Mumbai, Maharashtra 400077, India

S. S. Muthu (eds.), *Sustainable Packaging*, Environmental Footprints and Eco-design of Products and Processes, https://doi.org/10.1007/978-981-16-4609-6_3

USA in the packaging market, Asia–Pacific observes a rise in this sector. Indian packaging market is valued at 50.5 billion USD and is expected to have 26.7% CAGR during 2020–25 [6].

A report of The World Bank [7] raised concern about Carbon emission in India. Being the fastest emerging economy, India needs to be abreast the sustainable practices with a high growth rate [8]. It may be noted that there is a positive effect of sustainable packaging on Indian young consumers and they are willing to pay for the same [9]. Nowadays, awareness in customers of emerging economies toward purchase with sustainable packaging has been increased [10].

Though there are enormous benefits of using sustainable packaging and it directly influences the environmental performance, this concept has not penetrated considerably among the consumers and there are several hindrances for its effective adoption across the industries. Furthermore, very few studies were carried out in this segment in the context of developing economies such as India [9, 11]. It may be noted that the interrelationship between barriers in sustainable packaging in Indian firms is not discussed in the past studies. However, it is important to identify the barriers of sustainable packaging adoption and interconnection among them for its effective implementation. To deal with this research gap, the present paper tries to answer the following research objectives (ROs).

RO1: To shortlist the obstacles affecting the adoption of sustainable packaging in the firms of developing economies.

RO2: To investigate the interrelationship among these hindrances in the categories of cause and effect.

RO3: To identify the most significant factors.

The manuscript is arranged as follows: Theoretical background is given in Sect. 2, followed by Sect. 3, the research methodology and cause–effect model development. Section 4 covers the results and discussion and Sect. 5 details the conclusion.

2 Theoretical Background

In this segment, papers published on sustainable packaging have been discussed.

Dharmadhikari [12] revealed that a very high volume of packaging waste is produced in Indian supply chain and packaging cost is about 0.5–12% of the cost of the goods. This shows that there is a large scope for sustainable and economic packaging. Lee and Xu [13] emphasized about the product Life Cycle Assessment (LCA) of a product packaging display and stated that it is a rigid system and does not allow additional expenses of green packaging. In the area of sustainable packaging, the scope for disposal, biodegradability and compostable plastic packaging were highlighted by Lewis [14] with some key drivers like Eco-Efficiency, Environmental communication symbols and design for recycling and composting strategies. The strategy of eco-efficiency was broadly assessed by Ravi [15] for an electronic product packaging in a systematic framework, whereas Prakash and Pathak [9]created a framework that allowed to take decisions for removing the hurdles such as absence of awareness about environmental and sustainability problems and many other decisions by higher authorities of the organization. In order to investigate the reluctance toward the environmentally friendly packaging, the consumer behavior, purchase intentions and consumers' approaches were studied in detail by ref [4]. Also, a meaningful organized literature review was carried out by Popovic et al. [16] on the attributes of consumer food-buying behavior with a focus on sustainable packaging, and it was the assortment of several other studies carried out in the period 1994–2019. The consumer-centric study carried out by Nguyen et al. [17] revealed that consumers are aware of plastic pollution and the importance of green packaging and some of them are also more sensitive toward their discernments of the recyclable packaging.

In the domain of consumers' attitude toward green packaging, Steenis et al. [18] evaluated consumer response by analytically testing inclination, illative and attitudinal aspects toward packaging sustainability and consumers decision-making ability toward packaging material. Similarly, Prakash et al. [10] revealed the role of altruistic and egoistic values on optimistic impression of consumers' attitude toward green packaging. This attitude focuses on how the Indian youngsters decide the importance of environment and the health benefits of green packaging. In other study, Herbes et al. [19] revealed that how end of life, biodegradability and reusability became the effective decision criteria in the acceptance of biomethane-based packaging among the consumers. To define the role of consumers' interest in sustainable packaging, Boz et al. [20] conducted an articulated research, which showed how general attitude-related models, all-inclusive observations and analytic methods can be used to evaluate the influence of specific product design indications. Magnier et al. [21] explored the consumer reactions toward the environmental concern criterion including green packaging. This research analyzed how the purchase intentions of the consumer influence verbal package communication and environmental visual on the packaging.

Scott and Vigar-Ellis [22] stated that the consumers from a developing country are sparsely aware of the benefits of green packaging and they have been gradually learning. This research recommended that with the help of governmental and non-governmental organizations some eco-certification programs may be launched to create a belief among the consumers about the eco-friendly packaging. Martinho et al. [23] studied the effect of social and environmental factors on the purchase of products having green packaging. Herbes et al. [24] suggested that labels are important tools for generating ecological information and consciousness toward educating consumers to recognize eco-friendly packaging.

Nguyen et al. [25] instigated with the help of a case study about the use of PDCA for a variety of packaging issues. The case study revealed the process of PDCA cycle in speeding up new designs and reducing defects in sustainable packaging. Cheng [26] endorsed that it is the designer who has to play a vital role in the entire life cycle analysis of the packaging. A designer has to prepare questionnaire, which should showcase the importance of disposal of packaging materials in the environment.

Proper designing of packaging saves a lot of stuff, which otherwise gets dumped into the environment. Wandosell et al. [3] stated that researchers have keenly analyzed the strategies of green packaging to attend the social issues. Furthermore, with a focus on climate change, resource conservation and environmental protection, the scientific community should evaluate the green packaging issues with a global perspective.

Coelho et al. [27] suggested that single-use packaging system is harmful to the environment and introduced different ways of reusable packaging systems by focusing on design and its overall impact on sustainable packaging. This ultimately promotes the development of reusable packaging systems. Zhao et al. [28] experimentally revealed about the natural packaging material. The cellular extraction process produces high-quality cellulose from the durian rind. It also possesses a homogeneous structure, good appearance and excellent mechanical and thermal properties. It is a 'green' source of packaging due to low cost and a faster rate of biodegradation. Khalil et al. [2] emphasized the importance of sustainable packaging, which involves qualitative environmental management of the packaging material and suggests the possible use of cellulosic nanomaterials. Cellulose nanofibers are recognized as profusely available renewable polymeric materials. It was revealed that a proper design process of nanocellulose fibers has a great sustainability impact on packaging applications.

Siracusa and Rosa [29] focused on the use of bio-based materials with improved performance and concluded that the future trend with bio-nanocomposite materials may have a great demand in the food packaging industry for commercial applications. Helanto et al.[30] reviewed wood and fermentation-based materials having interesting barrier property and it can be used for the packaging purpose. It was suggested that wood-based products like hemicellulose-based barrier and hydrophilic hemicellulose have indistinct barrier properties along with good mechanical properties. Borgi et al. [31] suggested the first of its kind study on different crates used in food industry with a focus on its performance for delivery, materials and multiple uses of food products.

Boubeta et al. focused on the packaging strategy implemented for an agro-food company with the environmental and financial aspects. A sustainable model for boxes allocation has produced positive results. The case study also indicated that there was a reduction in 20% of total cost and the carbon emission was reduced by 31%. In another study, the retailer's point of view analysis was carried out by Gustavo et al.[32], which showed a more generous scope in redesigning product packaging. Also, the challenges pertaining to the same were highlighted.

Identification of the barriers to the sustainable packaging

Through literature review and using expert inputs, the following barriers have been identified (Table 1). These are the barriers pertaining to the organizational context and consume behavior.

Table 1 List of barriers to the adoption of sustainable packaging

S. N	Barrier	References
1	Customer resistance to design changes	[33]
2	Additional costs	[33]
3	Conflict with functional requirements	[34, 35]
4	Additional workload	[35–38]
5	Lack of cooperation among departments	[36]
6	Limited experience	[36, 38]
7	Organizational complexities	[36]
8	Supply chain complexities	[35, 36, 39]
9	Avoidance of sustainability ambition	[35, 40]
10	Lack of management support and commitment	[36, 41]
11	Attitude toward change	[42][33, 41]

3 Methodology and Cause–Effect Model Development

Decision Making Trial and Evaluation Laboratory (DEMATEL)

It is a mathematical approach used for resolving complex issues of engineering/management. This method helps to establish cause–effect relationship between the factors and presents the same into a visible structural model [43–45]. This method can also be used to determine contextual association and interdependence within factors under consideration [46]. Various steps involved in the DEMATEL approach are as follows [47–49]:

Step 1: Obtain average initial relation matrix (A) from initial relation matrix. Entries in the initial relation matrix were filled by the experts of the domain on the scale of 0–3, which represents the strength of interrelationship among identified factors, where 0 represents 'no influence', 1 represents 'slight influence', 2 represents 'higher influence' and 3 represents 'very high influence'. 15 response sheets were received from the experts and average of all the inputs was taken and final average initial relation matrix was formulated as shown in Table 2.

Step 2: Develop the normalized direct relation matrix (D) using equation D = A*S.

$$\text{Where, } S = \min\left[\frac{1}{\text{max value of column sum of A}}, \frac{1}{\text{max value of row sum of A}}\right]$$

$$T = D * (I - D)^{-1}$$

Step 3: Formulate the total relation matrix (T) (Table 3) using equation,

$$T = D * (I - D)^{-1}$$

Table 2 Average interrelationship matrix (A)

S. N	Factors	1	2	3	4	5	6	7	8	9	10	11	Row summation
1	Customer resistance to design changes	0	1	0	0	0	0	0	0	3	0	0	***4***
2	Additional costs	0	0	3	0	1	0	3	1	2	0	3	***13***
3	Conflict with functional requirements	0	3	0	3	3	3	3	2	2	1	3	***23***
4	Additional workload	0	2	3	0	2	0	2	1	0	0	1	***11***
5	Lack of cooperation among departments	0	2	3	3	0	2	3	3	3	0	2	***21***
6	Limited experience	0	2	3	2	3	0	3	3	2	0	2	***20***
7	Organizational complexities	0	3	3	3	3	0	2	3	3	0	2	***22***
8	Supply chain complexities	0	2	3	3	3	0	2	0	1	0	2	***16***
9	Avoidance of sustainability ambition	3	2	0	0	0	0	2	2	0	0	2	***11***
10	Lack of management support and commitment	1	3	3	3	3	2	3	3	3	0	3	***27***
11	Attitude toward change	2	2	2	1	2	0	3	2	2	0	0	***16***
Column summation		***6***	***22***	***23***	***18***	***20***	***7***	***26***	***20***	***21***	***1***	***20***	

Table 3 Total interrelationship matrix (T)

S. N	1	2	3	4	5	6	7	8	9	10	11	r_i
1	0.652	0.866	0.539	0.423	0.468	0.096	0.917	1.542	5.339	0.020	0.793	***11.654***
2	0.118	0.418	0.553	0.391	0.445	0.096	0.587	1.691	0.723	0.020	0.498	***5.540***
3	0.091	0.384	0.295	0.339	0.359	0.174	0.440	0.674	0.555	0.048	0.367	***3.725***
4	0.040	0.220	0.261	0.132	0.209	0.045	0.250	0.369	0.238	0.010	0.178	***1.951***
5	0.082	0.319	0.357	0.310	0.225	0.131	0.398	0.615	0.532	0.013	0.303	***3.285***
6	0.076	0.313	0.354	0.277	0.324	0.064	0.393	0.606	0.482	0.013	0.298	***3.200***
7	0.087	0.370	0.377	0.327	0.341	0.068	0.384	0.676	0.562	0.014	0.321	***3.525***
8	0.060	0.268	0.309	0.272	0.282	0.056	0.306	0.419	0.364	0.011	0.255	***2.601***
9	0.216	0.271	0.173	0.136	0.150	0.031	0.298	0.493	0.771	0.006	0.258	***2.804***
10	0.163	0.450	0.447	0.383	0.405	0.155	0.512	0.823	0.844	0.017	0.429	***4.628***
11	0.175	0.310	0.290	0.217	0.262	0.052	0.381	0.565	0.758	0.011	0.225	***3.247***
cj	***1.759***	***4.188***	***3.956***	***3.208***	***3.470***	***0.968***	***4.865***	***8.472***	***11.166***	***0.183***	***3.925***	

where *I* represents unit matrix.

Step 4: Formulate the inner dependency matrix (Table 4) by eliminating less than threshold value (α) entries from the matrix 'T'.

$$\text{Where, threshold value } (\alpha) = \frac{\text{sum of all entries of T}}{\text{number of all entries of T}}$$

$$\alpha = 46.161/121 = 0.381$$

Table 4 Inner dependency matrix (α = 0.381)

S. N	1	2	3	4	5	6	7	8	9	10	11
1	0.652	0.866	0.539	0.423	0.468		0.917	1.542	5.339		0.793
2		0.418	0.553	0.391	0.445		0.587	1.691	0.723		0.498
3		0.384					0.440	0.674	0.555		
4											
5							0.398	0.615	0.532		
6							0.393	0.606	0.482		
7							0.384	0.676	0.562		
8								0.419			
9								0.493	0.771		
10		0.450	0.447	0.383	0.405		0.512	0.823	0.844		0.429
11							0.381	0.565	0.758		

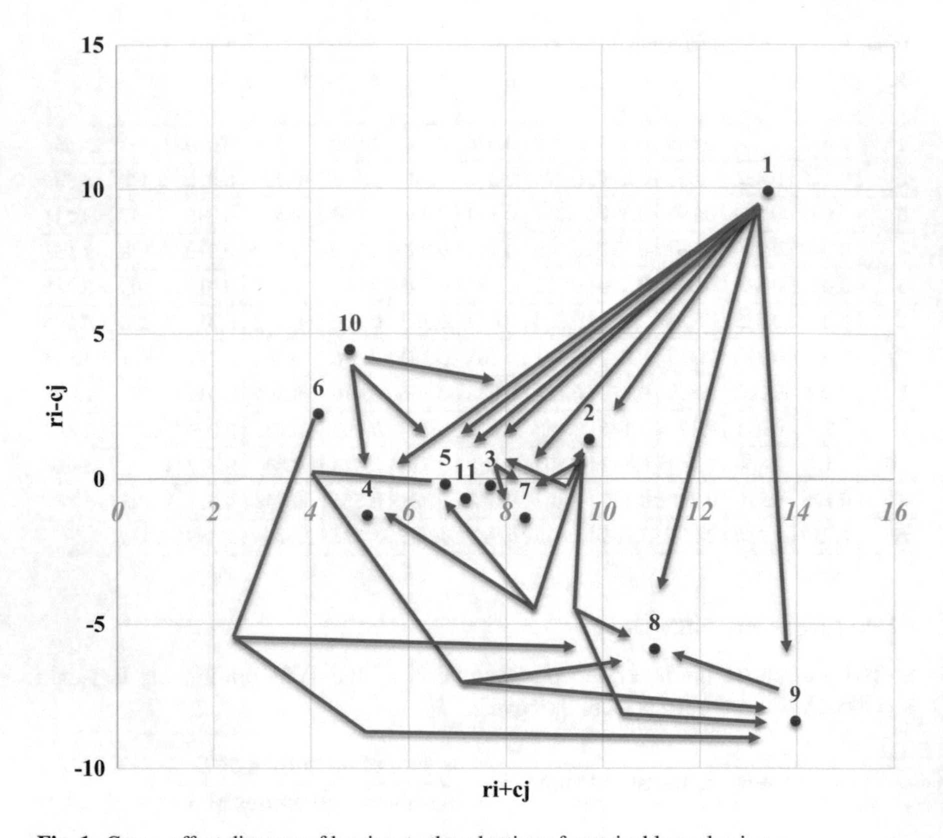

Fig. 1 Cause–effect diagram of barriers to the adoption of sustainable packaging

Step 5: Develop the cause–effect relationship diagram (Fig. 1) using rank and nature of factors (i.e., cause or effect), which is determined using '$r_i + c_j$' and '$r_i - c_j$' values (Table 5), respectively.

Where, r_i represents the sum of rows of matrix 'T' and
c_j represents the sum of column of matrix 'T'.

4 Results and Discussion

In this investigation, the hindrances to the adoption of sustainable packaging were identified and analyzed using DEMATEL methodology. The total relationship matrix (Table 3) highlights relationship between 11 barriers. Also, it indicates the strength of relationship. However, in inner dependency matrix (Table 4), the insignificant relations have been eliminated. The classification of factors into cause–effect categories is shown in Table 5. It may be noted that out of 11 factors, 4 factors belong to the cause category, whereas the remaining 7 were related to the effect cluster. The causal

Table 5 Classification of factors based on their nature

S. N	r_i	c_j	$r_i + c_j$	$r_i - c_j$	Nature
1	11.654	1.759	13.413	9.895	Cause
2	5.540	4.188	9.728	1.352	Cause
3	3.725	3.956	7.681	−0.231	Effect
4	1.951	3.208	5.159	−1.257	Effect
5	3.285	3.47	6.755	−0.185	Effect
6	3.200	0.968	4.168	2.232	Cause
7	3.525	4.865	8.390	−1.340	Effect
8	2.601	8.472	11.073	−5.871	Effect
9	2.804	11.166	13.970	−8.362	Effect
10	4.628	0.183	4.811	4.445	Cause
11	3.247	3.925	7.172	−0.678	Effect

factors are the factors that are very important from the decision-makers point of view. These four barriers, namely, 'Customer resistance to design changes', 'Additional costs', 'Limited experience' and 'Lack of management support and commitment' have the capability to influence the hindrances pertaining to the effect category, which are 'Conflict with functional requirements', 'Additional workload', 'Lack of cooperation among departments', 'Organizational complexities', 'Supply chain complexities', 'Avoidance of sustainability ambition' and 'Attitude toward change'.

It may be noted that sustainable packaging is a crucial topic, which requires considerable documentation and analysis for evaluating the design, selection and processing of materials and its life cycle assessment [2]. The sustainable packaging has a tremendous positive influence on the environment and there is a need to strictly follow the guidelines laid down by WHO and ISO for the safety of environment and mankind. The materials that best fit the purpose may be shortlisted and used for the packaging applications. While selecting the packaging material, its cost parameters may also be taken into account.

For improving the environmental performance index, the use of single-use packaging must be reduced and use of reusable packaging may be considered; and policies for the same should be formulated by the governmental organizations. The reusable packaging also helps in improving the financial dimension of the organization [27, 50]. For reducing the pollution caused by the packaging products, there is a need to align stakeholders with the strategies of the universal packaging along with the capabilities for sorting. Also, there is a need to standardize the labels and signage. Furthermore, consumer education on the sustainability aspects and environmental deterioration would definitely impact the usage of sustainable packaging [26].

5 Conclusion

In this study, 11 hindrances to the sustainable packaging practices in Indian organizations have been identified and the cause–effect relationship between the barriers has been established using DEMATEL approach. It may be noted that all the research objectives mentioned in Section 1 (Introduction) have been successfully achieved. Results revealed that a barrier, namely, 'Customer resistance to design changes' is the most significant one, whereas a dependent barrier, namely, 'Avoidance of sustainability ambition' is the least important one.

This investigation is useful for academicians and practitioners in understanding the barriers that influence the usage of sustainable packaging. Also, the organizational managers can develop effective guidelines for adopting sustainable packaging practices, which would improve the green image of the company and enhance the competitiveness in the global markets.

In this investigation, only 11 important factors were identified and analyzed. However, there may be other factors that may be influencing the adoption of sustainable packaging practices in the industries. Additionally, the inputs were considered from the experts who belong to Indian organizations; hence, findings may be applicable to the Indian context only. However, including responses from other geographical locations may generalize the model. Furthermore, the judgments used for the analysis could be subjective and they are not bias free. To overcome this limitation, the findings of the present study may be compared with the results obtained by employing other MCDM tools. In addition to this, quantitative analysis may be used for validating the findings.

In further studies, research activities in the direction of reusable packaging and its effect on the environmental and economic performance of the organization may be analyzed. The role of bio-based packaging materials on the environmental performance may be evaluated. Also, how sustainable packaging influences the waste management practices can be studied. Furthermore, the circular economy aspects of sustainable packaging may be explored. The innovation in packaging and its influence on sustainability can also be studied. In addition, cellulose nanofibers influence on the environment may be evaluated.

This work would guide the managers, Governmental organizations, policy and strategy makers to formulate the policies for the packaging sector for achieving sustainability and to enhance the performance of the entire supply chain.

References

1. Feng JC (2019) Integrated development of economic growth, energy consumption, and environment protection from different regions: based on city level. Energy Procedia 158:4268–4273
2. Khalil HA, Davoudpour Y, Saurabh CK, Hossain MS, Adnan AS, Dungani R, Paridah MT, Islam Sarker Md Z, Nurul Fazita MR, Syakir MI, Haafiz MKM (2016) A review on nanocellulosic

fibres as new material for sustainable packaging: process and applications. Renew Sustainable Energy Rev 64:823–836
3. Wandosell G (2021) Green packaging from consumer and business perspectives. Sustainability 13(3):1356
4. Orzan G (2018) Consumers' behavior concerning sustainable packaging: an exploratory study on Romanian consumers. Sustainability 10(6):1787
5. Grand View Research (2020) Green packaging market size, share & trends analysis report by type (Recycled Content, Reusable, Degradable), by application (Food & Beverages, Healthcare), by region, and segment forecasts, 2020–2027 https://www.grandviewresearch.com/industry-analysis/green-packaging-market(Accessed date: 17 April, 2021)
6. McKinsey and Company (2021) Sustainability in packaging: consumer views in emerging Asia. https://www.mckinsey.com/industries/paper-forest-products-and-packaging/our-insights/sustainability-in-packaging-consumer-views-in-emerging-asia(Accessed date: 17 April, 2021)
7. The World Bank, 2018. The World Bank in India. https://www.worldbank.org/en/country/india/overview (Accessed date: 16 April, 2021)
8. Nandy B (2015) Recovery of consumer waste in India–A mass flow analysis for paper, plastic and glass and the contribution of households and the informal sector. Resour Conserv Recycl 101:167–181
9. Prakash G (2017) Intention to buy eco-friendly packaged products among young consumers of India: a study on developing nation. J Clean Prod 141:385–393
10. Prakash G (2019) Do altruistic and egoistic values influence consumers' attitudes and purchase intentions towards eco-friendly packaged products? An empirical investigation. J Retail Consum Serv 50:163–169
11. Yadav R (2016) Altruistic or egoistic: which value promotes organic food consumption among young consumers? A study in the context of a developing nation. J Retail Consum Serv 33:92–97
12. Dharmadhikari S (2012) Eco-friendly packaging in supply chain. IUP J Supply Chain Manage 9(2):7
13. Lee SG (2005) Design for the environment: life cycle assessment and sustainable packaging issues. Int J Environ Technol Manage 5(1):14–41
14. Lewis H (2008) Eco-design of food packaging materials. In Environmentally compatible food packaging. Woodhead Publishing, pp 238–262
15. Ravi V (2015) Analysis of interactions among barriers of eco-efficiency in electronics packaging industry. J Clean Prod 101:16–25
16. Popovic I, Bossink BA, van der Sijde PC (2019) Factors influencing consumers' decision to purchase food in environmentally friendly packaging: what do we know and where do we go from here? Sustainability 11(24):7197
17. Nguyen AT, Parker L, Brennan L, Lockrey S (2020) A consumer definition of eco-friendly packaging. J Clean Prod 252:119792
18. van Steenis ND (2017) Consumer response to packaging design: the role of packaging materials and graphics in sustainability perceptions and product evaluations. J Clean Prod 162:286–298
19. Herbes C (2018) Consumer attitudes towards biobased packaging–A cross-cultural comparative study. J Clean Prod 194:203–218
20. Boz Z (2020) Consumer considerations for the implementation of sustainable packaging: a review. Sustainability 12(6):2192
21. Magnier L (2015) Consumer reactions to sustainable packaging: the interplay of visual appearance, verbal claim and environmental concern. J Environ Psychol 44:53–62
22. Scott L (2014) Consumer understanding, perceptions and behaviours with regard to environmentally friendly packaging in a developing nation. Int J Consum Stud 38(6):642–649
23. Martinho G (2015) Factors affecting consumers' choices concerning sustainable packaging during product purchase and recycling. Resour Conserv Recycl 103:58–68
24. Herbes C (2020) How green is your packaging—A comparative international study of cues consumers use to recognize environmentally friendly packaging. Int J Consum Stud 44(3):258–271

25. Nguyen V (2020) Practical application of plan–do–check–act cycle for quality improvement of sustainable packaging: a case study. Appl Sci 10(18):6332
26. Cheng, K. (2019). *Sustainable packaging approaches for current waste challenges* (Doctoral dissertation, Massachusetts Institute of Technology).
27. Coelho PM, Corona B, ten Klooster R, Worrell E (2020) Sustainability of reusable packaging-Current situation and trends. Resour Conserv Recycl: X 100037
28. Zhao G, Lyu X, Lee J, Cui X, Chen WN (2019) Biodegradable and transparent cellulose film prepared eco-friendly from durian rind for packaging application. Food Packag Shelf Life21:100345
29. Siracusa V, Rosa MD (2018) Sustainable packaging. In Sustainable food systems from agriculture to industry. Academic Press, pp 275–307
30. Helanto KE (2019) Bio-based polymers for sustainable packaging and biobarriers: a critical review. BioResources 14(2):4902–4951
31. Del Borghi A, Parodi S, Moreschi L, Gallo M (2020) Sustainable packaging: an evaluation of crates for food through a life cycle approach. Int J Life Cycle Assess 1–14
32. Gustavo JU Jr (2018) Drivers, opportunities and barriers for a retailer in the pursuit of more sustainable packaging redesign. J Clean Prod 187:18–28
33. Kassaye WW (1992) Balancing traditional packaging functions with the new "green" packaging concerns. SAM Adv Manag J 57(4):15
34. Dangelico RM (2010) Mainstreaming green product innovation: why and how companies integrate environmental sustainability. J Bus Ethics 95(3):471–486
35. Van Hemel C (2002) Barriers and stimuli for ecodesign in SMEs. J Clean Prod 10(5):439–453
36. Boks C (2006) The soft side of ecodesign. J Clean Prod 14(15–16):1346–1356
37. Martinez VG, English S (2015) Why designers won't save the world. In Proceedings of the 11th European academy of design conference. Boulonge-Billancourt, France, pp 22–24
38. Storaker A, Wever R, Dewulf K, Blankenburg D (2013) Sustainability in the front-end of innovation at design agencies. In EcoDesign 2013: 8th international symposium on environmentally conscious design and inverse manufacturing. Jeju Island, South Korea
39. Gardas BB (2018) Modelling the challenges to sustainability in the textile and apparel (T&A) sector: a Delphi-DEMATEL approach. Sustainable Production and Consumption 15:96–108
40. Edwards MG (2010) An integral metatheory for organisational sustainability: living with a crowded bottom line in chaotic times.
41. Raut RD (2017) To identify the critical success factors of sustainable supply chain management practices in the context of oil and gas industries: ISM approach. Renew Sustain Energy Rev 68:33–47
42. García-Arca J, Garrido ATGP, Prado-Prado JC (2017) Sustainable packaging logistics. The link between sustainability and competitiveness in supply chains. Sustainability 9(7):1098
43. Gardas BB (2019) A hybrid decision support system for analyzing challenges of the agricultural supply chain. Sustainable Prod Consumption 18:19–32
44. Narkhede BE (2020) Implementation barriers to lean-agile manufacturing systems for original equipment manufacturers: an integrated decision-making approach. Int J Adv Manuf Technol 108(9):3193–3206
45. Raut RD, Gardas BB, Narkhede BE, Narwane VS (2019) To investigate the determinants of cloud computing adoption in the manufacturing micro, small and medium enterprises. Benchmarking: Int J
46. Gardas BB (2019) Green talent management to unlock sustainability in the oil and gas sector. J Clean Prod 229:850–862
47. Du YW, Li XX (2021) Hierarchical DEMATEL method for complex systems. Expert Syst Appl 167:113871
48. Mao S, Han Y, Deng Y, Pelusi D (2020) A hybrid DEMATEL-FRACTAL method of handling dependent evidences. Eng Appl Artif Intell 91:103543
49. Petrovic I (2020) A hybridized IT2FS-DEMATEL-AHP-TOPSIS multicriteria decision making approach: case study of selection and evaluation of criteria for determination of air traffic control radar position. Decis Making: Appl Manage Eng 3(1):146–164

50. Shanmugam K, Doosthosseini H, Varanasi S, Garnier G, Batchelor W (2019) Nanocellulose films as air and water vapour barriers: a recyclable and biodegradable alternative to polyolefin packaging. Sustainable Mater Technol 22:e00115
51. Sustainable Packaging Coalition. (2011) Definition of sustainable packaging. Retrieved from https://sustainablepackaging.org/wp-content/uploads/2017/09/Definition-of-Sustainable-Packaging.pdf (Accessed on 16 April, 2021)
52. Boubeta IG (2018) Economic and environmental packaging sustainability: a case study. J Ind Eng Manage 11(2):229–238
53. Kumar A (2019) Exploring young adults'e-waste recycling behaviour using an extended theory of planned behaviour model: a cross-cultural study. Resour Conserv Recycl 141:378–389

Experimental Study for the Valorization of Polymeric Coffee Capsules Waste by Mechanical Recycling and Application on Contemporary Jewelry Design

Mariana Kuhl Cidade, Felipe Luis Palombini, Ana Paula Palhano, and Amanda Melchiors

Abstract Recent trends in the coffee market consumption in Brazil show the growth of polymeric capsules as one new preferred way of consuming the beverage, which contributes to the country's increasing problem of urban solid waste treatment, particularly related to plastic waste. One of the main causes of the low recycling and recovery rates of secondary plastic in Brazil is related to the low economic interest on work with this type of waste due to its intrinsically lower value when compared to the raw material. This study presents an experimental approach to increase the value of polymeric coffee capsule waste via a simplified recycling procedure and its application in contemporary jewelry. Several capsule samples were collected from urban solid waste and investigated via Fourier Transform Infrared (FT-IR) spectroscopy for resin identification and degradation analyses. Most polypropylene samples contained some level of degradation, mainly related to photodegradation and the presence of contaminants. Samples were then coarsely ground and melted in a silicone mold for the production of secondary plastic pieces as a replacement for gemstones. Following, the produced pieces were employed in the manufacturing of a jewelry collection, utilizing sterling silver in which the alloy was composed of scrap silver jewelry and copper recovered from electronic waste. Artisanal jewelry methods were followed for the manufacturing of a pendant and a ring. The findings support that secondary plastic can be valorized when applied on different and valued products, and even degraded polymers processed via a simplified recycling procedure can benefit from the material's aesthetic attributes for its application in contemporary jewelry design. Finally, a discussion is set for the need for a sustainable approach on the single-use plastic coffee capsules, regarding its recovery and social impacts, in developing countries such as Brazil.

M. K. Cidade (✉)
Department of Industrial Design, Federal University of Santa Maria — UFSM, Av. Roraima, nº 1000, Prédio 40, Sala 1136, Santa Maria, RS 97105-900, Brazil

F. L. Palombini
Design and Computer Simulation Group, School of Engineering, Federal University of Rio Grande do Sul — UFRGS, Av. Osvaldo Aranha 99/408, Porto Alegre, RS 90035-190, Brazil

A. P. Palhano · A. Melchiors
Industrial Design Undergraduate Course, Federal University of Santa Maria — UFSM, Av. Roraima, nº 1000, Prédio 40, Santa Maria, RS 97105-900, Brazil

S. S. Muthu (eds.), *Sustainable Packaging*, Environmental Footprints and Eco-design of Products and Processes, https://doi.org/10.1007/978-981-16-4609-6_4

Keywords Sustainability · Polypropylene · Recycling · Aesthetics · Jewelry · Plastic waste · Coffee · Trash-to-treasure

1 Introduction

Brazil is the current world leader in coffee consumption, with a retail volume of about 814 thousand metric tons in 2020 [27]. The country has a historic connection with the beverage, holding a major household presence [23], besides coffee having significant importance to the Brazilian economy, according to the Brazilian Ministry of Agriculture [46]. Coffee can be considered almost omnipresent in the daily life of Brazilians, being consumed during all meals of the day. Along with the known effects of caffeine, the consumption increase is being favored due to its health benefits. Despite the popularity and long-term relationship with the beverage, several methods of preparation and consumption are being developed, raising the interest on the public for new alternatives that combine freshness and practicality. Therefore, investments have been made on newer types of packaging solutions that could contribute to such aspects and then gain a larger portion of the market. Among different coffee categories, single-serve capsules present a relatively new consumption type in Brazil, although it has already increased its retail volume share from 0.03% in 2006, to 1.59% in 2020, or currently over 12.2 thousand metric tons [27]. Along with convenience, the recent growth in domestic demand for different varieties and premium coffee products has motivated consumers to invest more in this type of packaging, according to the Brazilian Association of the Coffee Industry [1]. Once single-serve coffee capsules can substantially increase the shelf life of the product, it also allows consumers to differ from different varieties without the need to consume a single ground coffee package at a time before the expiration date. For instance, in terms of retail value, capsules represent a higher share in coffee consumption in the country, climbing from 0.41%, in 2006, to 16.77%, in 2020 [27]. However, despite the many practical advantages of single-serve coffee capsules, this type of packaging also represents one of the main issues regarding its environmental impact.

In a recent Life Cycle Assessment (LCA) of multiple coffee preparation methods in Brazil, [23] found that single-serve capsules have the highest levels of energy consumption, mostly due to the specific type of packaging used. Particularly, the choice of materials used in the capsule manufacturing process is a key issue for the relatively poor LCA results [15, 35]. The majority of capsules are composed of polymeric resins, aluminum, or a mixture of them, which increases the difficulty of appropriate end-of-life disposal and management [34]. Multi-material packaging is known for its numerous advantages for manufacturing, such as time and cost reduction and increased product quality [71]. However, this type of packaging is also related to poor recycling capabilities, materials separation, and recovery [74]. When considering products with a great unitary consumption volume, such as the case of coffee capsules, this poses a major threat for public policies aiming at the reduction or recovery of this type of waste. For instance, it is estimated that approximately 59 billion capsules were produced globally in 2018, and about 56 billion of those were likely to be disposed of in landfills [32]. For a growing number of such a complex waste, coffee capsules pose a major issue for sustaining a circular economy, particularly in Brazil [4]. Most of all, for some waste to be acquired by recycling companies and individuals, it must be considered economically interesting to allow the gaining of profit, or at least favoring it somehow over the same type of raw material, in order to the circular economy to be feasible. It is noteworthy, though, that some materials have a higher success rate regarding their end-of-life recovery, including in the single-serve capsule market. For instance, aluminum capsules packaging benefits from the significantly higher price for obtaining the raw material, when compared to its secondary source via recycling [8, 31]. This scenario strengths the demand for aluminum packaging waste, making the recovery of this material much more achievable. On the other hand, even though aluminum coffee capsules waste is considered valorized for recycling and commercializing including in Brazil, due to the same reasons most polymeric capsules do not raise significant economic interest and therefore are frequently discarded. With the expressively lower cost of polyolefin-based commodities, the average value of secondary plastics is well below covering the very costs of collecting, sorting, cleaning, and recycling to be used in products with a similar purpose. Consequently, the valorization of polymeric coffee capsule waste is an urging matter for environmental and economic spheres in the country, mainly because of the current Brazilian waste treatment system and its difficulties.

When globally compared, waste treatment in Brazil is considered inefficient for the respective amount of waste generated, both in terms of the recycling and recovery of plastics. In the USA, for instance, about 35 million tonnes of plastics are generated annually, at which 8.4% is recycled, 15.8% is combusted with energy recovery, and 75.8% is landfilled, according to the US Environmental Protection Agency [25]. As for Europe, from about 29 million tonnes of plastic waste collected in 2018, about 32.5% was recycled, 42.6% was energy recovered, and 24.9% was landfilled [59]. On the other hand, in Brazil, for about 11 million tonnes of plastic generated, only 1.3% is estimated to be recycled indeed [73] and then reinserted into the industry. According to the Brazilian Plastic Industry Association [2], the country has a significant potential for the recovery of plastics, mainly due to the country's widespread solid waste collecting system. Following the report from the Brazilian Association of Special Waste and Public Cleaning Companies [3], over 92% of the municipal solid waste is collected. This shows that the collecting of waste is not enough for keeping a higher level of waste recovery, and, therefore, public policies should also address other key steps of waste treatment. Due to packaging design flaws and the industries' choices of materials selection, the interest in recycling polymers originated from waste becomes lesser evident in the country [17, 54]. In Brazil, urban solid waste (USW) treatment heavily relies on the work of individual collectors and cooperatives with multiple workers. Each municipality is responsible for the collection and transportation of USW from designated points to registered sorting facilities, called Sorting Units. These units are composed of low-income workers whose revenue depends on the commercialization of the sorted USW to private buyers. The Sorting Units are responsible for the separation and selection of dry waste, dividing them into different categories, including several types of polymers, glasses, paper, and metals. Often without any reprocessing method, each category of divided waste is later pressed into bales, which are then sold to industries as a secondary-sourced material for further recycling and use [29, 36, 42]. The main issue of this system is that the majority of the selected dry waste is only commercialized if it has enough interest from buyers. For instance, a potentially recyclable material must have its cost attenuated in the processes of collecting and recycling by the buying company, as a way to make the residue commercially competitive. Even though the consumer's willingness to pay more for recycled products [39] can be raised as a potential incentive to invest in the acquisition of plastic residue originated from sorted Urban Solid Waste, the final cost of a product that uses this type of material should be reasonable. On the other hand, the application choices for this type of secondary material should be revised in order to find alternatives that could be naturally more valuable, *i.e.*, that consumers would be even more inclined to buy. Another important issue regarding the premature lack of interest of buy secondary-sourced plastics is related to its final destination in the life cycle. In Brazil, the vast majority of waste that does not have economic interest ends up being disposed of by landfilling or even discarded in open dumps [54]. Therefore, the economic interest in certain types of residue is one of the main driving forces for USW in Brazil to be properly recovered. This situation aggravates the need for more holistic approaches addressing USW treatment. For instance, decreasing the number of processes a potential buyer should perform on recovering the plastic waste acquired, before its application, could make the investment as being considered worthy. For instance, some Sorting Units have included a mini polymer recycling plant in its facilities, generally comprising of a washer station, a knife mill for hard case plastics (*e.g.*, bottles, food containers), and an agglomerator for film-like plastics (*e.g.*, packages from snacks and general grocery items). Acquired by the workers of Sorting Units or donated by local government or private companies, such equipment allows the waste to be commercialized as granules, with an initial reprocessing procedure, facilitating its application on extrusion, and thermoforming during further recycling steps in the industry. This procedure significantly increases the interest of buyers, even allowing the waste to be sold by a much higher price per weight [26, 49], and consequently leading to a higher revenue for the workers of the sorting cooperatives. Another way to boost the recovery rate of secondary plastic could be to contribute with knowledge and subsidies for those workers to gain with the direct commercialization of products using secondary plastics. Beyond the selling of the residue as a raw material for further processing, workers could produce simplified goods with a much higher intrinsic value. This demonstrates that the recovery of complex polymeric packaging waste, such as the growing consumption of single-use coffee capsules, is an issue that embraces the tripod of sustainability. As a result, finding ways to increase the value of the waste is crucial for better recyclability, contributing to the environment, as well as the social and economic spheres [52, 54].

Concerns regarding plastic residues of USW have been of great interest and discussion in the literature, mainly regarding practices to reduce material waste and to increase its recovery by recycling or other means [26, 28, 62, 67]. One of the most important characteristics of thermoplastics is their capability to be reprocessed into new products via mechanical or chemical recycling [11, 13, 65]. Several processes can be followed to reintroduce polymeric waste into new products, even though its current condition may let it be found with some level of degradation which could impair some desirable mechanical properties, for instance [55, 63]. In this manner, linking the recyclability of this material with the issue of its economic interest as a residue, and even its social need to be valorized, multiple approaches can be followed to increase its value and thus to extend its life cycle [6, 8]. Commonly, recycled plastics can be used in the manufacturing of multiple products, varying from containers, packaging, and bins, to brooms, bicycle racks, and general utensils [49]. Polymers were also used as reinforcement in composites for multiple purposes. For instance, plastic residues were studied as a sand-substitution aggregate in concrete composites [48], as an additive for asphalt [5], in the manufacturing of plastic lumber [14], as well as in the fabrication of other building materials [64]. Polyethylene terephthalate (PET) bottles are also widely recycled into fibers for textiles and fabrics [56, 66]. For instance, in Brazil, multiple companies recycle PET bottles from soda and other soft drinks into thicker fibers for the manufacturing of brooms. Besides, different techniques can also be used to convert general polyolefins to fuels and other chemicals [43]. Regarding polymeric coffee capsules, Montagna et al. [51] showed that their mechanical recycling is possible, both using ground and injection, and extruded and injection methods. The authors also found satisfactory the performance of the recycled material regarding its mechanical and physical properties, by means of impact resistance and water absorption tests. Moreover, [24] evaluated different composition rates of polymeric coffee capsules with multiple characterization techniques and mechanical tests. Authors founds contribute to the potential of using the material as a source for newer products due to its overall good mechanical properties. Such an approach is also an important way to increase social awareness and to reduce the environmental impact [4]. On the other hand, despite its mechanical and physical properties, little has been studied regarding newer or different applications for recycled polymeric waste, including coffee capsules, which could increase its application capabilities making it a more interesting material.

Another way to increase the value of secondary plastic is through its aesthetic characteristics, which can imply different complexity levels and individuality to a product. One of the benefits of using recycled plastics as raw materials is its great variable and stimulating visual qualities [69]. Particularly, when recycling materials with different visual characteristics, as an example of working with unhomogenized residues with multiple colors, the obtained piece leads to both an attractive and unique final pattern. When properly used, these 'imperfect aesthetic qualities' of secondary materials, such as recycled plastics, can even contribute to the sense of craftsmanship and manual labor of the product, leading to increased added value [60]. Rognoli et al. [60] reviewed several cases of products made with roughly and unhomogenized recycled polymers as a key aesthetic attribute to increase the value of different types of waste. Authors claim that this category, called do-it-yourself (DIY) materials positively contribute to product design via material experimentation and are a result of self-production practices from individuals or collectives. Being an expression of creativity through materials, rough polymer recycling allows products to be unique, giving them personality and enriching the buyers' experience with an object they could feel more related to. Similarly, the irregular pattern of the wood grain can be explored with recycled plastics for the development of aesthetic timber panels [44]. Such type of application could replace wood with a much higher appreciation originated from its aesthetic value, closer to the original, in addition to its diverse and desirable physical properties. Additionally, [38] verified that products with a sustainable approach should have particular aesthetic features, like uniqueness and self-expression, that can be socially appreciated, and which could be obtained with recycled plastic. The perception of a sustainable product is not only linked to its aesthetics but relies on these intangible characteristics in order to be universally recognized. Besides, exploring inhomogeneous attributes, including those of rough secondary polymers, can be one important way to increase the value of a proposed product [37, 61]. Therefore, it is important to select not only the visual characteristics of residues used in a batch but also the recycling process through which they will be subject to and consequently the choice of machinery for this application. To obtain a more noticeable and representative aesthetic of some recycled plastic piece, for instance, extrusion prior to molding processes should be avoided, as a way to reduce homogenization. Naturally, the more homogenized a secondary material is, the more consistent its mechanical properties would became [11, 40, 49]. Then again, if the application of the recycled plastic is intended for a product with lower mechanical requirements, this should be less of an issue.

Multiple branches of product design can benefit from aesthetic qualities derived from the uniqueness of the materials employed [9]. More than just a visual perception of a product, aesthetic attributes are related to the perception by the senses, when the relationship of an object pleases and is appreciated by its beauty [45]. In product design, aesthetics is explored throughout different consumer goods, even though its surface is responsible for most of the effects [9]. Therefore, the choice of materials, manufacturing processes, and surface treatments play an important role in how a product is perceived and how does it interact with the user, particularly related to its sense of pleasure. For instance, one of the main areas of design where aesthetics has a major part is jewelry. With its origins related to symbolism, spiritualism, and expression of eighter individuality or collectiveness, jewelry aesthetics are connected to a variety of meanings that its user intends to show [33]. A relatively recent new field of jewelry design started to emerge in the 1960s, in a practice called contemporary jewelry [12]. As opposed to classic jewelry, in which precious and rare materials such as gold, silver, platinum, diamonds, and other gemstones are ever-present, in contemporary jewelry, new materials gained a place, such as woods, textiles, seeds, ornamental rocks, natural fibers, multiple metal alloys, rubber, plastics, among others [18]. New technologies were also a turning point for the development of contemporary jewelry, such as water jetting [21], laser machining [19], and additive manufacturing [50]. In this sense, Guerra and Cidade [30] presented the development of a piece of contemporary jewelry with the use of natural materials and essential oils aimed at the relief of allergies. By challenging preconceptions of traditional jewelry and its association to wealth, social status, and cultural positioning, contemporary jewelry uses new ways and techniques to emphasize the value of a piece, rather than simply relying on the costs of the materials used [70]. Likewise, even traditional jewelry manufacturing methods can be combined with new materials and technologies, in a way to enrich the meaning and value of a piece. For instance, Allen [7] describes the development of two pairs of wedding rings in which new materials and methods were used to increase their value, the first pair is made of unpolished bronze and wood from a Hawaiian native tree, and the second one is made of ferritic stainless steel with the one's fingerprint etched on the surface of the partner's ring. In this way, recycled plastic can be studied as a material for application in contemporary jewelry, benefiting from both its reprocessing properties and the distinctive aesthetic finish of each piece.

This chapter describes an experimental study of the recycling process of polymeric coffee capsules waste with aims at its valorization regarding the application of the material in a collection of contemporary jewelry. Collected samples of capsule residues were first washed and characterized by Fourier transform infrared spectroscopy (FT-IR) for resin identification and the assessment of potential degraded conditions. Even though the material application is not intended for having specific structural or physical requirements, mechanical tests were added to further validate the recycled material as a viable source for other applications. Therefore, the selected residue of polymeric coffee capsules was roughly ground, and a set of tensile tests was carried out with different ratios of concentration related to the addition or not of virgin resin, to assess its main mechanical properties. Later, the coarsely grounded samples were also applied in a simplified mechanical recycling procedure, using a developed silicone mold for the creation of gemstone-like pieces. The parts were then employed in an artisanal jewelry manufacturing process using recycled sterling silver alloyed with copper recovered from e-waste, for the development of a ring and a pendant. With the main purpose of exploring new, high-value applications for a type of waste that has little economic interest, we propose a workflow that could assist inexperienced workers with the manufacturing of innovative pieces. Finally, a discussion is presented aiming to approach the sustainable challenges of this type of packaging, particularly regarding developing countries with a still poorly established Urban Solid Waste Treatment.

2 Material and Methods

2.1 Coffee Capsules

Around 40 polymeric coffee capsules from the same brand were collected from domestic and commercial waste and separated according to multiple flavors. Capsules were manually disassembled, by opening the sealing that protects the coffee compound content, and cleaned with a neutral soap, followed by rinsing in water.

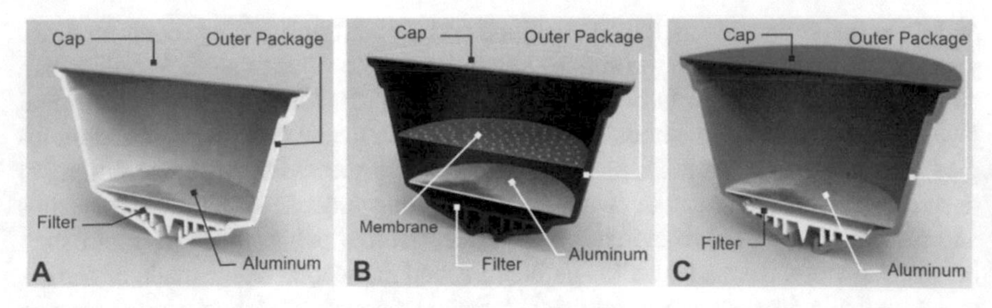

Fig. 1 Internal components of each polymeric coffee capsule: **A** milk mixture capsule; **B** roast coffee capsule; and **C** chocolate mixture capsule

Used capsules were originated from three types of beverages: milk mixture, roast coffee, and chocolate mixture. Figure 1 illustrates the internal components of each one. All flavors consist of an outer package, a sealer cap, and an aluminum foil which is perforated by the bottom filter. Some coffee capsules (Fig. 1B) also present a semi-permeable plastic membrane separated from the cap, which preserves their food content by isolating it from the exterior as well as allowing the passage of water when in use.

2.2 *Fourier Transform Infrared Spectroscopy (FT-IR)*

To verify the type of polymeric material in the coffee capsules, as well as to analyze possible degradation conditions, a Fourier transform infrared spectroscopy (FT-IR) analysis was conducted in different components of the collected samples. Disassembled samples were divided into the outer package, the cap, and the bottom filter, which represents the larger amount of volume available for further recycling and application. Samples were analyzed via Fourier transform infrared spectroscopy (FT-IR), for the identification of constituent polymer resins. Spectrum 100 (PerkinElmer®, Waltham, MA, USA) equipment was used in the FT-IR analyses by attenuated total reflectance mode (ART FT-IR), with a resolution of 4 cm^{-1} under 16 scans in the region between 4000 and 600 cm^{-1}, in combination with a polymer library. As expected, based on similar results from the literature [22, 24, 51], the outer package and internal filters were identified in FT-IR as polypropylene (PP). The cap was identified as a polyethylene terephthalate and polypropylene (PET/PP) laminated multifilm, and the membrane as PP. Potential degradation-associated bands were analyzed in the spectra.

2.3 Mechanical Analysis

The outer package and the internal filters, verified as polypropylene, from the samples were analyzed in a mechanical test. Due to typical high natural degradation in weathering conditions (such as heat and sunlight) of this type of resin [11, 20], polypropylene-based waste is an interesting candidate for assessing its mechanical behavior during recycling procedures [55]. In order to propose a simplified way to process and recycle such residue, a manual grinding was followed, leading to roughly pieces with approximately 3–5 mm size. The ground size is compatible with the virgin PP pellets H 503, a PP homopolymer with additives intended for general purposes, with a melt flow rate of 3.5 g/10 min (Braskem® S/A, Triunfo, RS, Brazil).

An electronic analytical balance AUW-220D (Shimadzu® Corp., Kyoto, Japan) was used for measuring different mixture concentrations of ground coffee capsules residues along with virgin PP. To analyze the different mechanical behavior of the ground capsule material (CM), three concentration ratios in weight of CM, being completed with virgin material when needed, were: 100%, 85%, and 70%. All materials were firstly dried in a hot-air oven at 80 °C for about 5 h, to avoid the presence of humidity and, consequently, the formation of air bubbles during the molding process. For each concentration mixture, eight ASTM D638-14 Type V specimens were prepared with Haake Minijet II (Thermo Fisher® Scientific Inc., Waltham, MA, USA) micro-injection molding machine. Besides the mixture samples, eight control specimens using 100% virgin material were also injected for comparison. The injection molding parameters were based on the previous work with recycled PP by the authors [55] and consist of: melting temperature of 220 °C; mold temperature of 60 °C; injection pressure and post pressure of 400 bar and 380 bar, respectively; and injection time and post pressure time of 2 s and 4 s, respectively. Finally, the mechanical performance of the injected specimens was evaluated and compared using a tensile test, via ASTM D638-14. Tensile testing machine EZ-LX series (Shimadzu® Corp., Kyoto, Japan) was used with a speed of 100 mm/min, along with Trapezium X (Shimadzu® Corp., Kyoto, Japan) proprietary software.

2.4 Mold Design and Recycling Procedures

With the objective of increase the market value of the capsule in this experimental study, a small volume of residue was required. Therefore, samples with the same characterized PP material were sorted by color and, in the same way as used in the mechanical tests, were also manually ground into approximately 3–5 mm size, in a coarsen way. As verified in the literature, polymeric coffee capsules can be recycled either by ground, extrusion, and injection, with a relatively small difference between their mechanical performance [51], besides showing reasonable weathering resistance [22]. Due to the intended application in contemporary jewelry with artisanal fabrication, and thus having no need for higher mechanical strength in the

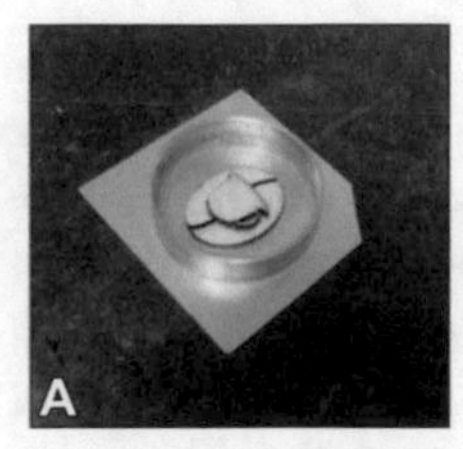

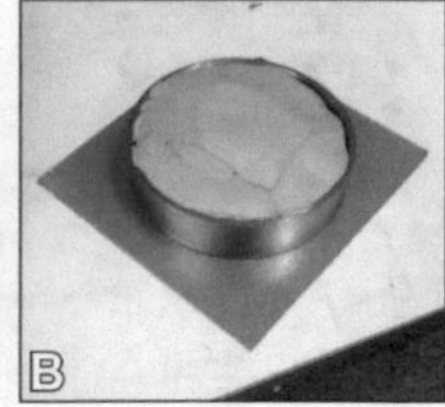

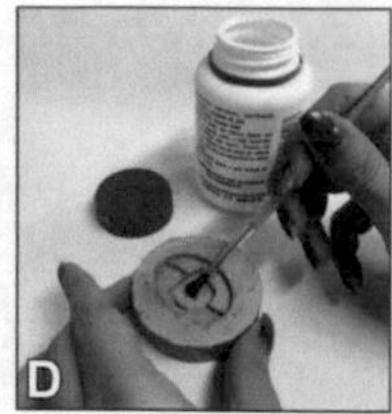

Fig. 2 Mold design procedure: **A** laser-cut piece in a stainless-steel container; **B** application of compound silicone rubber; **C** thermosetting in vulcanizer; **D** finished mold coating with releasing agent

recycled material, the recycling process was simplified by means to use equipment most commonly found in jewelry workshops. Likewise, the recycling procedure for this experimental jewelry application could be realized using different equipment, as long as following the required temperatures.

Two jewelry pieces were designed using the CAD software Rhinoceros® 3D (McNeel®, Seattle, WA, USA), consisting of a pendant and a ring, in which the recycled material was applying as a gemstone-like. The designed pieces were first cut in an MDF board using EXLAS-X4 1410 plotter CO_2 laser machining (Jinan XYZ Machinery® LLC., Jinan, SD, China), with a laser power of 40 W, with a speed of 15 mm/s, and details were then engraved at a speed of 95 mm/s. Figure 2 shows the molding procedure. Pieces were placed inside a stainless-steel container (Fig. 2A) and covered with compound silicone rubber (Fig. 2B) MA4000E (Kinner® Silicone Rubber Indústria e Comércio Ltda., Ribeirão Pires, SP, Brazil). After that, the silicone was thermoset in a vulcanizer (Zezimaq® Indústria e Comércio de Máquinas e Fornos Ltda., Belo Horizonte, MG, Brazil) for 1 h at 180 °C (Fig. 2C), and the MDF pieces were removed. The final mold was coated with a release agent (Fig. 2D) before the recycling procedures.

Giving the capsule material identified as polypropylene, the initial temperature of the recycling tests was based on its melting point of around 170 °C [75] to avoid overheating and damaging of the material. The manufacturing procedures of the pieces consisted of placing sequential thins layers of the ground secondary polymer inside the mold and heating it in a jewelry muffle oven (Zezimaq®, Belo Horizonte, MG, Brazil) until melting. Increasing temperatures were used at different tests, due to the characteristics of the oven being originally used as jewelry-aimed until a satisfactory fill of all the details from the jewels designed. After finished, the recycled pieces were included in the manufacturing of the jewels, using traditional jewelry techniques, that were detailed described.

3 Results and Discussion

3.1 FT-IR Characterization

Figure 3 shows the resulting FT-IR transmittance spectra for three different coffee capsule samples (#1, #2, and #3). Regions from analyzed samples were retrieved from the main polypropylene external body region due to having the most weathering exposure, in addition to the region having the largest volume for bulk usage in recycling procedures. The bottom spectrum corresponds to the standard reference library material (LM) for PP and shows the most typical bands associated with PP [57].

Overall, the three examples show different levels of degradation in particular bands. The band at 3319 cm^{-1}, which is observed in the three examples, can be assigned to some hydroxyl groups ($O - H$ bonds) that are related to the exposure of PP to ultraviolet rays, thus increasing the polymer degradation capacity [10, 22]. Though present in all samples, the peak cannot be seen in the standard PP library material. The formation of peaks in the regions at 1745 cm^{-1} and 1652 cm^{-1} is related to the formation of carbonyl groups ($C = H$ bonds) associated with PP chain degradation, also occurring by photooxidation of the material by weather conditions [47, 55]. Despite being noticed in all examples, sample #3 showed greater peaks in those regions that, once again, have no greater peaks in LM. As for smaller peaks at 730 cm^{-1}, which are not a PP characteristic band and are mostly showed in sample #3, may be related to the presence of polyamide as a contaminant in the analyzed material [22, 53]. Most of the analyzed samples presented degradation levels in agreement with those found in weathered coffee capsule samples analyzed by [22].

3.2 Mechanical Tests

Figure 4 presents the tensile test results for different concentration ratios recycled capsule material (CM) along with the virgin polypropylene (PP) control material, with the bars representing the average values and the error bars the standard deviation. The elastic modulus (Fig. 4A) of the 100% CM PP presented an average value around 21% lower than the control PP material. When using 85% and 70% of recycled capsule material the average values were approximately 9% and 4% lower, respectively. When compared to results in the literature [24], considering the tensile tests performed with 100% recycled PP from the body of coffee capsules, the values are around 18% higher, still, they both presented a high standard deviation. As expected, with recycled material the variation between specimens is higher due to the presence of multiple material contaminants and degradation levels, confirming the results of the FT-IR analyses. Also due to the coarse and manual grinding process, and the absence of an extrusion step prior to injection molding, the recycled pallets were

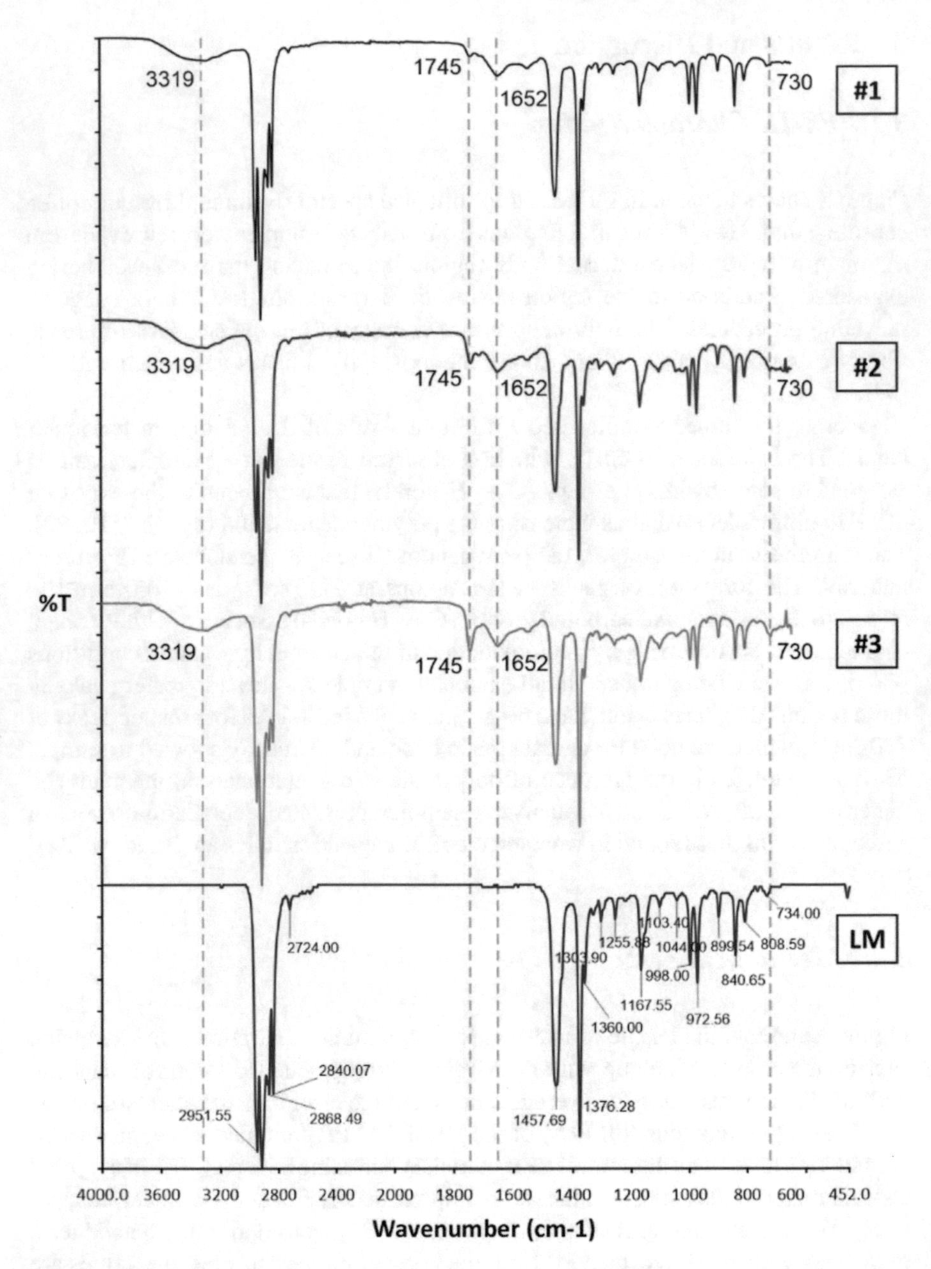

Fig. 3 FT-IR spectra of polypropylene coffee capsule samples (#1, #2, and #3) with different levels of degradation and a reference spectrum for the standard library material (LM)

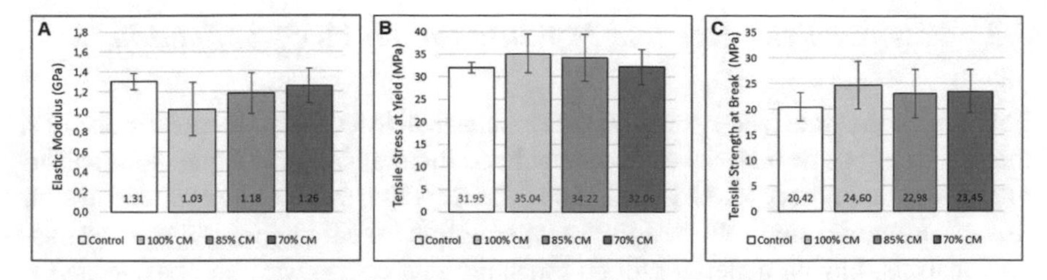

Fig. 4 Mechanical tensile test results for the control PP and recycled capsule material (CM) in different concentration ratios of 100%, 85%, and 70%: **A** elastic modulus; **B** tensile stress at yield; and **C** tensile strength at break. Values are presented as averages with error bars as the standard deviations for each ratio

not completed mixed. Nevertheless, their mechanical behavior is within acceptance when compared to the control material.

Similar to the elastic modulus analyses, the tensile stress at yield results (Fig. 4B), also show a different pattern with the variation of the concentration levels of capsule material. The 100% CM exhibit a 10% increase in tensile strength at yield when compared to the control material, resulting in a similar value reported by the comparable literature [24]. In the same way, increasingly adding more virgin PP in the mixture (85% CM and 70% CM) also leads to a reduction in the tensile stress at yield until similar results to the control material. The results represent an embrittlement performance in the specimen. As expected, the addition of degraded recycled polymers in different concentration ratios leads to a decrease in the overall molecular weight [55]. Consequently, this effect also halters an increase in crystallinity of the samples [40, 63] and induces a loss in the elastic phase properties of the polymer [11]. Similar behavior was seen in the tensile strength at break (Fig. 4C), where the 100% capsule material presented an approximately 20% higher value than the control PP; and the 85% and 70% mixtures showed values approximately 13% and 15% higher, respectively. Once again, when working with the recycling of degraded material, particularly PP, higher standard deviation values are expected [20, 55, 72]. In the tests realized, the 100% recycled capsule material showed standard deviation values for elastic modulus, tensile stress at yield, and tensile strength approximately 227%, 276%, and 71% higher than the control material, respectively. Overall, even if some noticeable performance losses were found when working with the recycled material, it still would allow it to be employed in different applications. Particularly, if the new intended use does not have a similar function and, therefore, does not require comparable mechanical properties, the material can be considered. Considering the purpose of this experimental research, still, the recycled material was also sough due to its visual characteristics, thus further reducing the need for a certain level of performance.

3.3 Polymer Recycling and Molding for Jewelry Application

Following some satisfactory results in the mechanical tests, for the intended purposes, the recycled capsule material can be considered for application as a gemstone substitute in a contemporary jewelry collection. The recycling and molding procedures for this experimental step followed the same simplified methods used in the mechanical analysis, having material ground manually in a coarse way, and basing on the melting temperature of PP for the fabrication of the pieces. However, instead of using some injection molding equipment as followed in the tensile tests, we propose a more simplified approach, which could be realized without specialized equipment. Figure 5 exemplifies the steps taken until finding the more appropriate parameters for successfully melting the recycled material into the desired molded shape. For the first test, a thin layer of ground plastic was placed on the mold (Fig. 5A), being heated at 185 °C in the preheated jewelry muffle oven (Fig. 5B) for about 15 min. A temperature lower than the that used for melting PP was chose to avoid damaging the material. After that, most of the material did not show significant melting (Fig. 5C), and therefore the temperature was raised to 200 °C, with the oven priorly preheated, resulting in a molted material (Fig. 5D). With the melting results showing no apparent damage on the material's surface, a second layer was placed into the mold (Fig. 5E). Following the same temperatures, the material melted completely (Fig. 5F), however,

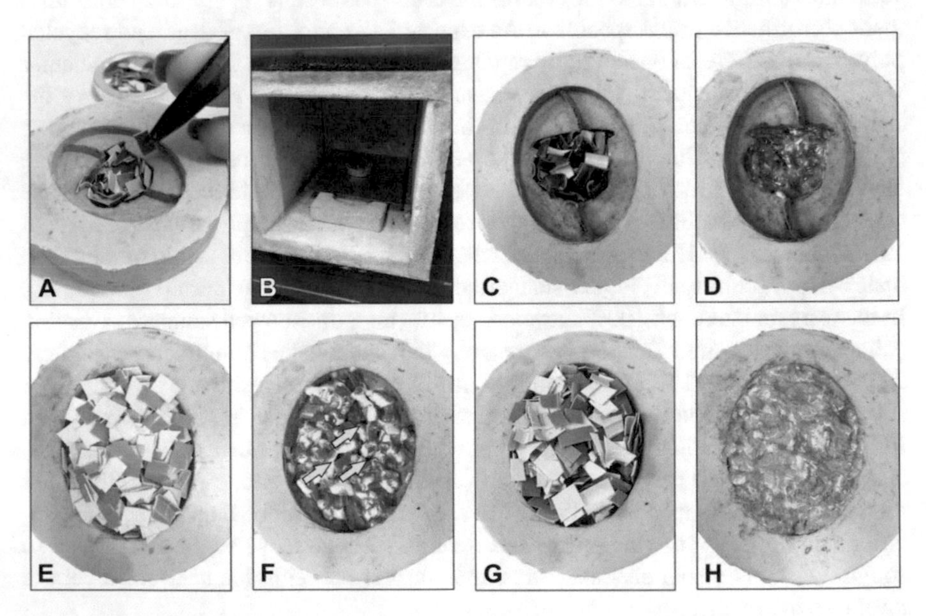

Fig. 5 Simplified material molding tests: **A** first layer of ground material placed into the mold; **B** heating procedure in a jewelry muffle oven; **C** material without melting at 185 °C; **D** first layer, melted at 200 °C; **E** second layer of material placed into the mold; **F** second layer melted with the presence of empty spaces (yellow arrows); **G** third layer placed into the mold; **H** third layer, melted with mechanical pressure applied

some empty spaces were noticed (yellow arrows). For the third and final layer placed (Fig. 5G), a mechanical pressure was applied on the top of the mold using a steel plate. The result of the third layer test was a more compacted melted material (Fig. 5H). Due to the lack of pressure typically obtained when using a specific injection molding equipment, for instance, as the specimens used in the tensile tests, the molten pieces lie only in the action of gravity and additional mechanical pressure to adapt into the mold cavity. Even if this could impair the mechanical performance of the applied material even further, this method was still preferred in the experiment because of being more feasible and replicable, thus requiring no specialized equipment. Naturally, if available, one could use jewelry vacuum furnaces to reduce the amount of air imprisoned inside the molten material.

The same procedures were followed for the manufacturing of the second piece, which was intended for being applied on a ring. The first produced pieces for jewelry application can be seen in Fig. 6, including the one used for the manufacturing of a pendant (Fig. 6A) and a ring (Fig. 6B). In the details A1 and B1, both silicone molds produced are shown, and their respective original designed pieces, made of CO_2 laser-engraved MDF, are presented in A2 and B2. The pieces were created using a coffee bean as a general design theme, with the inclusion of a coffee cup in the pendant. Details A3, A4, and B3 exemplify tested pieces with discontinuities and defects that did not receive the appropriate amount of material and mechanical pressure in the mold, by using a steel plate. Final pieces intended for the manufacturing of the

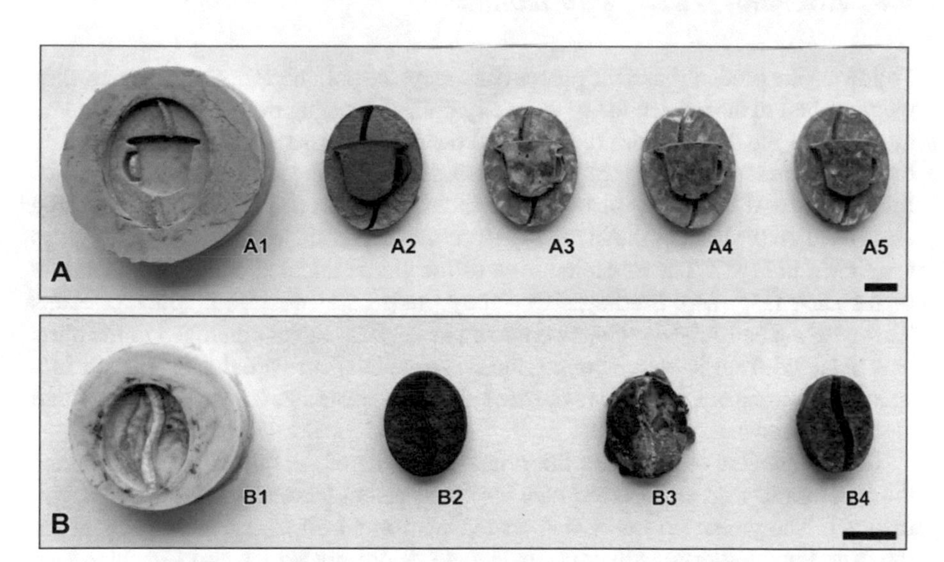

Fig. 6 Manufactured pieces for jewelry application. **A** Pendant piece production: (A1) silicone mold; (A2) original designed piece in laser-engraved MDF; (A3) first and (A4) second pieces produced with the presence of discontinuities and air bubbles; (A5) third piece without apparent defects. **B** Ring piece production: (B1) silicone mold; (B2) original designed piece in laser-engraved MDF; (B3) first piece produced with the presence of discontinuities and air bubbles; (B4) second piece without apparent external defects. Scale bars are 10 mm

pendant and the ring are shown in A5 and B4, respectively. External dimensions were approximately 36.5 × 28.5 mm for the pendant piece, and 17.0 × 10.3 mm for the ring one. No apparent defects or discontinuities were observed on the final pieces.

The experimental tests indicated that being identified as PP, the material could only be recycled and molded, without a proper injecting system, by using a higher temperature set in the jewelry muffle oven, at 200 °C. In a simplified way, using only the atmospheric pressure for the molding process, *i.e.,* without the use of injection molding techniques, some defects such as air spaces were noticed. Even though this would not make the piece significantly fragile for this application, considering that it was produced aiming at it being placed as a decorative gemstone-like in a jewelry collection, such defects could diminish its visual aesthetics by not filling the entire cavity of the design. Thus, the need for using more material and applying a top pressure with a steel plate was observed. The temperature could also be slightly increased with aims to decrease the viscosity of the melted material and hence favoring the filling of the mold, however, recycled PP with an indication of degraded phases has shown to be more sensitive to different temperatures [40, 55]; thus requiring a finer control of the heat source, which is somewhat difficult to be achieved with a conventional jewelry muffle oven though.

3.4 Artisanal Jewelry Fabrication

Following the production of the pieces from recycled polymeric coffee capsules, they were applied in the fabrication of a jewelry collection using traditional and artisanal methods. In the design, both the ring and pendant polymeric pieces were involved by plane silver structures for protection. For the ring application, a supporting piece for the base and a rim were also manufactured, as for the pendant, a chain coupling was added on the back of the involving structure. The main jewelry fabrication steps are shown in Fig. 7. The manufacturing of the pieces initiated with the preparation of the alloy (Fig. 7A). Sterling silver alloy graded 925 was used, which contains 92.5-wt% Ag and 7.5-wt% Cu, also known as Ag 925. Silver employed in the alloy was obtained from jewelry scrap, which was initially recovered and purified in a specialized company. Copper was extracted from e-waste sources, such as household electronics and uncoated wires.

Casting process started with the primary heating of the refractory crucible and the ingot mold with an oxyacetylene torch to prevent thermal shock in the molten material. The preheated mold was coated with mold oil as a releasing agent for the casted ingot. Sterling silver components (weight-graded Ag and Cu) were then placed in the heated crucible and melted with the torch (Fig. 7B), having its melting point of about 890 °C [68]. The alloy was poured in a horizontal, open jewelry ingot mold, shaped into a wire with a section of approximately 9 × 9 mm. The hot sterling silver ingot was released from the mold and submerged into a pickle solution of 10-wt% H_2SO_4 in water, for about 2 min to remove surface oxide. The piece was then

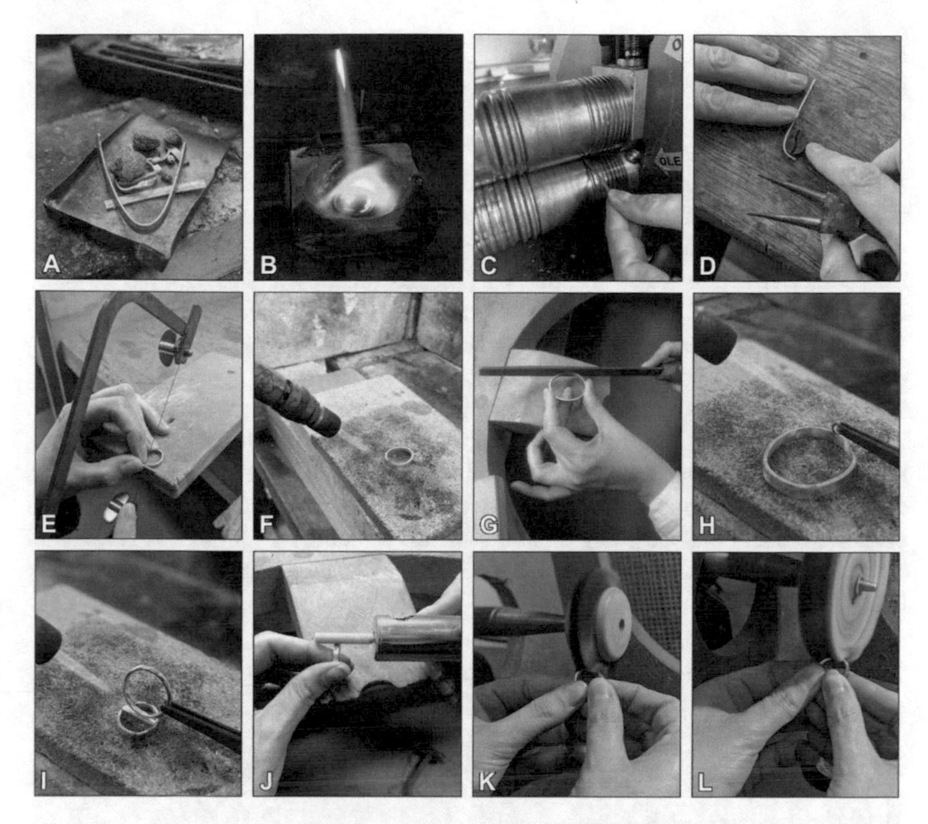

Fig. 7 Artisanal jewelry fabrication processes: **A** alloy preparation with copper recycled from e-waste and sterling silver recovered from other pieces (prior from purification process); **B** silver alloy melting with an oxyacetylene torch; **C** electric rolling milling; **D** manual shaping around the recycled pieces; **E** cutting pieces with jeweler's saw; **F** welding with liquified petroleum gas (LPG) torch; **G** surface smoothening with different coarseness files; **H** welding the chain coupling for the pendant; **I** welding the base support and rim of the ring; **J** sanding process with an abrasive paper in a flexible shaft; **K** polishing with a natural bristle wheel; **L** finishing with a muslin buffing wheel

submerged into a neutralizing bath with 5-wt% $NaHCO_3$ in water for about 2 min. Finally, the ingot is rinsed in water.

The shaping process of the produced ingot initiated with a rolling process (Fig. 7C), using a G3 130 electric rolling mill (Laminadores Rio Preto Indústria e Comércio Ltda., São José do Rio Preto, SP, Brazil). Initially, the ingot was rolled into a wire, as shown in Fig. 7C. Part of this initial wire was cut for the manufacturing of the ring rim, and the remaining part was once again roll-shaped into a sheet. The sheet part was then manually shaped in a way to involve the recycled polymeric pieces, with the use of jewelry pliers (Fig. 7D), with the excess being cut off with a jeweler's saw (Fig. 7E). Both involving parts (from ring and pendant) were welded using a liquefied petroleum gas (LPG) torch (Fig. 7F). A silver brazing alloy was manufactured, consisting of 65% Ag, 20% Cu, and 15% Zn, which has a melting point of around 670 °C [68]. No commercial brazing alloy was used in this experiment due to the possibility of having toxic Cadmium contamination in its composition [68].

After welding, the surfaces of the involving protective parts were processed with files of different coarseness (Fig. 7G). Files were used in a way to remove surface defects, edge burrs, and unwanted excess solder, to adjust the piece to the final desired dimensions, as well as for finishing and smoothing. A chain coupling was welded in the pendant structure (Fig. 7H). Likewise, the base support and the rim were welded into the ring (Fig. 7I). The welding of both parts also consisted of using the same silver brazing alloy and the LGP torch. Similar to the involving structures, the ring rim manufacturing process consisted of shaping the previously separated silver wire into a rim using pliers, and by cutting the excess material, welding, and filing it. Finally, surface polishing procedures were conducted in the ring and pendant for a smooth and shiny appearance. A manual sanding process was followed, using an abrasive paper attached to a flexible shaft by a split-shank mandrel (Fig. 7J). Both parts were also polished using a natural bristle wheel (Fig. 7K) in a bench-model polishing lathe with a dust collector and finally finished with a muslin buffing wheel (Fig. 7L).

Fig. 8 Final jewelry collection manufactured via artisanal processes: sterling silver **A** pendant and **B** ring, with recycled polymeric capsules as gemstones-like; and usage demonstration of the **C** pendant and **D** ring

Figure 8 presents the final manufactured pieces. The jewelry collection proposed consists of a pendant (Fig. 8A) and a ring (Fig. 8B) made of sterling silver with the application of the recycled polymeric coffee capsules as an alternative for gemstones. Figure 8C and D show a usage demonstration of the pendant and ring, respectively.

The produced pieces explored the heterogeneous surface design of secondary plastics to highlight the individuality of each piece, thus contributing to valorize the often-neglected material. Instead of using the recycled polymeric capsules as a component with structural properties or as an aggregate or reinforcement in composites, the material was employed as an aesthetic element. This application demonstrated that a simplified recycling approach on degraded plastics, *i.e.,* using coarsely ground material, contributes to an irregular surface pattern and consequently to a unique design for each piece manufactured. Regarding its application in jewelry, the inclusion of recycled material in a central role in a pendant or ring could generate favorable consumer attitudes [58]. Particularly when considering the consumption of luxury-related products, where both quality and exclusivity are important attributes [16], the unique potential of secondary plastic as a decorative element should be explored.

4 Implications for Waste Valorization

Coffee plays a major role in the Brazilian consumption market, reaching almost a complete household presence [46]. This raises a significant issue regarding how the beverage is being packaged, *i.e.*, if whether or not in an environmentally sound way. Data shows a recent increasing demand for premium varieties of coffee in the country [1], encompassing a growing trend in both retail volume and value data in consuming single-serve coffee capsules [27]. In countries with a still incipient system of Urban Solid Waste treatment, defining the materials and processes used in the manufacturing of packaging with a rapidly growing demand in addition to such an ephemeral usage is of great importance [4]. Considering a global production estimative of about 59 billion capsules annually, when the vast majority is likely to be disposed of in landfills [32], this matter is even more pronounced.

Single-serve capsules are known for having one of the highest levels of energy consumption among coffee preparation methods [23], primarily due to the materials selection [15, 35]. For instance, in the coffee capsule market, two main types of base materials can be highlighted: aluminum- or polymer-based containers, or even a mixture of them, and that increases the complexity of managing its end of life once as a residue [34]. If the first one benefits from a significantly greater demand as a secondary source from waste, the second one struggles to find potentially interested buyers. Once the market price for primary aluminum is high due to its extremely energy-demanding process [8, 31], the need for recycled sources contributes to its high demand. Consequently, even if the production of capsules with aluminum only leads to higher energy consumption compared to polymer-based ones [23], the choice for the metal-based option increases its recyclability, which is necessary for the current Brazilian USW treatment [54]. On the other hand, polymer residues generally cannot repeat the same success, partially because of its commodity price when a virgin resin. In addition to the need for the buyer to reprocess the plastic residue, by cleaning, pelletizing, and recycling, which all increases the investment required, the physical and mechanical properties of the material naturally decrease, thus making it difficult for it to be applied on a product similar to what it was originally. Furthermore, when some waste does not find interested most of it ends up being disposed into landfills or, even worst, into open-air dumps. With a decomposing time of centuries [28, 40] and an urging menace to marine life [41] each incorrectly discarded plastic packaging poses a major hazard on all ecosystems. Considering that just over 1% of the plastic waste is estimated to be properly recycled in Brazil [73], improving the country's general perception of this type of material may be a turning point. By making the residue more attractive to potential buyers the material end-of-life environmental impact diminishes significantly. Besides the environmental threat of single-serve coffee packaging solutions, whom the waste treatment depends on also must be taken into consideration as an urgent matter in Brazil. Virtually all the USW in the country's major cities, where the majority of people live, hinges on the commercial success of selling them to interested buyers [17, 54]. And these waste trades are almost entirely performed by the workers of Sorting Units, which receive and sort dry waste in facilities all over the country. Moreover, those workers are mainly low-income people who heavily depend on the sale of the waste for their revenue. As a result, increasing the secondary market value of plastics not only presents benefits to the environment but also the social and the economic spheres of a sustainable economy.

This research aimed to explore the hidden value of secondary plastic sourced from single-serve coffee capsules. Using a simplified recycling approach, without employing sophisticated techniques and equipment for the reprocessing of the material, we evaluate the possibility of reusing it with a different purpose. Firstly, our FT-IR analysis showed that some level of degradation is expected to be found on the material identified as polypropylene since it was retrieved from domestic and commercial waste and may have passed through the action of weathering. By recycling it using a manual and coarse grinding for the fabrication of the test specimen, we verified in a preliminary mechanical tensile test that even if the material would not present the same elastic and plastic properties as the compared virgin material, as expected, it could still be used on less mechanically demanding applications, such as in a jewelry collection. One of the least mentioned attributes of secondary plastic is its potential aesthetics attributes [60]. By exploring the heterogeneity and uniqueness of a piece produced with waste materials of diverse colors, we could achieve some interesting surface patterns. As a way to increase the perceived value of the recycled coffee capsule material, we propose its use as a gemstone-like substitute in the manufacturing of a pendant and a ring, also using secondary-sourced silver and copper, in the metal alloy. Even if some specialized knowledge may be required for the manufacturing of the jewelry pieces themselves, the suggested method for the "polymeric gemstones" can be replicated with no need for a particular set of equipment. Therefore, this suggests that more people could employ this material in newer applications aiming at delivering a product with a higher added value.

The bottom-line from this experimental research is to demonstrate that polymer packaging waste should not be explored only by means of its value as secondary material. With little investment, one can still benefit from its properties—even if reduced—for the manufacturing of simplified and high-valued goods. The known willingness to pay more for a product containing some level of recycled material [39] confirms that the general public is open to newer approaches following this trend. Polymer waste valorization should be stimulated as a way to reduce the environmental impact as well as to improve the quality of life of low-income workers related to waste treatment, particularly in developing countries. Despite little efforts are put into research and development regarding aesthetic attributes in recycled material, this has the potential of turning into one of the most significant characteristics of secondary plastics. Having such a deep connection to the perception of a sustainable product [38], more could be invested in visual attributes of these types of materials. The inhomogeneity of colors and a rough surface finish becomes desirable features that emphasize the perception of a sustainable product. Finally, contemporary jewelry rises as an important field for embracing newer materials and processes, by both gaining and giving them value as they are combined with traditional techniques. Henceforth, jewelry design must also be seen as a way to approach sustainable issues of dealing with problematic packaging materials.

5 Conclusions

Recent trends in the Brazilian coffee market led to the growth of a new type of residue, polymeric coffee capsules, which contributes to highlighting the country's already poor capacity to deal with plastic waste. Among the main reasons for the low recyclability and recovery of plastic waste is the material's intrinsic low value, hence decreasing the economic interest of stakeholders to acquire it. For this reason, new approaches to valorize secondary plastics can be a way to minimize its waste, given that its reprocessing is encouraged by the creation of different, high-value products. One way to apply recycled plastics is in contemporary jewelry, where new materials and methods are combined with traditional techniques for the creation of innovative pieces with high added value.

This study presented an experimental procedure of valorizing polymeric coffee capsule wastes employing a simplified recycling approach for the application of the secondary material as gemstone-like pieces in contemporary jewelry. Regarding the over 30 polymeric coffee capsules collected from USW, most of them presented different levels of degradation, as analyzed via FT-IR, mainly related to photodegradation and the presence of contaminants. PP parts were selected for a simplified mechanical recycling process, in which the material was coarsely ground and melted in a jewelry oven, to produce new parts using a developed silicone mold. Produced secondary plastic parts were used as a gemstone alternative for the development of a coffee-themed, jewelry collection.

Sterling silver was used for the manufacturing of the jewelry collection, in which the alloy was produced using silver obtained from jewelry scrap and copper recovered from electronic waste. Different artisanal jewelry manufacturing methods were followed, including rolling milling, manual shaping and cutting, welding, sanding, and polishing a pendant and a ring. The produced jewelry pieces took advantage of the nonhomogeneous aesthetic attributes of the secondary plastic, placing the polymeric parts in plain sight, as opposed to using them as a structural or reinforcement material. The findings of this experimental study support the suggestions that (1) new applications for polymeric coffee capsule waste can contribute to its valorization, with aims to increase the economic interest in such residue; (2) degraded secondary plastics can still be employed as a decorative element, such as a replacement for gemstones in contemporary jewelry, when no particular structural function is required; and (3) simplified recycling procedures with roughly ground materials can be explored in ways to provide unique visual characteristics to products made of secondary plastics.

Acknowledgements Authors thank the "National Council for Scientific and Technological Development—CNPq" for supporting this study through the project "Chamada Universal MCTIC/CNPq 2018". This study was financed in part by the Coordenação de Aperfeiçoamento de Pessoal de Nível Superior—Brasil (CAPES)—Finance Code 001. The authors also thank the Design and Materials Selection Lab (LDSM) from the Federal University of Rio Grande do Sul for the usage of the FT-IR equipment.

References

1. ABIC (2020) Tendências do mercado de café [Coffee market trends]
2. ABIPLAST (2019) Perfil 2018 da Indústria Brasileira de Transformação de Material Plástico [2018 Profile of the Brazilian Plastics Processing Industry]
3. ABRELPE (2020) Panorama dos resíduos sólidos no Brasil 2018/2019 [Panorama of solid waste in Brazil 2018/2019]
4. Abuabara L, Paucar-Caceres A, Burrowes-Cromwell T (2019) Consumers' values and behaviour in the Brazilian coffee-in-capsules market: promoting circular economy. Int J Prod Res 57:7269–7288. https://doi.org/10.1080/00207543.2019.1629664
5. Ahmadinia E, Zargar M, Karim MR, Abdelaziz M, Shafigh P (2011) Using waste plastic bottles as additive for stone mastic asphalt. Mater Des 32:4844–4849. https://doi.org/10.1016/j.matdes.2011.06.016
6. Al-Salem SM, Lettieri P, Baeyens J (2010) The valorization of plastic solid waste (PSW) by primary to quaternary routes: from re-use to energy and chemicals. Prog Energy Combust Sci 36:103–129
7. Allen J (2014) The immaterial of materials. In: Materials experience. Elsevier, pp 63–72
8. Ashby MF (2013) Materials and the environment: eco-informed material choice, 2nd edn. Butterworth-Heinemann, Oxford
9. Ashby MF, Johnson K (2013) Materials and design: the art and science of material selection in product design. Elsevier Science & Technology
10. Auta HS, Emenike CU, Jayanthi B, Fauziah SH (2018) Growth kinetics and biodeterioration of polypropylene microplastics by Bacillus sp. and Rhodococcus sp. isolated from mangrove sediment. Mar Pollut Bull 127:15–21. https://doi.org/10.1016/j.marpolbul.2017.11.036

11. Azapagic A, Emsley A, Hamerton L (2003) Polymers, the environment and sustainable development. John Wiley & Sons, Ltd, Chichester, UK
12. Ba'ai NM, Hashim HZ (2015) Waste to wealth: the innovation of areca catechu as a biomaterial in esthetics seed-based jewelry. In: Proceedings of the international symposium on research of arts, design and humanities (ISRADH 2014). Springer Singapore, Singapore, pp 373–381
13. Barlow CY, Morgan DC (2013) Polymer film packaging for food: an environmental assessment. Resour Conserv Recycl 78:74–80. https://doi.org/10.1016/j.resconrec.2013.07.003
14. Breslin VT, Senturk U, Berndt CC (1998) Long-term engineering properties of recycled plastic lumber used in pier construction. Resour Conserv Recycl 23:243–258. https://doi.org/10.1016/S0921-3449(98)00024-X
15. Brommer E, Stratmann B, Quack D (2011) Environmental impacts of different methods of coffee preparation. Int J Consum Stud 35:212–220. https://doi.org/10.1111/j.1470-6431.2010.00971.x
16. Buenstorf G, Cordes C (2008) Can sustainable consumption be learned? A model of cultural evolution. Ecol Econ 67:646–657. https://doi.org/10.1016/j.ecolecon.2008.01.028
17. Campos HKT (2014) Recycling in Brazil: challenges and prospects. Resour Conserv Recycl 85:130–138. https://doi.org/10.1016/j.resconrec.2013.10.017
18. Cappellieri A, Tenuta L, Testa S (2020) Jewellery between product and experience: luxury in the twenty-first century. pp 1–23
19. Cidade MK, Palombini FL, Duarte L da C, Paciornik S (2018) Investigation of the thermal microstructural effects of CO_2 laser engraving on agate via X-ray microtomography. Opt Laser Technol 104:56–64. https://doi.org/10.1016/j.optlastec.2018.02.002
20. Crawford CB, Quinn B (2017) Physiochemical properties and degradation. In: Microplastic pollutants. Elsevier, pp 57–100
21. de Abreu e Lima CE, Lebrón R, de Souza AJ, Ferreira NF, Neis PD (2016) Study of influence of traverse speed and abrasive mass flowrate in abrasive water jet machining of gemstones. Int J Adv Manuf Technol 83:77–87. https://doi.org/10.1007/s00170-015-7529-9
22. de Bomfim ASC, Maciel MMÁD, Voorwald HJC, Benini KCC de C, de Oliveira DM, Cioffi MOH (2019) Effect of different degradation types on properties of plastic waste obtained from espresso coffee capsules. Waste Manage 83:123–130. https://doi.org/10.1016/j.wasman.2018.11.006
23. de Figueiredo Tavares MP, Mourad AL (2020) Coffee beverage preparation by different methods from an environmental perspective. Int J Life Cycle Assess 25:1356–1367. https://doi.org/10.1007/s11367-019-01719-2
24. Domingues MLB, Bocca JR, Fávaro SL, Radovanovic E (2020) Disposable coffee capsules as a source of recycled polypropylene. Polímeros 30. https://doi.org/10.1590/0104-1428.05518
25. EPA (2019) Advancing sustainable materials management: 2017 fact sheet
26. Eriksson O, Reich MC, Frostell B, Björklund A, Assefa G, Sundqvist JO, Granath J, Baky A, Thyselius L (2005) Municipal solid waste management from a systems perspective. J Clean Prod. Elsevier Ltd, pp 241–252
27. Euromonitor International (2021) Coffee in Brazil. 10
28. Eyerer P (ed) (2010) Polymers - opportunities and risks I. Springer, Berlin Heidelberg, Berlin, Heidelberg
29. Fuss M, Vasconcelos Barros RT, Poganietz WR (2018) Designing a framework for municipal solid waste management towards sustainability in emerging economy countries - an application to a case study in Belo Horizonte (Brazil). J Clean Prod 178:655–664. https://doi.org/10.1016/j.jclepro.2018.01.051
30. Guerra AL, Cidade MK (2019) Design e Joalheria: desenvolvimento de uma joia com óleos essenciais para o alívio de sintomas alérgicos [Design and jewelry: development of a jewelry with essential oils for the relief of allergic symptoms]. Des e Tecnol 9:115–130. https://doi.org/10.23972/det2019iss18pp115-130
31. Haggar S El (2010) Sustainable industrial design and waste management: cradle-to-cradle for sustainable development. Academic Press

32. Halo Coffee (2020) Is recycling aluminium coffee capsules really the way forward? https://halo.coffee/blogs/blog/is-recycling-aluminium-coffee-capsules-really-the-way-forward. Accessed 12 May 2020
33. Hesse RW (2007) Jewelrymaking through history: an encyclopedia. Greenwood Press, Westport, Connecticut, USA
34. Hicks AL (2018) Environmental implications of consumer convenience: coffee as a case study. J Ind Ecol 22:79–91. https://doi.org/10.1111/jiec.12487
35. Humbert S, Loerincik Y, Rossi V, Margni M, Jolliet O (2009) Life cycle assessment of spray dried soluble coffee and comparison with alternatives (drip filter and capsule espresso). J Clean Prod 17:1351–1358. https://doi.org/10.1016/j.jclepro.2009.04.011
36. Ibáñez-Forés V, Bovea MD, Coutinho-Nóbrega C, de Medeiros-García HR, Barreto-Lins R (2018) Temporal evolution of the environmental performance of implementing selective collection in municipal waste management systems in developing countries: A Brazilian case study. Waste Manage 72:65–77. https://doi.org/10.1016/j.wasman.2017.10.027
37. Karana E (2012) Characterization of "natural" and "high-quality" materials to improve perception of bio-plastics. J Clean Prod 37:316–325. https://doi.org/10.1016/j.jclepro.2012.07.034
38. Karana E, Nijkamp N (2014) Fiberness, reflectiveness and roughness in the characterization of natural and high quality materials. J Clean Prod 68:252–260. https://doi.org/10.1016/j.jclepro.2014.01.001
39. Klaiman K, Ortega DL, Garnache C (2016) Consumer preferences and demand for packaging material and recyclability. Resour Conserv Recycl 115:1–8. https://doi.org/10.1016/j.resconrec.2016.08.021
40. La Mantia FP (2002) Handbook of plastics recycling. Rapra Technology Ltd, Shrewsbury
41. Law KL, Thompson RC (2014) Microplastics in the seas. Science (80) 345:144–145. https://doi.org/10.1126/science.1254065
42. Liikanen M, Havukainen J, Viana E, Horttanainen M (2018) Steps towards more environmentally sustainable municipal solid waste management – a life cycle assessment study of São Paulo, Brazil. J Clean Prod 196:150–162. https://doi.org/10.1016/j.jclepro.2018.06.005
43. Lopez G, Artetxe M, Amutio M, Bilbao J, Olazar M (2017) Thermochemical routes for the valorization of waste polyolefinic plastics to produce fuels and chemicals. A review. Renew Sustain Energy Rev 73:346–368
44. Löschke SK, Mai J, Proust G, Brambilla A (2019) Microtimber: the development of a 3d printed composite panel made from waste wood and recycled plastics. In: Lecture notes in civil engineering. Springer, pp 827–848
45. Macnab M (2011) Design by nature: using universal forms and principles in design. New Riders, Berkeley, USA
46. MAPA (2020) Exportação de café deve bater novo recorde em 2020, projeta setor [Coffee exports should hit a new record in 2020, sector projects]
47. Marsich L, Ferluga A, Cozzarini L, Caniato M, Sbaizero O, Schmid C (2017) The effect of artificial weathering on PP coextruded tape and laminate. Compos Part A Appl Sci Manuf 95:370–376. https://doi.org/10.1016/j.compositesa.2017.01.016
48. Marzouk OY, Dheilly RM, Queneudec M (2007) Valorization of post-consumer waste plastic in cementitious concrete composites. Waste Manage 27:310–318. https://doi.org/10.1016/j.wasman.2006.03.012
49. Merrington A (2017) Recycling of plastics. In: Applied plastics engineering handbook: processing, materials, and applications, 2nd edn. Elsevier Inc., pp 167–189
50. Milewski JO (2017) Additive manufacturing metal, the art of the possible. In: Springer series in materials science. Springer Verlag, pp 7–33
51. Montagna LS, Catto AL, Oliveira JB, Griebeler A, Santana RMC (2017) Reciclagem Mecânica de Cápsulas Plásticas de Café [Mechanical recycling of plastic coffee capsules]. In: 14° Brazilian congress of polymers [Congresso Brasileiro de Polímeros – CBPol]. Águas de Lindóis (SP)

52. Murakami F, Sulzbach A, Pereira GM, Borchardt M, Sellitto MA (2015) How the Brazilian government can use public policies to induce recycling and still save money? J Clean Prod 96:94–101. https://doi.org/10.1016/j.jclepro.2014.03.083
53. Oliver-Ortega H, Méndez J, Mutjé P, Tarrés Q, Espinach F, Ardanuy M (2017) Evaluation of thermal and thermomechanical behaviour of bio-based polyamide 11 based composites reinforced with lignocellulosic fibres. Polymers (Basel) 9:522. https://doi.org/10.3390/polym9100522
54. Palombini FL, Cidade MK, de Jacques JJ (2017) How sustainable is organic packaging? A design method for recyclability assessment via a social perspective: a case study of Porto Alegre city (Brazil). J Clean Prod 142:2593–2605. https://doi.org/10.1016/j.jclepro.2016.11.016
55. Palombini FL, Demori R, Cidade MK, Kindlein W, de Jacques JJ (2018) Occurrence and recovery of small-sized plastic debris from a Brazilian beach: characterization, recycling, and mechanical analysis. Environ Sci Pollut Res 25:26218–26227. https://doi.org/10.1007/s11356-018-2678-7
56. Park SH, Kim SH (2014) Poly (ethylene terephthalate) recycling for high value added textiles. Fash Text 1:1
57. Părpăriţă E, Nistor MT, Popescu M-C, Vasile C (2014) TG/FT–IR/MS study on thermal decomposition of polypropylene/biomass composites. Polym Degrad Stab 109:13–20. https://doi.org/10.1016/j.polymdegradstab.2014.06.001
58. Perez D, Stockheim I, Tevet D, Matan Rubin M (2020) Consumers value manufacturer sincerity: the effect of central eco-friendly attributes on luxury product evaluations. J Clean Prod 267:122132. https://doi.org/10.1016/j.jclepro.2020.122132
59. Plastics Europe (2019) Plastics – The facts 2019: an analysis of European latest plastics production, demand and waste data. 42
60. Rognoli V, Bianchini M, Maffei S, Karana E (2015) DIY materials. Mater Des 86:692–702. https://doi.org/10.1016/j.matdes.2015.07.020
61. Rognoli V, Karana E (2014) Toward a new materials aesthetic based on imperfection and graceful aging. In: Materials experience. Elsevier, pp 145–154
62. Ross S, Evans D (2003) The environmental effect of reusing and recycling a plastic-based packaging system. J Clean Prod 11:561–571. https://doi.org/10.1016/S0959-6526(02)00089-6
63. Scott G (1999) Polymers and the environment. Royal Society of Chemistry, Cambridge, UK
64. Singh N, Hui D, Singh R, Ahuja IPS, Feo L, Fraternali F (2017) Recycling of plastic solid waste: a state of art review and future applications. Compos Part B Eng 115:409–422. https://doi.org/10.1016/j.compositesb.2016.09.013
65. Teegarden DM (2004) Polymer chemistry: introduction to an indispensable science. NSTA Press, Arlington
66. Telli A, Özdil N (2015) Effect of recycled PET fibers on the performance properties of knitted fabrics. J Eng Fiber Fabr 10:155892501501000. https://doi.org/10.1177/155892501501000206
67. Thompson RC, Moore CJ, vom Saal FS, Swan SH (2009) Plastics, the environment and human health: current consensus and future trends. Philos Trans R Soc B Biol Sci 364:2153–2166. https://doi.org/10.1098/rstb.2009.0053
68. Untracht O (2011) Jewelry concepts & technology. Doubleday, New York
69. Walker S (1995) The Environment, Product Aesthetics and Surface. Des Issues 11:15. https://doi.org/10.2307/1511767
70. Wallace J, Dearden A (2005) Digital jewellery as experience. Future interaction design. Springer-Verlag, London, pp 193–216
71. Wargnier H, Kromm FX, Danis M, Brechet Y (2014) Proposal for a multi-material design procedure. Mater Des 56:44–49. https://doi.org/10.1016/j.matdes.2013.11.004
72. Weinstein JE, Crocker BK, Gray AD (2016) From macroplastic to microplastic: degradation of high-density polyethylene, polypropylene, and polystyrene in a salt marsh habitat. Environ Toxicol Chem 35:1632–1640. https://doi.org/10.1002/etc.3432

73. Wit W de, Hamilton A, Scheer R, Stakes T, Allan S (2019) solving plastic pollution through accountability, a report for WWF. Gland, Switzerland
74. Worrell E, Reuter MA (eds) (2014) Handbook of recycling. Elsevier
75. Xiong J, Liao X, Zhu J, An Z, Yang Q, Huang Y, Li G (2017) Natural weathering mechanism of isotatic polypropylene under different outdoor climates in China. Polym Degrad Stab 146:212–222. https://doi.org/10.1016/j.polymdegradstab.2017.10.012

Biobased Materials as a Sustainable Potential for Edible Packaging

Anka Trajkovska Petkoska, Davor Daniloski, Nishant Kumar, Pratibha, and Anita T. Broach

Abstract Edible packaging obtained from natural biopolymers as an alternative to synthetic food packaging have become very attractive for food engineers, scientists and consumers due to their edibility, functionality, biodegradability and compostability. Development of edible films and coatings with improved physical, mechanical, functional, barrier and sensory properties is a key for applications in different food sectors as wrapping and packaging materials. Biobased and biodegradable materials are ideal candidates as components to edible packaging. In addition, bioactive compounds and nutraceuticals, especially compounds extracted from by-products, are ideal addition to edible materials in context of extending their role towards active and intelligent packaging options. In this context, nanotechnology tools enable proper inclusions of the bioactives and nutraceuticals into edible materials; increase the stability of bioactive compounds, utilisation and delivery mechanisms from the packaging to food item or human body, but also to extend their usage, improve the quality and safety of packed food items.

A. Trajkovska Petkoska (✉)
Faculty of Technology and Technical Sciences, St. Clement of Ohrid University of Bitola, Dimitar Vlahov, 1400 Veles, Republic of North Macedonia
e-mail: anka.trajkovska@uklo.edu.mk

A. Trajkovska Petkoska · A. T. Broach
CSI: Create.Solve.Innovate. LLC, 2020 Kraft Dr., Suite 3007, VA 24060 Blacksburg, USA

D. Daniloski
Advanced Food Systems Research Unit, Institute for Sustainable Industries and Liveable Cities and College of Health and Biomedicine, Victoria University, Victoria 8001, Melbourne, Australia

Food Chemistry and Technology Department, Teagasc Food Research Centre, Moorepark, Fermoy, P61 C996, Cork, Ireland

N. Kumar
National Institute of Food Technology Entrepreneurship and Management, Haryana Kundli-131028, Sonipat, India

Pratibha
National Institute Of Technology, Haryana Kurukshetra-136119, India

S. S. Muthu (eds.), *Sustainable Packaging*, Environmental Footprints and Eco-design of Products and Processes, https://doi.org/10.1007/978-981-16-4609-6_5

Keywords Edible packaging · Bioactive compounds · Nutraceuticals · Food preservation · Nanotechnology

1 Introduction

Packaging as an integral part of the food product provides several functions to the packed items, protection from external contamination and spoilage, maintaining food quality and safety, increase of shelf life, convenience, easier handling, storage and transportation, promotion of the product, as well as providing information to the consumer. All of them affect the socio-economic perspective of food packaging, the sale, purchase and consumption, with the main aim to deliver the food in the best way to customers and improve their satisfaction [95, 140]. Figure 1 summarises the main functions of packaging and its extension towards novel active and intelligent packaging options. Synthetic or petroleum-derived polymers are heavily used packaging materials owing to their features, such as lightweight, good barrier against gases and water, heat stability, mechanical performance, easy formability and low cost; they can be easily tailored to the product application. Their annual production continues to increase; the packaging sector accounts for ≈40% of the total worldwide plastic consumption, mainly because of the wide range of properties of the plastics. In general, polyethylene (PE), polyethylene terephthalate (PET), polypropylene (PP), polyamide (PA) and polystyrene (PS) are the most used plastics for packaging

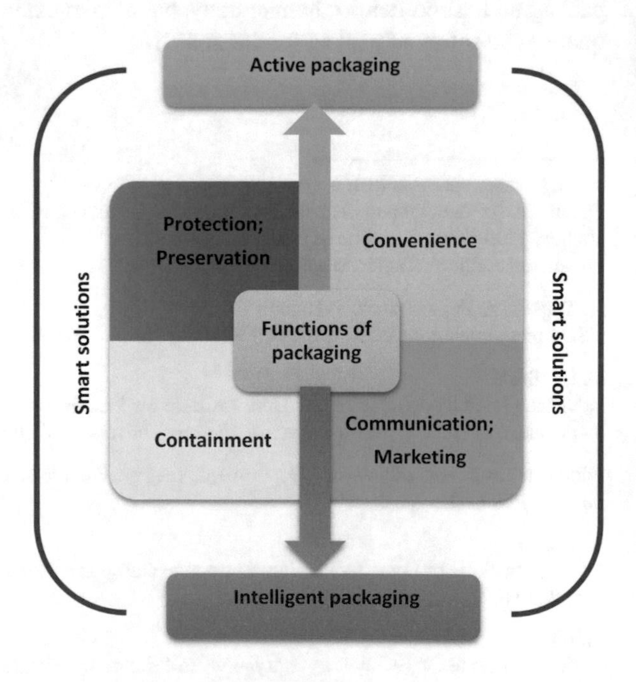

Fig. 1 Main functions of packaging and their extension towards novel active and intelligent options of packaging

purposes; however, the solution for waste generation derived from them—responsible for landfills zones and serious environmental pollution around the globe—still remains obscure and undeveloped. In most cases, plastic packaging contains residues from the food or biological substances, so their recycling is impractical and economically inconvenient. Nowadays, the total amount of plastic waste exceeds 200 million tonnes, with an annual increase of ≈5%[12, 19, 65, 108, 146].

Since petroleum-based polymers are not biodegradable and can cause serious environmental problems like pollution and biodiversity issues, there has been shown a great interest in the development of biobased and biodegradable packaging materials that are sustainable and environmental-friendly options. In this context, edible materials have a great potential in the food packaging because they are non-toxic, non-polluting, biobased and biodegradable. They can be designed to improve the food quality and safety, and at the same time to extend the shelf life of packaged foodstuffs by controlling water transfer, gas exchange, and inhibiting the oxidation processes. However, edible materials in form of films and coatings rarely exhibit good mechanical and barrier properties as petroleum-based plastics, and therefore there is still an open field of research to improve their structural and functional properties and achieve better competencies [19, 105, 146]. Hence, the natural biopolymers, particularly proteins, polysaccharides, lipids and their composites, are good candidates for film formation or coatings in context of edible materials. The combined forms of these biopolymers are often used to create edible packaging forms with desirable properties for customised applications. The functional properties of edible materials can be modulated by chemical modifications of the biopolymers by utilising emulsifiers, crosslinking agents, plasticizers and an addition of functional agents that will provide an active role of packaging [24, 68, 105].

The chapter considers overview of biobased materials as potential for edible packaging in forms of films and coatings, and bioactive compounds especially those from by-products added to edible materials in context to create active role of edible packaging.

2 Biobased and Biodegradable Materials: Bioplastics

There is a growing demand for the production of plastics using renewable resources, renowned as sustainable and eco-friendly alternatives to petroleum-based materials. The term bioplastics refers to biobased and biodegradable materials (Fig. 2); they are organised into a family of plastics that are either: (i) materials from biomass source: biobased, (ii) materials that are metabolised into organic biomass at the end of life: biodegradable, or (iii) materials that belong to both categories: biobased and biodegradable [88, 159, 47, 50, 109]. Usually, a material made of renewable source (plant or animal feedstock) is designated with terms like biobased, renewable and biotic; they are an interchangeable. These biopolymers can be either chemically synthesised from bio-derived monomers, directly extracted from biomass or industrial wastes.

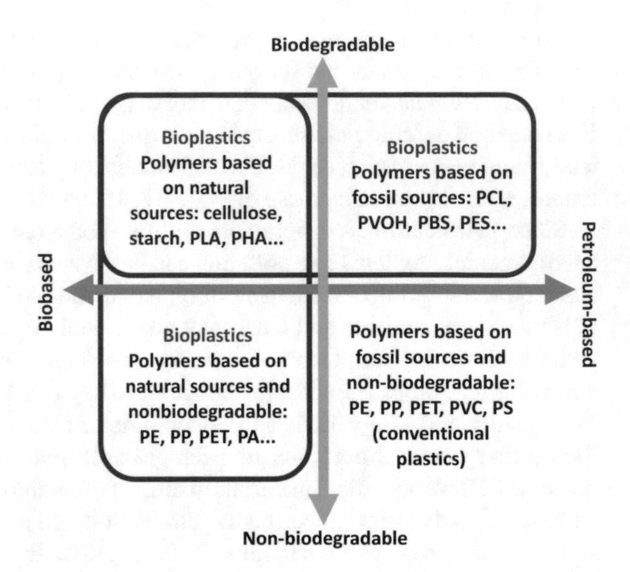

Fig. 2 Different types of plastics depending on their origin and biodegradability, where PLA = Polylactic acid, PHA = polyhydroxy alkanoate, PCL = polycapro lactone, PBS = polybutylene succinate, PE = polyethylene, PP = polypropylene, PET = polyethylene terephtalate, PA = polyamide, PVC = polyvinyl chloride, PS = polystyrene, PVOH = polyvinyl alcohol

In addition, another reason that development and utilisation of bioplastics increasingly gaining multidisciplinary attention is due to the fact that their disintegration and composting can be used as fertilisers and soil conditioners [38]. The term compostable refers to a material's ability to biodegrade within a certain period in composting conditions. Nevertheless, not all biodegradable products are compostable. In general, the diversity of biodegradable plastics is found in the variation in biodegradation routes; they could be microbially degraded, allowing for alternative end-of-life management, such as industrial or home composting, anaerobic digestion depending on the plastics' type. On the other hand, bioplastics can be recycled or incinerated as other synthetic plastics; nonetheless, they are not widely recycled since they are considered as contaminants in the current recycling system. Finally, in the family of non-biodegradable fossil-based plastics also belong the non-biodegradable biobased polymers, such as bio-polyethylene terephthalate (bio-PET) and bio-polyethylene (bio-PE) (Fig. 2) [88].

Currently, bioplastics represent only ~1% of the total annually produced plastic. The bioplastic market is growing and diversifying as a result of discovering more sophisticated biopolymers, applications and products [159]; new and innovative biopolymers, such as PLA (Polylactic Acid), biobased PP (Polypropylene) and many others are with increasing production rates. In addition, biobased polyesters, polyhydroxy alkanoates (PHA) and polyhydroxy butyrate (PHB), are from microbial origin and are 100% biobased and biodegradable in different environments [78, 150, 155, 92].

Many of the food manufacturers and scientists are trying to reduce the non-renewable plastics with the novel sustainable options for food packaging, and therefore to develop innovative smart packaging options. In this context, good representatives are edible films and coatings, both with a renewable origin and renowned for providing suitable food safety and quality. Usually, they are constructed from food components, proteins, polysaccharides, lipids or their combination, but the biopolymer-derived materials often lack the performance features of conventional plastics, such as strength, flexibility and barrier properties, even though a lot of efforts are made to improve them by using different approaches. In most of the cases, their functional properties can possibly be enhanced with active ingredients, e.g. antioxidants and antimicrobials, and consequently they can fit to applications with required mechanical, barrier and structural properties. Simultaneously, their properties are continuously improving in order to protect a variety of food products in an efficient and functional way [19, 106, 108, 111].

3 Bioactive Ingredients and Nutraceuticals

In recent years, there is a tendency to incorporate bioactive additives into packaging materials, especially those of a natural origin. Those include polyphenols, essential oils, carotenoids and many others, renowned as a frequent case of promotion of active packaging options [2, 38, 93]. These ingredients are used due to their antioxidant, antimicrobial, colouring, flavouring and other nutritional efficiency, but their incorporation could also impact the structural and functional properties of edible materials, specifically optical, mechanical, barrier, antimicrobial and antioxidant [19, 57, 86, 104, 31, 74]. In general, the plants and their parts (leaves, flowers, seeds, grains, roots, etc.) are good sources of bioactive compounds (phenols, essential oils, terpenoids, flavonoids) (Fig. 3) [93].

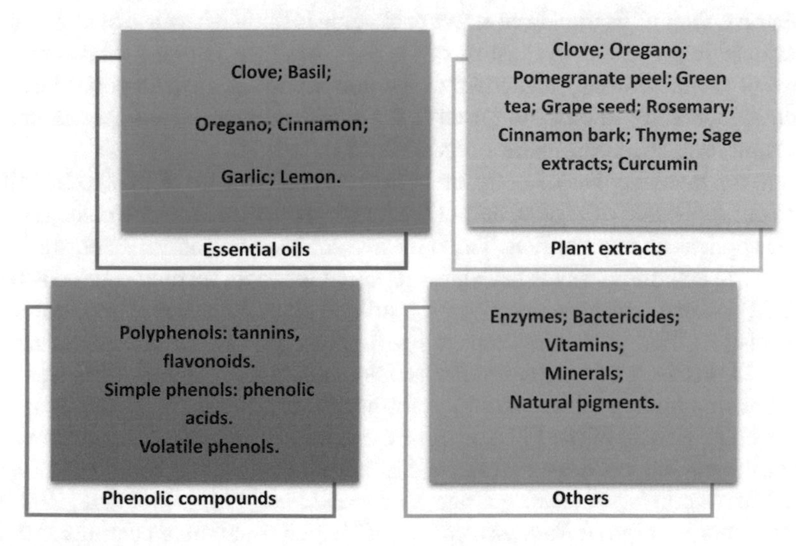

Fig. 3 Natural bioactive compounds usually used in edible packaging

Nutraceuticals are compounds with "possible beneficial effect" for human health; the word is coined from the words "nutrition" and "pharmaceutical". They have been defined as "*the phytocomplex if they derive from a food of vegetal origin, and as the pool of the secondary metabolites if they derive from a food of animal origin, concentrated and administered in the more suitable pharmaceutical form*". Examples of substances that have nutritional and nutraceutical effect are antioxidants, vitamins, polyunsaturated fatty acids, dietary fibres, prebiotics and probiotics [33].

In many cases, edible materials can be fortified by adding single or multiple bioactive ingredients in order to obtain the desirable set of functional attributes and protect the food from spoilage and deterioration. However, in such cases, the formulation and fabrication methods should be optimised depending on the nature of the active ingredients, film-forming materials and the functional effects that are required [19, 38]. The strategy of mixing the active (functional) ingredients into edible films and coatings depends on the miscibility of the active ingredients with the edible materials. It could be direct mixing or an addition in different encapsulated or other functionalised forms that will dictate the migration and release mechanisms of bioactive ingredients within edible films/coatings, and their potential effects on the packed food. One of the best options to protect the sensitive bioactive compounds is by enclosing them in a solid matrix encapsulation; it can also help for bioavailability enhancement, masking astringent flavours and controlled release in the gastrointestinal tract (GIT). Some of the techniques for bulk encapsulation are spray drying, spray chilling, freeze-drying and ultrasound, to name a few [21, 43, 55, 99].

Essential oils (EOs) are active compounds extracted from plants (e.g. fruits, seeds, nuts, leaves, roots) [61, 7, 8]. The data available explains the incorporation of EOs within the structure of edible materials, proving antioxidant and antimicrobial effects; however, they are proven to modify both mechanical and barrier properties [105]. On the contrary, their utilisation in an active packaging is limited because of their volatile and oxidisable properties [57]. Most of the EOs, or other natural extracts, required the use of manufacturing methods that are carried out at specific conditions, e.g. at room temperature in order to preserve their original properties (electrospinning, electrospraying, microcapsulation, etc.) [10, 141].

There are many studies of usage and effective contribution of EOs to inhibit the growth of foodborne pathogens, the antibacterial mechanism on these pathogens and specific application cases of EOs. The extensive antibacterial, antioxidant, antifungal and antiviral activities of EOs have been reported for food, medicine and cosmetics [122], 62]. A few instances described the utilisation of EOs from lemongrass [39] or basil [4] in food packaging applications; ginger EO with antibacterial activity against *Staphylococcus aureus* and *Escherichia coli* [145]. Rockrose EO was more active towards gram-positive than to gram-negative bacteria [76]; another case of antimicrobial activity of the EO extract of *Ocimum gratissimum L.* against multiresistant microorganisms in planktonic and biofilm form; a significant reduction was verified for *Staphylococcus aureus* and *Escherichia coli* [145]. [46] and [103] have reported incorporation of *Ruta graveolens EO* in chitosan edible coatings and their antimicrobial effect that have been used to pack fruits and vegetables. The EOs derived from thyme, citrus limonia [25] and other citrus fruits (lemon, mandarin and orange) [130], have been reported as antimicrobial agent for meat [110]. On the other hand, nettle EO has been investigated for its potential and inhibitory effect against *Escherichia coli, Listeria monocytogenes* and *Staphylococcus aureus* [77], while clove and oregano EOs have shown a protective effect against microbial growth and the oxidation of lipids in salmon burgers [59], etc.

Phenolic compounds are other type of bioactive compounds that contain at least one phenol unit; they are usually present in most of the plants, and are responsible for the plant's defense against external stimuli, such as radiation, predators and microorganisms. They possess an antioxidant and antimicrobial characteristics and are of major interest not only in food, but also in cosmetic and pharmaceutical industries [12, 16, 151]. Phenolic compounds such as polyphenols (tannins, flavonoids), simple phenols (phenolic acids, coumarins) and volatiles phenols (carvacrol, eugenol, thymol) are the most common representatives (Fig. 3) [15, 93]. They are an active addition with antioxidant role in edible materials and can minimise the lipid oxidation and other deterioration effects that help for extending the shelf life of food products [143].

Therefore, the bioactive compounds come from variety of sources; there are many cases of their usage. Some examples are reported for walnut and walnut shell [143], green tea extracts or other tea polyphenols in active packaging—they affect the physical, structural and antioxidant properties of edible films as well [32, 94, 102]. Another study has revealed the antioxidant activity of phenolic compounds extracted from hawthorn species [2], while curcuma ethanol extract in gelatine-based films has affected the antioxidant and physical properties of edible films [16]. In addition, propolis addition to edible materials has performed antimicrobial and antioxidant activities, and therefore it could be considered for novel active packaging solutions [151].

Furthermore, the use of Mediterranean herbs is reported as addition to edible packaging options. They have polyphenols, terpenoids and flavonoids with antimicrobial and antioxidant activity that can be used in different forms in active systems, including sachets, multilayers, labels and coatings. Moreover, the use of EOs from Mediterranean herbs in edible materials can improve the properties of packaging system and consequently could extend the shelf life of variety of food items (e.g. muscle products, fruits, vegetables) [104].

3.1 Bioactive Compounds Extracted from By-Products

Plants and particularly fruit by-products are the excellent source of active compounds, specifically phenolic compounds, vitamins, carotenoids, among others; they are usually cheap or with a low economic value [30, 129]. Agro-waste is defined as "*plant or animal residues that are not (nor further processed into) food or feed, and create additional environmental and economic issues in the farming and primary processing sectors*", while food waste is defined as "*food losses of quality and quantity through the process of the supply chain taking place at production, post-harvest, and processing stages*". Also, food waste means a substantial loss of other resources like land, water, energy and labour. According to FAO, more than 30% of the world's food production is lost or wasted (it is ≈1.3 billion tonnes) [6, 47, 139].

There are many cases reported on utilisation of agro-waste materials worldwide, i.e. pineapple peel [75], lemon peel [26, 136], Persian lime peel extract [63], walnut shell [143], sugarcane bagasse [144], mango puree and peel [134] pomegranate peel extract [68], apple peel polyphenols [113], cocoa and soya by-products [41, 81], oregano waste extract [138], aloe vera [131], black chokeberry [49], aronia [71], melanin that is isolated from watermelon seeds [74], pumpkin peel [128], chlorophylls from cucumber peel [45], onion skins and whey [117, 83], grapefruit seed extract 118, tomato pomace [3], anthocyanin extracted from jabuticaba fruit, purple sweet potato peels or red cabbage (potential for manufacture colorimetric indicator films) [18, 72, 152]. Moreover, there are reported cases of utilisation of by-products from animal origin, e.g. by-products from the processing of fish that are a good source of gelatine [77].

Noteworthy, a good biocompatibility of fruit puree film with food products has been found; when an edible, fruit-based film has been applied to a food product, part of it could be absorbed. A fruit-based film also could incorporate bioactive compounds into the product, providing an additional nutritional and bioactive benefit (antioxidants, functional carbohydrates) beyond the expected technological attributes; in addition, the inclusion of natural antimicrobial additives improves the microbiological quality, advantages in terms of the safety and shelf life of products [134, 79].

All these data encourage the valorisation of the plant by-products as powerful sources of natural antioxidants and antimicrobials compounds. They can be used as food additives or ingredients for developing functional food, active edible packaging and extending the shelf life of the foodstuffs, but also novel packaging options with potential health benefits. The utilisation of the bio-waste creates new food market and technology perspectives that belong within the concept of a circular economy, too [12, 6, 37].

4 Edible Materials as Candidates for Sustainable Food Packaging

Edible packaging is made from ingredients that can be consumed alongside the contained food or beverage; it is an option towards sustainable and eco-friendly alternatives in packaging sector and they do not require any additional processing after using as food packaging. Actually, the edible packaging is defined as a thin layer that covers the food surface as a coating and forms a cohesive system; in case if it is not consumed, it degrades very fast and reduces the landfill demands [82, 140].

Generally, materials used for the preparation of edible films are classified into following categories: proteins, polysaccharides, lipids and composites that possess

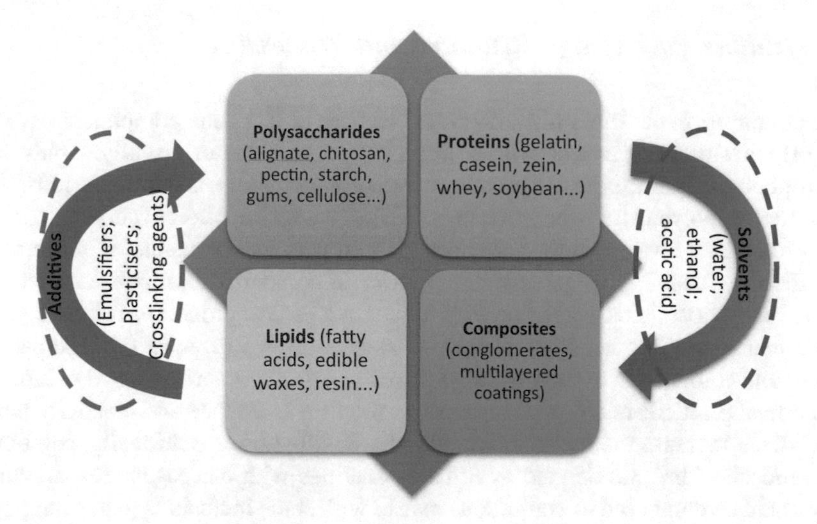

Fig. 4 Edible materials used for edible food packaging

film-forming properties or could be applied as coatings on a variety of food products [14, 105] (Fig. 4).

4.1 History of Edible Materials Development

The first use of edible materials for food packaging exists for a long time, however not in the same form or chemical composition as today's edible materials. Over the centuries, people have developed different kinds of films and coatings to protect the food from spoilage or to keep it for a longer period of time [13]. Namely, food products in the past were packed in natural materials such as casings, leaves, wood or baskets; the first recorded use of biopolymer was in China when the citrus fruits were preserved from the water loss by molten wax coating (twelfth century). Later, in Japan (fifteenth century), a protein edible film and the first freestanding film was used to improve preservation of food. Whereas, in the sixteenth century, in other part of the world (England) coating with lard was used to prevent/prolong the shelf life of meat, though this resulted in some loss of taste and texture. Use of gelatine and oil to preserve meat was patented in US for the first time. Moreover, nuts were coated with sucrose to prevent rancidity and therefore to enhance the quality, while emulsion of oil or waxes in water was used for protection of the fruits in order to prevent moisture loss, furnish shine or perform brighter appearance [52, 83, 140].

4.2 Edible Packaging: Structure and Processing

The composition of edible films/coatings requires at least one component capable to form a structural matrix with a sufficient cohesiveness. Usually, lipids and hydrophobic substances, resins, waxes or some non-soluble proteins are the most efficient for the moisture transfer protection. In contrast, hydrocolloids (polysaccharides and proteins) are water-soluble; they are low-efficient barrier against water transfer and their permeability to gases is often lower in comparison to traditional plastic films, but hydrocolloids usually provide higher mechanical properties to edible packaging than lipids or other hydrophobic substances. However, most of these natural and edible compounds usually need additives in order to improve the film-forming properties: plasticizers are used to improve the film resistance and elasticity while emulsifiers increase the hydrophobic globule distribution. Additionally, combined or composite films are defined as films or coatings with heterogeneous structure, they could be composed of continuous matrix with some inclusions (e.g. solid particles, lipid globules, etc.) or is composed of several layers originating from the same or different biomaterials. Figure 5 provides information of the main properties that edible materials could ensure when they are used as food packaging in form of films or coatings [29, 40, 140].

The formation of edible coatings and films can be done by wet and dry processing methods (direct casting, coating, dipping, layer-by-layer assembly and extrusion). For wet processing, films can be produced from edible materials with film-forming ability: film materials should be dispersed and dissolved in solvents (e.g. water,

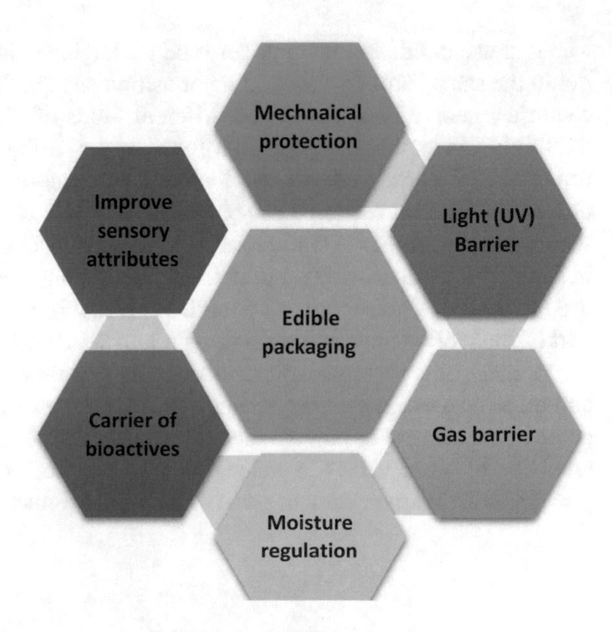

Fig. 5 Main functions of edible packaging for packed food

alcohol or their mixture); then film solution is casted and dried at a desired temperature and relative humidity to obtain freestanding films. In contrast, for coatings, the film solutions can be applied directly to the food item by methods known as dipping, spraying, brushing and panning, and later followed by drying process. Plasticisers, emulsifiers as well as other agents like crosslinking agents, antimicrobial agents, colours or flavours can therefore be added into the film-forming solutions [133, 104, 85, 13, 67].

4.3 Characteristics of Main Edible Components Used in Edible Packaging

The hydrocolloids are a general term of polysaccharides and proteins [57]. Polysaccharides used for edible films and coatings are usually cellulose, starch [89], dextran, inulin, alginate [75, 81, 120], carrageenan [154], pectin [112], chitosan [137], mucilage [44] and their derivatives. Polysaccharide films/coatings can be from plant, marine organisms and microbial origin [90]; they serve as good oxygen, odour and oil barriers with good mechanical properties, but their major drawback is moisture permeability, due to their hydrophilic nature [27, 105, 114].

The cellulose and its derivatives (methylcellulose, hydroxypropyl methylcellulose, carboxy methylcellulose) are the most abundant natural organic compounds that are odourless and tasteless and therefore are widely used as raw materials for edible films. However, due to the hydrophilic nature, the cellulose derivative films show poor barrier properties to water vapour. In addition, carboxymethyl cellulose and pectin are important polysaccharides with great potential in making edible coatings; they are not only odourless and tasteless, but also non-toxic, non-allergic, water-soluble, transparent and resistant to oil and fats; considering their advantageous characteristics, including abundance, biodegradability and renewability; they can be good candidates to deliver active additives [96, 117, 93, 160].

Starch is a polysaccharide composed of two macromolecules, amylose and amylopectin; it is used in its native or modified form. Different ratio of amylose/amylopectin present in different starch sources allow production of variety of edible films. The most commercially used are corn, potato, rice and wheat starches [93].

Chitosan is frequently used in food industry due to its antioxidant, antimicrobial and chelation properties, as a clarifying agent, and an inhibitor of enzymatic browning. Its ability of film formation, and decent mechanical strength, stands out compared to other biopolymer counterparts; it presents a big potential as a food packaging material. Numerous research advancements have been performed in the last years; the modification of chitosan films by using different additives improves not only the basic barrier properties, but also imparts many functionalities, ability to extend the food shelf life, while maintaining food safety and quality [107, 115, 132, 137].

Gums are another group of polysaccharides that are produced naturally by some botanical (trees and shrubs, seeds and tubers), algae or microbial sources [70, 93]. They have been used as film-forming materials and some examples are guar gum [121], Arabic gum [100], Persian gum [127], tragacanth gum [63], Moringa oleifera gum [9, 69], etc.

Proteins could be from plant (corn, soy, pea and quinoa) and animal (whey, casein, collagen, gelatine, egg white) origin. Proteins used for edible films and coatings are gelatine [22], corn zein [101], wheat gluten, soy protein, canola [11], whey and casein [24]. They are generally formed by wet processes, from solutions or dispersions of the protein as the solvent (water, ethanol or their mixture) evaporates. Proteins should be denatured by heat, acid, base or solvent in order to form more extended structure that is required for the film formation. However, protein films exhibit poor moisture resistance, but better mechanical and barrier properties than polysaccharides [19, 43, 64].

Plant proteins exhibit glossy appearance, toughness, low water solubility, resistance to microbial attack and high hydrophobicity [101]. In contrast, milk proteins are classified as caseins and whey proteins. They have the ability to form transparent and tasteless films, for example, casein films are stable at different pH, temperature and salt levels, but also whey proteins possess film-forming properties [24]. In addition, gelatine is obtained from the hydrolysis of collagen; fish, pork or bovine collagen is good representative. Gelatine can form film with high transparency and good tensile strength; blends with different hydrocolloids are widely used. For example, bovine gelatine-based films possess a hydrophobic nature, and the inclusion of chitin increases its hydrophobicity, while a fish gelatine film with inclusion of anthocyanins has shown good mechanical properties and water resistance, but also an antioxidant activity [82, 93, 143].

Lipids used for edible packaging include natural waxes, vegetable oils, fatty acids and resins. They exhibit certain disadvantages like mechanical and chemical instabilities as well as decreased organoleptic quality with a time. It is for this reason the lipids that are usually combined with other film-forming materials, polysaccharides and proteins increase the moisture resistance of the films [58, 20, 48].

Edible films and coatings could also be heterogeneous by blending more than one compound during the manufacturing process, e.g. a blend of polysaccharides, proteins and/or lipids. These films and coatings are defined as composite materials. By combining the materials into composite films and coatings, one can adjust the properties of the composition to suit a specific application. In general, each individual coating material possesses some unique properties such as renewability, abundance, biodegradability, high adsorption capabilities, ease of functionalization, mechanical, barrier, physical and others but in most cases some limited functions, and therefore a combination of different materials can be more effective to overcome the drawbacks of a single substance [19, 105]. Layer-by-layer deposition [149] or blending method [124] is the usually used processing techniques to obtain composite edible films/coatings. Furthermore, composites can be prepared as binary films/coatings, such as lipid-based films, protein–carbohydrate, carbohydrate–carbohydrate or protein–protein films; nevertheless, they could be prepared as ternary and quadruple edible films and coatings as well [28, 84].

4.4 Bionanocomposites

Bionanocomposites as a relatively new generation of edible packaging materials is a combination of biobased polymer that acts as a matrix and fillers that have at least one dimension in nanometer scale. In nanocomposites, the fillers are dispersed in polymer matrix with a function to improve the functional, mechanical, thermal, barrier or physical properties, without hindering their biodegradable and non-toxic characters. These properties do not occur naturally in the biopolymers by themselves, and therefore there is a big interest for different types of "fillers" or "layers" or "nanomaterials" that can bring a novel set of properties in the nanocomposites like antimicrobial, antioxidant or both benefits [56, 57, 146]. They are usually prepared as: (i) laminated composites—composed of layers of materials held together by the matrix binder; (ii) fibrous composites—prepared of reinforcing fibres in a matrix and (iii) particulate composites—consist of particles dispersed in a matrix [126, 142, 153].

Nanomaterial defined by the European Commission is a "*natural, incidental or manufactured material containing particles, in an unbound state or as an aggregate or as an agglomerate and where, for 50% or more of the particles in the number size distribution, one or more external dimensions is in the size range 1 to 100 nm*" [51]. There are many reports on a variety of forms and shapes of nanomaterials, specifically spherical, fibrous, tubes and flakes, to name a few [126]. Furthermore, a nanofiller or nano-reinforcement according to dimensions can be (1) nanolayers, nanosheets or nanoflakes with one nanoscale dimension (layered silicates—nanoclay and nanolayered double hydroxides, starch nanocrystals; (2) nanotubes that have two nanoscale dimensions (nanofibers, nanorods and nanowhiskers: cellulose, chitin nanofibers) and (3) nanoparticles (NPs) known also as 3D nanoparticles, isodimensional nanoparticles, nanogranules, nanocrystals or nanospheres with three dimensions in the nanometer scale (silver NPs, and metal oxide NPs, TiO_2, ZnO, Al_2O_3, SiO_2, chitosan NPs, etc.) [141, 147].

According to the origin, nanofillers are divided as organic or inorganic, and metallic or non-metallic. The most used nanofillers include metal NPs [56], e.g. gold or silver (AuNPs, AgNPs), zinc oxide (ZnONPs) [125], copper sulphide (CuSNP) [120], then titanium dioxide (TiO_2-NPs) [157], graphene nanoplatelets [78], layered silicate nanoclays, such as montmorillonite [126], kaolinite, nano-silicon dioxide [123]. Moreover, forms like nanofibers, nanosphere and nanotubes could also be met in practice. Nanoparticles, particularly iron, silver, zinc oxides and carbon may serve as carriers for enzymes, antioxidants, anti-browning agents, flavours and other bioactive materials to improve the shelf life even after the package is opened [80, 91].

Silver nanoparticles (AgNPs) are the most important nanoparticle used in active food packaging due to its antimicrobial property to improving the stability of the materials and prolong the shelf life of food products [65]. The other nanoparticles, such as TiO_2-NPs, ZnONPs, are also widely used in food packaging sectors [5, 122, 158]. Nevertheless, not only metallic NPs, but also non-metallic nanoforms are of a big interest in novel types of active packaging. Hence, whey protein nanofibrils can be used in edible packaging—their antioxidant effects combined with TiO_2 nanotubes as antimicrobial agents have been reported [36]. Another report was conducted on nisin-loaded pectin NPs that exhibited antimicrobial activity dependent on the biopolymer [66]. In addition, there are also reports that were conducted on different polymers, nanochitin with enhanced functional and structural properties [60]; whey protein nanofibrils [36], but also cellulose nanocrystals [148] or flax cellulose nanocrystals [87], nanocellulose [89], cellulose nanofibers [155] or cellulose nanocrystal in chitosan forms [148], nano-chitosan [135], chitosan nanoparticles [97], melanin nanoparticles in chitosan-based nanocomposite films [119], etc.

In most cases, nanoformulations in the food industry are confirmed for improving the bioavailability, which is very important to prepare novel types of edible products, functional foods, to protect active ingredients against degradation or to reduce their side effects. Nanocarriers could be in forms of nanogels, core-shell NPs, nanofibers, cyclodextrin complexes, mesoporosous silica NP, nanoemulsions, micelles, liposomes, etc. [54]. In the recent time, a particular interest is shown for nanocoatings made by biopolymers as well composite laminated films obtained by layer-by-layer deposition or by spin coating process [1, 73].

Finally, bionanocomposites offer an inspiring route for creating new and multi-beneficial innovative packaging solutions. Inserted nanomaterials into packaging materials as nanosensors could alert the customer if a food has gone to spoilage or some deterioration processes, due to intrinsic or extrinsic factors that affect the food quality and stability. In general, intrinsic factors from the packed food are water activity, pH, microbes, enzymes and the level of reactive compounds—they can be regulated and monitored by using specific materials and ingredients incorporated in packaging material. In contrast, temperature, total pressure, partial pressure of various gases, relative humidity, mechanical stress or light are common extrinsic factors that influence the rate of degradation reactions during the food storage. There are attempts to reach their monitoring through the packaging system by implanting indicators or nanosensors [50, 91].

4.5 Utilisation of (Nano)technologies in Designing of Edible Packaging

Besides the opportunities for improved mechanical, thermal, barrier properties of edible materials, active forms of packaging could provide also antibacterial, antifungal or antioxidant behaviour since they are carriers of important active substances. Consequently, the active packaging ensures longer shelf life, enhanced quality and safety of packaged foodstuffs. In the most cases, enhanced functional properties are owing to inclusion of active (nano)materials into biopolymer composites as packaging systems [23, 153, 34]. They could be included as microcapsules or nanoencapsulated forms through encapsulation processes. Microcapsules are capsules with a diameter in between 3 and 800 μm, while nanocapsules are usually in the range between 10 and 1000 nm; both microcapsules and nanocarriers can protect a bioactive compound from environmental conditions, e.g. oxidation, heat, light, pH and enzyme degradation. Noteworthy, nanoencapsulation is one of the most used methods for embedding bioactive compounds; it consists of coating target compounds (core substance) with different external materials (shell, wall or carrier material) to generate small particles that exert positive effects [53, 116]. The incorporation of bioactive compounds and nutraceuticals into edible materials can happen by using different techniques, like extrusion method, fluidized bed coating, spray cooling or (nano)spray drying, (nano)emulsion [42, 156], and result in different forms from nano- to micro-dimensions: liposomes [35], nanoparticles [38], micelles [80], microcapsules [17], nancocapsules [91], etc.

Nanotechnologies' applications enhance the food bioavailability, taste, texture and consistency that can usually be achieved through modification of particle size (functionalisation), surface charge of nanomaterials, synergistic action of nanomaterials in food preparation and protection or application of nanosensors in smart food packaging [91]. They offer passive, active and/or bioactive properties (barrier, antimicrobial, antioxidant, oxygen scavenging), and may control the liberation of functional ingredients exerted either by the intended or non-intended migration of the nanomaterials or by the active substances they may carry [91, 141].

The main benefits of nanotechnology for incorporation of bioactive substances in food products and packaging are improved stability and shelf life, preserving them against degradation during processing, distribution, storage; improved mechanical and thermal properties; integration of functional additives (fatty acids, vitamins, minerals) in biomaterials; mask undesirable tastes; protect nutraceuticals from extreme conditions (low gastric pH); optimise their release during the digestion process; obtain higher activity levels of the encapsulated functional ingredients and induce better homogeneity of the food system [98].

5 Conclusion: Future Perspectives and Limitations

The production of plastics in food packaging has increased over the last decades. Plastics have many benefits as packaging materials; they are designed to be strong, low in weight and possess suitable protective functions. Nevertheless, when they are produced from fossil feedstock they can be big pollutants for the planet and society. Namely, those plastics can heavily contribute to carbon emission and climate changes; even after their end of life and disposal when these materials enter ecosystems (lands, rivers, oceans); thus, they are again harmful to the environment. Currently, the majority of plastic packaging are single use and are not part of any recycle or biodegradable cycles. They are known to generate environmental, ecological and socio-economic problems. As a consequence, the eco-friendly and sustainable packaging designs are more than needed—they should be focussed to reduce carbon footprints during their production processes, serving as protective functional materials and considering their after use life. Eradication of the waste disposal is the only smart way to decrease landfill zones, leading to a full satisfaction of the most of the global sustainability goals (SDGs). Thus, usage of packaging solutions that could be recycled, reused or reformulated to other valuable products is also the way to accomplish most of those SDGs for cleaner and less polluted world. The combination of these steps can direct to a high utilisation of biobased materials from sustainable sources, designed in a way to be easily recycled, reused or be biodegradable and compostable. Edible materials from the single or combined version of biopolymers are good examples of this focus as smart and novel types of eco-friendly trends in food packaging. More particularly, edible films and coatings containing natural bioactive ingredients (antioxidant, antimicrobial) have been the focus of many studies, owing to their great potential in food packaging. Hence, bionanocomposites still need scientific research and improvement in order to increase the shelf life, quality, functionality and marketability of variety of packaging options that they can offer. There are still many hurdles that need to be addressed, namely, preparation methods with better yield, more functional properties, release properties, consumer acceptance and policy regulation of the novel packaging options. The main concerns among consumers are awareness and safety issues of these packaging options; however, the cost and large-scale preparation of edible films and coatings remain the main concerns for professionals in the food industry.

Sustainable utilisation and application of edible films and coatings in food industry require a synergistic effort of many food and polymer scientists and engineers, industrial partners and governmental institutions to develop suitable packaging options that can overcome the drawbacks of natural-based materials. Despite being biobased and biodegradable, sustainable with natural origin, the next-generation packaging materials need to possess strength, plasticity and most of the properties of the synthetic-based polymers. The use of inexpensive, underutilised food processing by-products or agro-waste is a promising strategy for producing cheaper packaging materials. Those materials may still present the sensory and physicochemical characteristics that differ them from films made up of conventional polymers, particularly "greener" and economic extraction methods should be developed. In this regard, edible films based on by-products with novel health-promoting functionalities may also be developed; there are already examples of probiotic/prebiotic films, increasing their market appeal as healthy food components with desirable sensory properties.

Proper toxicology studies and safety measures should be undertaken before the application of nanomaterials in both the food products and the packaging. Namely, the special category of novel packaging—intelligent type that utilise natural colourant-based, pH-responsive indicators—has tremendous potential for future sustainable smart packaging applications and an increased interest in intelligent packaging by consumers and producers. Accordingly, the nanomaterials and nanosensors would help the consumers by providing information on the state of the packed food, the food's nutritional status and enhanced security through the microbial detection. There is still a knowledge gap of the migration, toxicity, consumer acceptance and recyclability of nanoreinforced or all-natural type of plastic packaging; implementation of nanotechnology in food packaging applications still provides ample of issues that should be solved, not only their safety, but also the industrial scale-up, end-of use status or recyclability solution. Furthermore, the regulatory bodies such as FDA and EU require more information to develop and implement safety regulations for the application of nanomaterials in food packaging systems.

References

1. Acevedo-Fani A, Soliva-Fortuny R, Martín-Belloso O (2017) Nanoemulsions as edible coatings. Curr Opin Food Sci 15:43–49
2. Alirezalu A, Ahmadi N, Salehi P, Sonboli A, Alirezalu K, Mousavi Khaneghah A, Barba FJ, Munekata PE, Lorenzo JM (2020) Physicochemical characterization, antioxidant activity, and phenolic compounds of Hawthorn (Crataegus spp.) fruits species for potential use in food applications. Foods 9:436
3. Aloui H, Baraket K, Sendon R, Silva AS, Khwaldia K (2019) Development and characterization of novel composite glycerol-plasticized films based on sodium caseinate and lipid fraction of tomato pomace by-product. Int J Biol Macromol 139:128–138
4. Amor G, Sabbah M, Caputo L, Idbella M, de Feo V, Porta R, Fechtali T, Mauriello G (2021) Basil essential oil: composition, antimicrobial properties, and microencapsulation to produce active chitosan films for food packaging. Foods 10:121
5. Anaya-Esparza LM, Ruvalcaba-Gómez JM, Maytorena-Verdugo CI, González-Silva N, Romero-Toledo R, Aguilera-Aguirre S, Pérez-Larios A, Montalvo-González E (2020) Chitosan-TiO_2: a versatile hybrid composite. Materials 13:811
6. Andrade MA, Lima V, Silva AS, Vilarinho F, Castilho MC, Khwaldia K, Ramos F (2019) Pomegranate and grape by-products and their active compounds: are they a valuable source for food applications? Trends Food Sci Technol 86:68–84
7. Anis A, Pal K, Al-Zahrani SM (2021) Essential oil-containing polysaccharide-based edible films and coatings for food security applications. Polymers 13:575
8. Atarés L, Chiralt A (2016) Essential oils as additives in biodegradable films and coatings for active food packaging. Trends Food Sci Technol 48:51–62
9. Badwaik HR, Al Hoque A, Kumari L, Sakure K, Baghel M, Giri TK (2020) Moringa gum and its modified form as a potential green polymer used in biomedical field. Carbohydr Polym 116893
10. Bakry AM, Abbas S, Ali B, Majeed H, Abouelwafa MY, Mousa A, Liang L (2016) Microencapsulation of oils: a comprehensive review of benefits, techniques, and applications. Compr Rev Food Sci Food Saf 15:143–182
11. Bandara N, Akbari A, Esparza Y, Wu J (2018) Canola protein: a promising protein source for delivery, adhesive, and material applications. J Am Oil Chem Soc 95:1075–1090
12. Barbosa CH, Andrade MA, Séndon R, Silva AS, Ramos F, Vilarinho F, Khwaldia K, Barbosa-Pereira L (2021a) Industrial fruits by-products and their antioxidant profile: can they be exploited for industrial food applications? Foods 10:272
13. Barbosa CH, Andrade MA, Vilarinho F, Fernando AL, Silva AS (2021b) Active edible packaging. Encyclopedia 1:360-370
14. Basumatary IB, Mukherjee A, Katiyar V, Kumar S (2020) Biopolymer-based nanocomposite films and coatings: recent advances in shelf-life improvement of fruits and vegetables. Crit Rev Food Sci Nutr 1–24
15. Benbettaïeb N, Debeaufort F, Karbowiak T (2019) Bioactive edible films for food applications: mechanisms of antimicrobial and antioxidant activity. Crit Rev Food Sci Nutr 59:3431–3455

16. Bitencourt C, Fávaro-Trindade C, Sobral PDA, Carvalho R (2014) Gelatin-based films additivated with curcuma ethanol extract: antioxidant activity and physical properties of films. Food Hydrocolloids 40:145–152
17. Cai C, Ma R, Duan M, Deng Y, Liu T, Lu D (2020) Effect of starch film containing thyme essential oil microcapsules on physicochemical activity of mango. LWT 131:109700
18. Capello C, Trevisol TC, Pelicioli J, Terrazas MB, Monteiro AR, Valencia GA (2020) Preparation and characterization of colorimetric indicator films based on chitosan/polyvinyl alcohol and anthocyanins from agri-food wastes. J Polym Environ 1–14
19. Chen W, Ma S, Wang Q, Mcclements DJ, Liu X, Ngai T, Liu F (2021) Fortification of edible films with bioactive agents: a review of their formation, properties, and application in food preservation. Crit Rev Food Sci Nutr 1–27
20. Chevalier E, Chaabani A, Assezat G, Prochazka F, Oulahal N (2018) Casein/wax blend extrusion for production of edible films as carriers of potassium sorbate—a comparative study of waxes and potassium sorbate effect. Food Packag Shelf Life 16:41–50
21. Chopde S, Datir R, Deshmukh G, Dhotre A, Patil M (2020) Nanoparticle formation by nanospray drying & its application in nanoencapsulation of food bioactive ingredients. J Agric Food Res 2:100085
22. Chung D (2020) Fish gelatin: molecular interactions and applications. Elsevier, Biopolymer-Based Formulations
23. Daniloski D, Gjorgjijoski D, Petkoska AT (2020) Advances in active packaging: perspectives in packaging of meat and dairy products. Adv Mater Lett 11:1–10
24. Daniloski D, Petkoska AT, Lee NA, Bekhit AE-D, Carne A, Vaskoska R, Vasiljevic T (2021) Active edible packaging based on milk proteins: a route to carry and deliver nutraceuticals. Trends Food Sci Technol 111:688–705
25. de Oliveira Filho JG, de Deus IPB, Valadares ACF, Fernandes CC, Estevam EBB, Egea MB (2020) Chitosan film with citrus limonia essential oil: physical and morphological properties and antibacterial activity. Colloids Interfaces 4:18
26. Deb Majumder S, Sarathi Ganguly S (2020) Effect of a chitosan edible-coating enriched with citrus limon peel extracts and Ocimum tenuiflorum leaf extracts on the shelf-life of bananas. Biosurface Biotribology 6:124–128
27. del sol González-Forte, L., Amalvy, J. I. & Bertola, N. (2019) Corn starch-based coating enriched with natamycin as an active compound to control mold contamination on semi-hard cheese during ripening. Heliyon 5:1–8
28. Dhumal CV, Sarkar P (2018) Composite edible films and coatings from food-grade biopolymers. J Food Sci Technol 55:4369–4383
29. Díaz-Montes E, Castro-Muñoz R (2021) Edible films and coatings as food-quality preservers: an overview. Foods 10:249
30. Dilucia F, Lacivita V, Conte A, Nobile MAD (2020) Sustainable use of fruit and vegetable by-products to enhance food packaging performance. Foods 9:857
31. Domínguez R, Barba FJ, Gómez B, Putnik P, Kovačević DB, Pateiro M, Santos EM, Lorenzo JM (2018) Active packaging films with natural antioxidants to be used in meat industry: a review. Food Res Int 113:93–101
32. Dou L, Li B, Zhang K, Chu X, Hou H (2018) Physical properties and antioxidant activity of gelatin-sodium alginate edible films with tea polyphenols. Int J Biol Macromol 118:1377–1383
33. Durazzo A, Nazhand A, Lucarini M, Atanasov AG, Souto EB, Novellino E, Capasso R, Santini A (2020) An updated overview on nanonutraceuticals: focus on nanoprebiotics and nanoprobiotics. Int J Mol Sci 21:2285
34. Emamhadi MA, Sarafraz M, Akbari M, Fakhri Y, Linh NTT, Khaneghah AM (2020) Nanomaterials for food packaging applications: a systematic review. Food Chem Toxicol 1–9
35. Esposto BS, Jauregi P, Tapia-Blácido DR, Martelli-Tosi M (2021) Liposomes vs. chitosomes: cncapsulating food bioactives. Trends Food Sci Technol 108:40–48

36. Feng Z, Li L, Wang Q, Wu G, Liu C, Jiang B, Xu J (2019) Effect of antioxidant and antimicrobial coating based on whey protein nanofibrils with TiO_2 nanotubes on the quality and shelf life of chilled meat. Int J Mol Sci 20:1184
37. Fierascu RC, Sieniawska E, Ortan A, Fierascu I, Xiao J (2020) Fruits by-products–a source of valuable active principles. A short review. Front Bioeng Biotechnol 8:319
38. Fleming E, Luo Y (2021) Co-delivery of synergistic antioxidants from food sources for the prevention of oxidative stress. J Agric Food Res 3:100107
39. Gago C, Antão R, Dores C, Guerreiro A, Miguel MG, Faleiro ML, Figueiredo AC, Antunes MD (2020) The effect of nanocoatings enriched with essential oils on 'rocha' pear long storage. Foods 9:240
40. Galus S, Arik Kibar EA, Gniewosz M, Kraśniewska K (2020) Novel materials in the preparation of edible films and coatings—a review. Coatings 10:674
41. Garrido T, Etxabide A, Leceta I, Cabezudo S, de la Caba K, Guerrero P (2014) Valorization of soya by-products for sustainable packaging. J Clean Prod 64:228–233
42. Geranpour M, Assadpour E, Jafari SM (2020) Recent advances in the spray drying encapsulation of essential fatty acids and functional oils. Trends Food Sci Technol 102:71–90
43. Gharibzahedi SMT, Smith B (2021) Legume proteins are smart carriers to encapsulate hydrophilic and hydrophobic bioactive compounds and probiotic bacteria: a review. Compr Rev Food Sci Food Saf 20:1250–1279
44. Gheribi R, Khwaldia K (2019) Cactus mucilage for food packaging applications. Coatings 9:655
45. Goularte, AC, Capello C, Valencia GA (2020) Recovery of chlorophylls from cucumber (Cucumis sativus L.) peel using a synthetic layered silicate. Recent Adv Food Sci 5:6
46. Grande Tovar CD, Delgado-Ospina J, Navia Porras DP, Peralta-Ruiz Y, Cordero AP, Castro JI, Chaur Valencia MN, Mina JH, Chaves López C (2019) Colletotrichum gloesporioides inhibition in situ by chitosan-ruta graveolens essential oil coatings: effect on microbiological, physicochemical, and organoleptic properties of guava (Psidium guajava L.) during room temperature storage. Biomolecules 9:399
47. Guillard V, Gaucel S, Fornaciari C, Angellier-Coussy H, Buche P, Gontard N (2018) The next generation of sustainable food packaging to preserve our environment in a circular economy context. Front Nutr 5:121
48. Gutiérrez-Pacheco MM, Ortega-Ramírez LA, Silva-Espinoza BA, Cruz-Valenzuela MR, González-Aguilar GA, Lizardi-Mendoza J, Miranda R, Ayala-Zavala JF (2020) Individual and combined coatings of chitosan and carnauba wax with oregano essential oil to avoid water loss and microbial decay of fresh cucumber. Coatings 10:1–16
49. Halász K, Csóka L (2018) Black chokeberry (Aronia melanocarpa) pomace extract immobilized in chitosan for colorimetric pH indicator film application. Food Packag Shelf Llife 16:185–193
50. Halonen NJ, Pálvölgyi PS, Bassani A, Fiorentini C, Nair R, Spigno G, Kordas K (2020) Bio-based smart materials for food packaging and sensors–a review. Front Mater 7:82
51. Hannon JC, Kerry J, Cruz-Romero M, Morris M, Cummins E (2015) Advances and challenges for the use of engineered nanoparticles in food contact materials. Trends Food Sci Technol 43:43–62
52. Hardenburg RE (1967) Wax and related coatings for horticultural products: a bibliography. Agric Res Bull 15–51
53. Hosseini H, Jafari SM (2020) Introducing nano/microencapsulated bioactive ingredients for extending the shelf-life of food products. Adv Colloid Interface Sci 102210
54. Ishkeh SR, Shirzad H, Asghari MR, Alirezalu A, Pateiro M, Lorenzo JM (2021) Effect of chitosan nanoemulsion on enhancing the phytochemical contents, health-promoting components, and shelf life of raspberry (Rubus sanctus Schreber). Appl Sci 11:2224
55. Jafari SM, Arpagaus C, Cerqueira MA, Samborska K (2021) Nano spray drying of food ingredients; materials, processing and applications. Trends Food Sci Technol 109:632–646
56. Jafarzadeh S, Jafari SM (2020) Impact of metal nanoparticles on the mechanical, barrier, optical and thermal properties of biodegradable food packaging materials. Crit Rev Food Sci Nutr 1–19

57. Jamróz E, Kopel P (2020) Polysaccharide and protein films with antimicrobial/antioxidant activity in the food industry: a review. Polymers 12:1289
58. Jeya Jeevahan J, Chandrasekaran M, Venkatesan SP, Sriram V, Britto Joseph G, Mageshwaran G, Durairaj RB (2020) Scaling up difficulties and commercial aspects of edible films for food packaging: a review. Trends Food Sci Technol 100:210–222
59. Jonušaite K, Venskutonis PR, Martínez-Hernández GB, Taboada-Rodríguez A, Nieto G, López-Gómez A, Marín-Iniesta F (2021) Antioxidant and antimicrobial effect of plant essential oils and sambucus nigra extract in salmon burgers. Foods 10:776
60. Joseph B, Mavelil Sam R, Balakrishnan P, Maria J, H., Gopi, S., Volova, T., CM Fernandes, S. & Thomas, S. (2020) Extraction of nanochitin from marine resources and fabrication of polymer nanocomposites: recent advances. Polymers 12:1664
61. Ju J, Xie Y, Guo Y, Cheng Y, Qian H, Yao W (2019a) Application of edible coating with essential oil in food preservation. Crit Rev Food Sci Nutr 59:2467–2480
62. Ju J, Xie Y, Guo Y, Cheng Y, Qian H, Yao W (2019b) The inhibitory effect of plant essential oils on foodborne pathogenic bacteria in food. Crit Rev Food Sci Nutr 59:3281–3292
63. Khaledian S, Basiri S, Shekarforoush SS (2021) Shelf-life extension of pacific white shrimp using tragacanth gum-based coatings containing Persian lime peel (Citrus latifolia) extract. LWT 141:110937
64. Kornet R, Shek C, Venema P, van der Goot AJ, Meinders M, van der Linden E (2021) Substitution of whey protein by pea protein is facilitated by specific fractionation routes. Food Hydrocoll 117:1–11
65. Kraśniewska K, Galus S, Gniewosz M (2020) Biopolymers-based materials containing silver nanoparticles as active packaging for food applications–a review. Int J Mol Sci 21:1–18
66. Krivorotova T, Cirkovas A, Maciulyte S, Staneviciene R, Budriene S, Serviene E, Sereikaite J (2016) Nisin-loaded pectin nanoparticles for food preservation. Food Hydrocoll 54:49–56
67. Kumar N (2019) Polysaccharide-based component and their relevance in edible film/coating: a review. Nutr Food Sci 49:793–823
68. Kumar N, Neeraj P, Trajkovska Petkoska A (2021) Improved shelf life and quality of tomato (Solanum Lycopersicum L.) by using chitosan-pullulan composite edible coating enriched with pomegranate peel extract. ACS Food Sci Technol 1–11
69. Kumar N, Pratibha PS (2020) Bioactive compounds of Moringa (Moringa Species). In: Murthy HN, Paek KY (eds) Bioactive compounds in underutilized vegetables and legumes. Cham: Springer International Publishing
70. Lee H, Rukmanikrishnan B, Lee J (2019) Rheological, morphological, mechanical, and water-barrier properties of agar/gellan gum/montmorillonite clay composite films. Int J Biol Macromol 141:538–544
71. Lee KH, Chun Y, Jang YW, Lee SK, Kim HR, Lee JH, Kim SW, Park C, Yoo HY (2020) Fabrication of functional bioelastomer for food packaging from aronia (Aronia melanocarpa) juice processing by-products. Foods 9:1565
72. Liang T, Sun G, Cao L, Li J, Wang L (2019) A pH and NH_3 sensing intelligent film based on Artemisia sphaerocephala Krasch. gum and red cabbage anthocyanins anchored by carboxymethyl cellulose sodium added as a host complex. Food Hydrocolloids 87:858–868
73. López-Rubio A, Blanco-Padilla A, Oksman K, Mendoza S (2020) Strategies to improve the properties of amaranth protein isolate-based thin films for food packaging applications: nano-layering through spin-coating and incorporation of cellulose nanocrystals. Nanomaterials 10:2564
74. Łopusiewicz Ł, Drozłowska E, Trocer P, Kostek M, Śliwiński M, Henriques MH, Bartkowiak A, Sobolewski P (2020) Whey protein concentrate/isolate biofunctional films modified with melanin from watermelon (Citrullus lanatus) seeds. Materials 13:3876
75. Lourenço SC, Fraqueza MJ, Fernandes MH, Moldão-Martins M, Alves VD (2020) Application of edible alginate films with pineapple peel active compounds on beef meat preservation. Antioxidants 9:667
76. Luís Â, Ramos A, Domingues F (2020) Pullulan films containing rockrose essential oil for potential food packaging applications. Antibiotics 9:681

77. Mahjoorian A, Jafarian S, Fazeli F (2021) Nettle (Utrica dioica) Essential oil incorporation in edible film from caspian whitefish (Rutilus frisii kutum) scale: physical, antimicrobial, and morphological characterization. J Aquat Food Prod Technol 1–11
78. Manikandan NA, Pakshirajan K, Pugazhenthi G (2020) Preparation and characterization of environmentally safe and highly biodegradable microbial polyhydroxybutyrate (PHB) based graphene nanocomposites for potential food packaging applications. Int J Biol Macromol 154:866–877
79. Matheus JRV, Miyahira RF, Fai AEC (2020) Biodegradable films based on fruit puree: a brief review. Crit Rev Food Sci Nutr 1–8
80. Maurya VK, Shakya A, Aggarwal M, Gothandam KM, Bohn T, Pareek S (2021) Fate of β-carotene within loaded delivery systems in food: state of knowledge. Antioxidants 10:426
81. Medina-Jaramillo C, Quintero-Pimiento C, Gómez-Hoyos C, Zuluaga-Gallego R, López-Córdoba A (2020) Alginate-edible coatings for application on wild andean blueberries (Vaccinium meridionale swartz): effect of the addition of nanofibrils isolated from cocoa by-products. Polymers 12:824
82. Mihalca V, Kerezsi AD, Weber A, Gruber-Traub C, Schmucker J, Vodnar DC, Dulf FV, Socaci SA, Fărcaș A, Mureșan CI (2021) Protein-based films and coatings for food industry applications. Polymers 13:769
83. Milea ȘA, Aprodu I, Enachi E, Barbu V, Râpeanu G, Bahrim GE, Stănciuc N (2021) β-lactoglobulin and its thermolysin derived hydrolysates on regulating selected biological functions of onion skin flavonoids through microencapsulation. CYTA-J Food 19:127–136
84. Mohanty AK, Vivekanandhan S, Pin J-M, Misra M (2018) Composites from renewable and sustainable resources: challenges and innovations. Science 362:536–542
85. Moradi M, Kousheh SA, Razavi R, Rasouli Y, Ghorbani M, Divsalar E, Tajik H, Guimarães JT, Ibrahim SA (2021) Review of microbiological methods for testing protein and carbohydrate-based antimicrobial food packaging. Trends Food Sci Technol 111:595–609
86. Motelica L, Ficai D, Ficai A, Oprea OC, Kaya DA, Andronescu E (2020) Biodegradable antimicrobial food packaging: trends and perspectives. Foods 9:1438
87. Mujtaba M, Salaberria AM, Andres MA, Kaya M, Gunyakti A, Labidi J (2017) Utilization of flax (Linum usitatissimum) cellulose nanocrystals as reinforcing material for chitosan films. Int J Biol Macromol 104:944–952
88. Narancic T, Cerrone F, Beagan N, O'Connor KE (2020) Recent advances in bioplastics: application and biodegradation. Polymers 12:920
89. Nazrin A, Sapuan S, Zuhri M, Ilyas R, Syafiq R, Sherwani S (2020) Nanocellulose reinforced thermoplastic starch (TPS), polylactic acid (PLA), and polybutylene succinate (PBS) for food packaging applications. Front Chem 8:1–12
90. Nešić A, Cabrera-Barjas G, Dimitrijević-Branković S, Davidović S, Radovanović N, Delattre C (2020) Prospect of polysaccharide-based materials as advanced food packaging. Molecules 25:135
91. Nile SH, Baskar V, Selvaraj D, Nile A, Xiao J, Kai G (2020) Nanotechnologies in food science: applications, recent trends, and future perspectives. Nano-Micro Lett 12:1–34
92. Nilsen-Nygaard J, Fernández EN, Radusin T, Rotabakk BT, Sarfraz J, Sharmin N, Sivertsvik M, Sone I, Pettersen MK (2021) Current status of biobased and biodegradable food packaging materials: impact on food quality and effect of innovative processing technologies. Compr Rev Food Sci Food Saf 20:1333–1380
93. Nogueira GF, Oliveira RAD, Velasco JI, Fakhouri FM (2020) Methods of incorporating plant-derived bioactive compounds into films made with agro-based polymers for application as food packaging: a brief review. Polymers 12:2518
94. Nunes JC, Melo PTS, Lorevice MV, Aouada FA, de Moura MR (2020) Effect of green tea extract on gelatin-based films incorporated with lemon essential oil. J Food Sci Technol 1–8
95. Oliveira Filho JG, Braga ARC, de Oliveira BR, Gomes FP, Moreira VL, Pereira VAC, Egea MB (2021) The potential of anthocyanins in smart, active, and bioactive eco-friendly polymer-based films: a review. Food Res Int 110202

96. Omran AAB, Mohammed AA, Sapuan S, Ilyas R, Asyraf M, Rahimian Koloor SS, Petrů M (2021) Micro-and nanocellulose in polymer composite materials: a review. Polymers 13:231
97. Othman SH, Othman NFL, Shapi'i, R. A., Ariffin, S. H. & Yunos, K. F. M. (2021) Corn starch/chitosan nanoparticles/thymol bio-nanocomposite films for potential food packaging applications. Polymers 13:390
98. Pateiro M, Gómez B, Munekata PE, Barba FJ, Putnik P, Kovačević DB, Lorenzo JM (2021) Nanoencapsulation of promising bioactive compounds to improve their absorption, stability, functionality and the appearance of the final food products. Molecules 26:1547
99. Pattnaik M, Pandey P, Martin GJ, Mishra HN, Ashokkumar M (2021) Innovative technologies for extraction and microencapsulation of bioactives from plant-based food waste and their applications in functional food development. Foods 10:279
100. Pech-Canul ADLC, Ortega D, García-Triana A, González-Silva N, Solis-Oviedo RL (2020) A brief review of edible coating materials for the microencapsulation of probiotics. Coatings 10:197
101. Pena-Serna C, Penna ALB, Lopes Filho JF (2016) Zein-based blend coatings: impact on the quality of a model cheese of short ripening period. J Food Eng 171:208–213
102. Peng Y, Wu Y, Li Y (2013) Development of tea extracts and chitosan composite films for active packaging materials. Int J Biol Macromol 59:282–289
103. Peralta-Ruiz Y, Grande-Tovar CD, Navia Porras D.P, Sinning-Mangonez A, Delgado-Ospina J, González-Locarno M, Maza Pautt Y, Chaves-López C (2021) Packham's triumph pears (Pyrus communis L.) post-harvest treatment during cold storage based on chitosan and rue essential oil. Molecules 26:725
104. Pérez-Santaescolástica C, Munekata PE, Feng X, Liu Y, Bastianello Campagnol PC, Lorenzo JM (2020) Active edible coatings and films with Mediterranean herbs to improve food shelf-life. Crit Rev Food Sci Nutr 1–13
105. Petkoska AT, Daniloski D, D'Cunha NM, Naumovski N, Broach AT (2021) Edible packaging: sustainable solutions and novel trends in food packaging. Food Res Int 140:109981
106. Porta R, Sabbah M, di Pierro P (2020) Biopolym Food Packag Mater 21:4942
107. Priyadarshi R, Rhim J-W (2020) Chitosan-based biodegradable functional films for food packaging applications. Innov Food Sci Emerg Technol 62:102346
108. Puscaselu R, Gutt G, Amariei S (2019) Rethinking the future of food packaging: biobased edible films for powdered food and drinks. Molecules 24:3136
109. Qamar SA, Asgher M, Bilal M, Iqbal HM (2020) Bio-based active food packaging materials: Sustainable alternative to conventional petrochemical-based packaging materials. Food Res Int 109625
110. Quesada J, Sendra E, Navarro C, Sayas-Barberá E (2016) Antimicrobial active packaging including chitosan films with Thymus vulgaris L. essential oil for ready-to-eat meat. Foods 5:57
111. Qureshi D, Nayak SK, Anis A, Ray SS, Kim D, Nguyen TTH, Pal K (2020) Introduction of biopolymers: food and biomedical applications. Elsevier, Biopolymer-Based Formulations
112. Reichembach LH, de Oliveira Petkowicz CL (2021) Pectins from alternative sources and uses beyond sweets and jellies: an overview. Food Hydrocoll 106824
113. Riaz A, Aadil RM, Amoussa AMO, Bashari M, Abid M, Hashim MM (2021) Application of chitosan-based apple peel polyphenols edible coating on the preservation of strawberry (Fragaria ananassa cv Hongyan) fruit. J Food Process Preserv 45:1–10
114. Rodríguez MC, Yépez CV, González JHG, Ortega-Toro R (2020) Effect of a multifunctional edible coating based on cassava starch on the shelf life of Andean blackberry. Heliyon 6:1–8
115. Romanazzi G, Feliziani E, Baños SB, Sivakumar D (2017) Shelf life extension of fresh fruit and vegetables by chitosan treatment. Crit Rev Food Sci Nutr 57:579–601
116. Rostamabadi H, Falsafi SR, Boostani S, Katouzian I, Rezaei A, Assadpour E, Jafari S M (2021) Design and formulation of nano/micro-encapsulated natural bioactive compounds for food applications. In: Application of nano/microencapsulated ingredients in food products. Elsevier

117. Roy S, Kim H-J, Rhim J-W (2021aa) Synthesis of carboxymethyl cellulose and agar-based multifunctional films reinforced with cellulose nanocrystals and shikonin. ACS Appl Polym Mater 3:1060–1069
118. Roy S, Kim HC, Panicker PS, Rhim J-W, Kim J (2021b) Cellulose nanofiber-based nanocomposite films reinforced with zinc oxide nanorods and grapefruit seed extract. Nanomaterials 11:877
119. Roy S, Kim HC, Zhai L, Kim J (2020) Preparation and characterization of synthetic melanin-like nanoparticles reinforced chitosan nanocomposite films. Carbohydr Polym 231: 115729
120. Roy S, Rhim J-W (2020) Effect of CuS reinforcement on the mechanical, water vapor barrier, UV-light barrier, and antibacterial properties of alginate-based composite films. Int J Biol Macromol 164:37–44
121. Ruelas-Chacon X, Aguilar-González A, de la Luz Reyes-Vega M, Peralta-Rodríguez RD, Corona-Flores J, Rebolloso-Padilla ON, Aguilera-Carbo AF (2020) Bioactive protecting coating of guar gum with thyme oil to extend shelf life of tilapia (Oreoschromis niloticus) fillets. Polymers 12:3019
122. Sahraee S, Milani JM, Ghanbarzadeh B, Hamishehkar H (2020) Development of emulsion films based on bovine gelatin-nano chitin-nano ZnO for cake packaging. Food Sci Nutr 8:1303–1312
123. Sami R, Almatrafi M, Elhakem A, Alharbi M, Benajiba N, Helal M (2021) Effect of nano silicon dioxide coating films on the quality characteristics of fresh-cut cantaloupe. Membranes 11:140
124. Samsi MS, Kamari A, Din SM, Lazar G (2019) Synthesis, characterization and application of gelatin–carboxymethyl cellulose blend films for preservation of cherry tomatoes and grapes. J Food Sci Technol 56:3099–3108
125. Sani IK, Marand SA, Alizadeh M, Amiri S, Asdagh A (2021) Thermal, mechanical, microstructural and inhibitory characteristics of sodium caseinate based bioactive films reinforced by ZnONPs/encapsulated Melissa Officinalis essential oil. J Inorg Organomet Polym Mater 31:261–271
126. Sarfraz J, Gulin-Sarfraz T, Nilsen-Nygaard J, Pettersen MK (2021) Nanocomposites for food packaging applications: an overview. Nanomaterials 11:10
127. Sharif N, Falcó I, Martínez-Abad A, Sánchez G, López-Rubio A, Fabra MJ (2021) On the use of persian gum for the development of antiviral edible coatings against murine norovirus of interest in blueberries. Polymers 13:224
128. Sharma M, Bhat R (2021) Extraction of carotenoids from pumpkin peel and pulp: comparison between innovative green extraction technologies (ultrasonic and microwave-assisted extractions using corn oil). Foods 10:787
129. Sharma S, Singh RK (2020) Cold plasma treatment of dairy proteins in relation to functionality enhancement. Trends Food Sci Technol 102:30–36
130. Shehata SA, Abdeldaym EA, Ali MR, Mohamed RM, Bob RI, Abdelgawad KF (2020) Effect of some citrus essential oils on post-harvest shelf life and physicochemical quality of strawberries during cold storage. Agronomy 10:1466
131. Solaberrieta I, Jiménez A, Cacciotti I, Garrigós MC (2020) Encapsulation of bioactive compounds from aloe vera agrowastes in electrospun poly (ethylene oxide) nanofibers. Polymers 12:1323
132. Souza VG, Pires JR, Rodrigues C, Coelhoso IM, Fernando AL (2020) Chitosan composites in packaging industry—current trends and future challenges. Polymers 12:417
133. Suhag R, Kumar N, Petkoska AT, Upadhyay A (2020) Film formation and deposition methods of edible coating on food products: a review. Food Res Int 136:109582
134. Susmitha A, Sasikumar K, Rajan D, Nampoothiri KM (2021) Development and characterization of corn starch-gelatin based edible films incorporated with mango and pineapple for active packaging. Food Biosci 41:100977
135. Tapilatu Y, Nugraheni PS, Ginzel T, Latumahina M, Limmon GV, Budhijanto W (2016) Nano-chitosan utilization for fresh yellowfin tuna preservation. Aquatic Procedia 7:285–295

136. Terzioğlu P, Parin FN (2020) Polyvinyl alcohol-corn starch-lemon peel biocomposite films as potential food packaging. Celal Bayar Üniversitesi Fen Bilimleri Dergisi 16:373–378
137. Tian B, Liu Y (2020) Chitosan-based biomaterials: from discovery to food application. Polym Adv Technol 31:2408–2421
138. Tran TN, Mai BT, Setti C, Athanassiou A (2020) Transparent bioplastic derived from CO_2-based polymer functionalized with oregano waste extract toward active food packaging. ACS Appl Mater Interfaces 12:46667–46677
139. Tsang YF, Kumar V, Samadar P, Yang Y, Lee J, Ok YS, Song H, Kim K-H, Kwon EE, Jeon YJ (2019) Production of bioplastic through food waste valorization. Environ Int 127:625–644
140. Umaraw P, Verma AK (2017) Comprehensive review on application of edible film on meat and meat products: an eco-friendly approach. Crit Rev Food Sci Nutr 57:1270–1279
141. Vasile C, Baican M (2021) Progresses in food packaging, food quality, and safety—controlled-release antioxidant and/or antimicrobial packaging. Molecules 26:1263
142. Vilarinho F, Vaz MF, Silva AS (2020) The use of montmorillonite (MMT) in food nanocomposites: methods of incorporation, characterization of MMT/Polymer nanocomposites and main consequences in the properties. Recent Pat Food Nutr Agric 11:13–26
143. Villasante J, Martin-Lujano A, Almajano MP (2020) Characterization and application of gelatin films with pecan walnut and shell extract (Carya illinoiensis). Polymers 12:1424
144. Waghmare NK, Khan, S (2021) Extraction and characterization of nano-cellulose fibrils from indian sugarcane bagasse-an agro waste. J Nat Fibers 1–9
145. Wang X, Shen Y, Thakur K, Han J, Zhang J-G, Hu F, Wei Z-J (2020) Antibacterial activity and mechanism of ginger essential oil against Escherichia coli and Staphylococcus aureus. Molecules 25:1–17
146. Xavier M, Parente I, Rodrigues P, Cerqueira MA, Pastrana L, Gonçalves C (2021) Safety and fate of nanomaterials in food: the role of in vitro tests. Trends Food Sci Technol 109:593–607
147. Xie F, Pollet E, Halley PJ, Averous L (2014) Advanced nano-biocomposites based on starch. In: Ramawat KG, Mérillon J-M (eds) Polysaccharides. Springer International Publishing, pp 1–75
148. Yadav M, Behera K, Chang Y-H, Chiu F-C (2020) Cellulose nanocrystal reinforced chitosan based uv barrier composite films for sustainable packaging. Polymers 12:202
149. Yang M, Liang Z, Wang L, Qi M, Luo Z, Li L (2020) Microencapsulation delivery system in food industry—challenge and the way forward. Adv Polym Technol 2020:1–14
150. Yeo JCC, Muiruri JK, Thitsartarn W, Li Z, He C (2018) Recent advances in the development of biodegradable PHB-based toughening materials: approaches, advantages and applications. Mater Sci Eng, C 92:1092–1116
151. Yong H, Liu J (2021) Active packaging films and edible coatings based on polyphenol-rich propolis extract: a review. Compr Rev Food Sci Food Saf 20:2106–2145
152. Yong H, Wang X, Bai R, Miao Z, Zhang X, Liu J (2019) Development of antioxidant and intelligent pH-sensing packaging films by incorporating purple-fleshed sweet potato extract into chitosan matrix. Food Hydrocolloids 90:216–224
153. Youssef AM, El-Sayed SM (2018) Bionanocomposites materials for food packaging applications: concepts and future outlook. Carbohyd Polym 193:19–27
154. Zakuwan SZ, Ahmad I (2019) Effects of hybridized organically modified montmorillonite and cellulose nanocrystals on rheological properties and thermal stability of k-carrageenan bio-nanocomposite. Nanomaterials 9:1547
155. Zhang B, Huang C, Zhao H, Wang J, Yin C, Zhang L, Zhao Y (2019) Effects of cellulose nanocrystals and cellulose nanofibers on the structure and properties of polyhydroxybutyrate nanocomposites. Polymers 11:2063
156. Zhang R., Zhang Z, Mcclements DJ (2020) Nanoemulsions: an emerging platform for increasing the efficacy of nutraceuticals in foods. Colloids Surf B Biointerfaces 194:111202
157. Zhang W, Chen J, Chen Y, Xia W, Xiong YL, Wang H (2016) Enhanced physicochemical properties of chitosan/whey protein isolate composite film by sodium laurate-modified TiO_2 nanoparticles. Carbohyd Polym 138:59–65

158. Zhang X, Xiao G, Wang Y, Zhao Y, Su H, Tan T (2017) Preparation of chitosan-TiO_2 composite film with efficient antimicrobial activities under visible light for food packaging applications. Carbohyd Polym 169:101–107
159. Zhao X, Cornish K, Vodovotz Y (2020a) Narrowing the gap for bioplastic use in food packaging: an update. Environ Sci Technol 54:4712–4732
160. Zhao Y, Sun H, Yang B, Weng Y (2020b) Hemicellulose-based film: potential green films for food packaging. Polymers 12:1–14

The Wicked Problem of Packaging and Consumers: Innovative Approaches for Sustainability Research

Linda Brennan, Lukas Parker, Simon Lockrey, Karli Verghese, Shinyi Chin, Sophie Langley, Allister Hill, Nhat Tram Phan-Le, Caroline Francis, Maddison Ryder, Michaela Jackson, Anouk Sherman, Ella Chorazy, and Natalia Alessi

Abstract This chapter presents the methodology and consumer insights from a highly impactful research and design project, conducted over two years by the Fight Food Waste Cooperative Research Centre and RMIT University. The project is designed to inform the Australian packaging industry in developing products and services that will reduce food waste. Food waste in Australia is a wicked problem, replete with paradoxes: people hate plastic but plastic saves food. The deep insights gained from the project have already resulted in significant changes to industry practices and new guidelines for packaging design. The methodology developed especially for this project is based on design thinking and ethnographic approaches, combined with quantifiable validation procedures. The study ensures industry actors have the right tools to make packaging decisions that reduce food waste.

Keywords Reducing food waste · Design thinking · Co-creation · Food packaging · Save food packaging · Multi-disciplinary research · Food industry · Consumer consultation

1 Introduction

Food waste is a wicked problem affecting communities and the environment globally. This chapter presents the methodology and insights of a highly impactful research and design project conducted over two years by the Fight Food Waste Cooperative Research Centre and RMIT University. Based on the consumer stage of the food supply chain, the project focussed on food packaging, exploring both design and consumer perception complexities and their impact on food waste. This chapter

L. Brennan · L. Parker (✉) · S. Chin · N. T. Phan-Le · M. Jackson · A. Sherman · E. Chorazy · N. Alessi
School of Media and Communication, RMIT University, 124 La Trobe St, Melbourne, Vic 3000, Australia
e-mail: Lukas.parker@rmit.edu.au

S. Lockrey · K. Verghese · S. Langley · A. Hill · C. Francis · M. Ryder
School of Design, RMIT University, Melbourne, Australia

S. S. Muthu (eds.), *Sustainable Packaging*, Environmental Footprints and Eco-design of Products and Processes, https://doi.org/10.1007/978-981-16-4609-6_6

provides a background to the issues surrounding food waste, details the design thinking and ethnographic methodology developed for this project, and includes a thorough evaluation of the project's processes and methods, including the pros and cons of such methods. We conclude by outlining how the project has already resulted in significant changes to industry practices and in new guidelines for packaging design.

1.1 The Problem of Food Waste Globally

Food waste is a significant environmental, economic, and social issue [29]. Reducing food waste is a significant way to lower production costs and increase the efficiency of the food system, improve food security and nutrition, and contribute towards a more environmentally sustainable food system [43]. Managing demand for food, in part by reducing food waste, is a key part of creating sustainable food systems to meet the world's growing population [104]. Addressing food waste, therefore, has a range of positive implications for global society and the biosphere.

Food loss and waste (FLW) occurs along the entire food supply chain (see Fig. 1). There are multiple definitions of FLW [13]. There are variations in these definitions about the inclusion of various stages of the food supply chain (and whether a distinction is made between food loss and food waste), in which different end-of-life options are considered as FLW, and the inclusion or exclusion of inedible parts of the food product [109]. There is also a differentiation between the loss of quantity and the loss of quality when determining FLW [43]. Most definitions consider 'food' to mean foodstuffs intended for human consumption. This project's definition of FLW is limited to food intended for human consumption that is wasted by consumers in the last stages of the food supply chain (see Fig. 1).

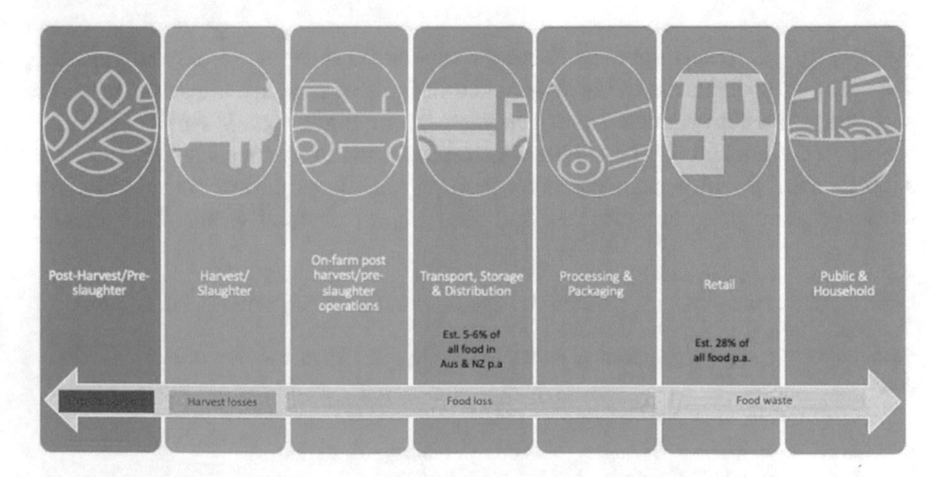

Fig. 1 Stages of the food system and estimated food loss and food waste. Based on [62], p. 22

The United Nations Food and Agriculture Organisation (UNFAO) estimates that each year, about one-third of all food produced for human consumption in the world is lost or wasted [43]. Globally, the volume of edible food waste is estimated to be 1.3 billion tonnes, with 13.8% of food produced being wasted at the upstream stages of the food supply chain: retail and the consumer [43]. Food waste at the consumer level is often caused by poor purchasing habits, confusion over labels, excess buying, and poor storage [43].

1.2 Current Approaches to Reducing Food Waste

Existing FLW solutions are usually aimed at reducing, recovering, or recycling waste. Food waste reduction is seen as the highest value solution [109]. Interventions aiming to reduce food waste in households have included consumer education [137] and technological interventions in household appliances, packaging, and technologies for food planning and sharing.

Consumer education has focussed on aspects of consumer behaviour, such as how products are used within the household [88, 94], and informing consumers about food waste issues, for example working with retailers and local governments on providing consumer information [6, 86] or using social media to share stories about food waste and reduction [136]. Skills training on how to reduce waste in the household has also been applied to the problem [95, 101]. However, much of this education is also applied to the context of health and weight management and consequently, the conflation of the issue of food waste with health means that the efficacy of interventions cannot be traced to a single causal factor.

Available packaging technologies are rapidly changing [107]. However, the uptake of technologies by the food and drink industry is still slow [19]. Some of this is ascribed to consumers' unwillingness to adopt and demand new packaging innovations [2]. Active and intelligent packaging can save food but is not well understood by consumers [7]. Digital applications can be built that allow consumers to use the food that is already available in their homes instead of buying new food [27], and there are apps that facilitate food sharing rather than wasting food [77]. Smart technologies that message consumers about looming expiry dates or refrigerator contents are also available [119]. In addition to smart communication technologies, there are smart appliances that are designed to ensure that food is kept at its optimum without the consumer even being aware that this is occurring. Table 1 provides a brief overview of existing approaches to reducing food loss waste.

Additionally, Spang and colleagues [109] argue that existing interventions tend to focus on pressure points for FLW, rather than addressing broader systemic causes and effects. Their review of existing interventions at the consumer level found that most interventions aim to change consumer behaviour and increase process efficiency, but few involve policy change. They also argue that it is necessary to consider the health and nutritional needs of consumers and to ensure that interventions aimed at

Table 1 Approaches to food loss waste interventions

Intervention type	Description
Consumer education	This type of intervention includes the following: • providing reference and communications material for local authorities to use in promoting reduced food wastage in their local area • advertising campaigns within supermarkets and other food retail stores • supermarket-led social media information campaigns (e.g. on company Facebook pages) • supermarket-led email newsletter information campaigns • intensive consumer training within households to learn skills and practices that prevent waste • changes to consumer 'environments', such as reducing plate or portion sizes
Technological	• This type of intervention includes the following: • consumer tools such as food sharing apps and active or intelligent packaging • refrigerators with cameras inside that are linked to a consumer's phone, so they can check what food they already have when they are shopping • refrigerators that send alerts to consumers via text message or email when food in the fridge is about to expire • colour-coding within the refrigerator assigning colours to particular food types so that consumers are more aware at a glance of what they have in their fridge • increasing the number of temperature-controlled compartments in a refrigerator to account for different refrigeration needs for different foods

(*Developed based on* [16, 38, 45, 93, 109, 121, 137, 136])

reducing waste by reducing purchases of certain foods—for instance, fresh fruits and vegetables—also aim to increase consumption of those foods.

1.3 The Scope and Nature of Food Waste in Australia

In the Australian context, it has been estimated that the value of food waste in Australia is $AUD20 billion [35]. According to the National Food Waste Baseline, Australia generated 11.8 million tonnes of food waste in 2016/17, of which four million was diverted to food rescue and animal feed [124]. This equates to 298 kg of waste per person [124]. The most significant stages of the supply chain at which the remaining 7.3 million tonnes of waste was generated include primary production (34%), households (34%), and manufacturing (24%) [124]. The role the consumer plays in reducing food waste is the subject of this research project. We know that consumers perceive plastic and food packaging as 'bad', but we do not yet know how much they are able to engage with the idea that packaging can save food and therefore simultaneously be more environmentally sustainable and cut household budgets if food is not wasted.

The top five food categories that are wasted in Australia are as follows:

1. Fresh fruit and vegetables.
2. Meat and seafood.
3. Dairy and eggs.
4. Bakery items.
5. Packaged and processed foods.

1.4 Consumer's Perceptions and the Role of Packaging in Reducing Food Waste

Packaging is often viewed as having a negative impact on the environment [56]. Packaging is leftover once the product is consumed, and the customer has to dispose of it either in the bin or through recycling [31]. However, in many cases, packaging protects food and prolongs its shelf life, reducing the overall environmental impact by reducing food waste [68]. Food packaging can help reduce household food waste by extending the shelf life of food products via innovative packaging design [65], being available in numerous sizes for different sized households [9], communicating on-pack the best way to use and store food items [121], assisting households to use date labels to better manage their food, and slowing the degradation of minimally processed vegetables [91]. Understanding, perception, and use of packaging by consumers also play a role in household food waste generation [13]. The negative perception of packaging and the lack of understanding of its purpose by households are likely to contribute to less-than-ideal packaging use amongst households. There is also a growing body of literature that identifies and examines food packaging functions and technologies that are specifically designed to reduce food waste [13, 127, 128].

Furthermore, new packaging technologies could reduce food waste by addressing some of the broader systemic causes of FLW—such as extending shelf life [25] and maintaining food quality [76], providing appropriate portion sizes [39], or using date labelling effectively [48, 82]—without discouraging the consumption of healthy foods (refer to Table 1).

Food packaging also plays a vital role in food waste reduction through functional measures [66]. Food packaging is continually advancing in shelf-life extension and waste reduction. Existing designs and integrated technologies include physical [121], chemical [10], sensory [122], and microbiological protection innovations [89, 127]. Consequently, there is well-established research on packaging features that extend the shelf life of food by using physical, chemical, and microbiological protection. However, research specific to packaging functions that save food from waste (also called 'save food packaging') is an under-developed field. Furthermore, food waste occurs within a behavioural ecosystem [3, 12], in which the consumer's interaction with the food and its packaging is different depending on the context of use [84], such as at work, at home, while out, and with friends or family or alone. Consequently, [65] asserts that the focus should be on the way the product and packaging work together as a system, rather than just the packaging. This approach aims to expand consumers' awareness that packaging is an actor in the broader food system and is not the only determining factor relating to environmental impact. Research on contemporary consumers' specific behaviours in relation to packaging is required to stay up to date, as demands for relevant future design are constantly shifting [7]. Understanding the consumer's potential responses to food packaging technologies is essential if consumer food waste is to be reduced.

Packaging designers also need to identify the aspects of packaging design (e.g. portion size or the ability to empty the packaging) that would reduce waste for specific products [127]. Beyond identifying aspects of form and function, designers need to be cognisant of stakeholders' attitudes and behaviours towards innovations in packaging design so that the uptake of the designs is optimised [73]. Packaging design can also be used as a medium through which to inform consumers of best practices for food waste reduction—for instance, to communicate portions [39] or inform consumers when the product has expired [116]. These factors in packaging design need to be based on the identification of food protection issues for particular products and an understanding of consumer behaviours that contribute to reduced food waste [127].

Consumer knowledge and levels of awareness, interest, and appreciation are major factors in consumers' refusal or acceptance of emerging packaging techniques (whether those technologies are specifically directed at reducing food waste or not). The complex relationship consumers have with food packaging creates a barrier to efficient food saving practices [124]. Consumer education on the benefits of packaging technologies is a repeated recommendation from both academics and industry [2, 7, 65, 123, 129]. However, as technologies improve, consumers may not have the capacity to assess the value of packaging due to the technical complexity of the products [135]. It is much easier to demonise 'plastic', as it presents a simple narrative, supported by high media weight [57, 74, 96, 117]. Consequently, educating consumers about food packaging must also include simple and easy-to-convey counter-messaging against the 'demon' plastic [32] if plastic is the solution to reducing food waste [123].

2 The Fight Food Waste Cooperative Research Centre

The Fight Food Waste Cooperative Research Centre (FFWCRC) is a partnership between 10 research and 48 industry partners. The collective aims are to improve the productivity, competitiveness, and sustainability of the Australian food industry. Progress towards these aims is via three main programs: Reduce, Transform, and Engage.

The Reduce program involves reducing supply chain losses by addressing resource flows and waste, function and perceptions of packaging and processing, opportunity identification, and food donation for social impact. The Transform program aims to transform waste resources by identifying valuable waste products, assessing waste transformation gaps and limitations, optimising opportunities, and assessing policy. The Engage program involves education and behavioural change via industry and skills training, future industry professionals' education, and behaviour change facilitation. Project 1.2.2 seeks to contribute mainly to the Reduce program but will also inform the Transform and Engage programs.

3 FFWCRC Project 1.2.2: Consumer Perceptions of Food Packaging

3.1 Consumer Perceptions Are Important in Reducing Food Waste

There are a wide variety of factors that influence consumer behaviour with food that is wasted. These broadly relate to the relationship between consumers' willingness to consume (WTC) [116] and their willingness to waste (WTW) [131]. There are three broad categories of factors that drive food waste in households: values, the challenges of everyday life, and managing stock in households [51]. These form a framework that has been used in all stages of Project 1.2.2 to understand the range of factors that contribute to consumer behaviour with food. Important to this work is the idea of a behavioural infrastructure in which consumption behaviours are embedded [12]. If the aim is to shift consumer behaviours at the individual level, then consideration must be given to the behavioural ecosystem in which the behaviours occur. For example, reducing household food waste through composting is only feasible if there are composting facilities available. How consumers respond to calls to reduce food waste will also depend on their perceptions of the importance of the issue [4], how they engage with food [5], as well as how they feel about food in general [97]. While much is already known about 'the green consumer' [78], we do not yet know how 'green' values and attitudes directly relate to food packaging and the issue of reducing food waste via improved packaging. Given that consumers can react negatively to complex technological solutions [72], there is a need to understand their concerns as well as their lived experiences in relation to food and waste.

3.2 Design Thinking Methodologies Provide the Opportunity to Deep Dive into the Consumer Journey

Design thinking is a process in which problems are reconsidered through a framework that engages those most connected to the problem in a process of co-creation, challenging assumptions, and creating positive, focussed solutions [120]. This process is a user-centred approach for tackling complex wicked problems [17] that do not have clearly identifiable solutions and where the user is instrumental in resolving the issue. The term 'wicked problem' refers to a problem that is complex and difficult or impossible to solve. [98] contrasted 'wicked' problems with 'tame' problems, clarifying that the former are less easily defined and often socially complex, multi-causal, and highly resistant to resolution, such as that of obesity prevalence in many developed nations. Design thinking represents a collaborative co-creation process of knowledge creation and iterative learning towards innovation [49]. The process involves gathering together a range of different people and perspectives, particularly those who will be the 'end-user' of the outcome in the brainstorming process and exploring and experimenting with possible ideas and solutions [15]. Design thinking involves three steps: inspiration, ideation, and implementation. Given the complex nature of the environment, particularly the nature of sustainability, and the need to use a scientific evidence base for business decision-making, our team added an additional element to the inspiration phase: information gathering (see Fig. 2).

The design thinking process (Fig. 2) is a form of co-creation that includes co-design methods where affected stakeholders are included in the development of solutions to the problem. Co-creation is the process where two or more people create something together collaboratively and in agreement with each other about desired outcomes [53]. Co-design, by comparison, is a form of design that actively involves

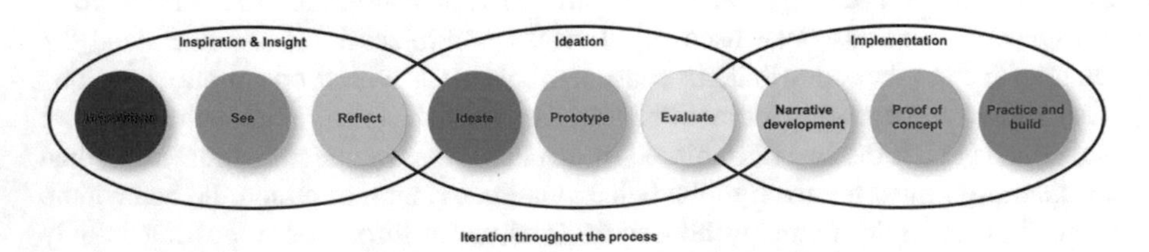

Fig. 2 Generic design thinking approach

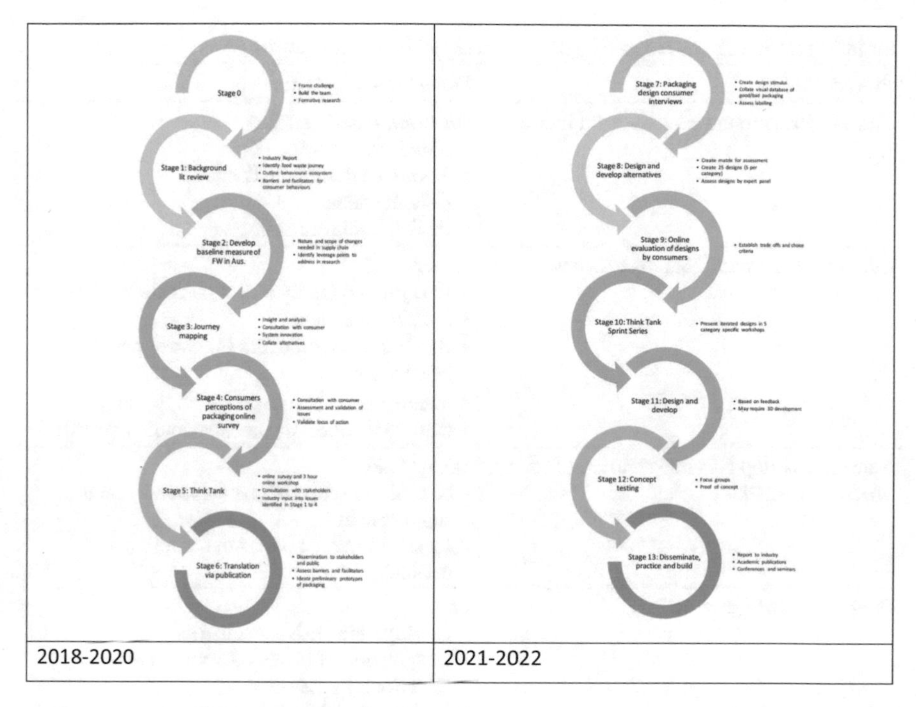

Fig. 3 Project 1.2.2 study protocol map

a variety of stakeholders in the design process; its roots are embedded in theories of participatory design [61].

3.3 The Design Thinking Approach to Project 1.2.2

The full study protocol for Project 1.2.2 was mapped out in the years prior to the inception of the FFWCRC. A series of industry round tables helped establish the key priorities of the first research projects to flow from the research centre. Project 1.2.2 is also integrally linked with other projects in the FFWCRC, such as Project 1.2.1: Save Food Packaging and Criteria. The protocol map is depicted in Fig. 3 and explanatory text is contained in Table 2.

3.4 Stages of Project 1.2.2

Project 1.2.2 adopted a multi-stage multi-method approach to understanding consumer perceptions of food packaging with a view to developing new packaging solutions that could be implemented in order to save food. A design thinking systems

Table 2 Protocol stages mapped against the design thinking framework

Protocol stage	Design thinking stage
Stage 0: Background research with industry	***Inspiration and insight*** Understand • Frame the design challenge • Build the team • Background formative research
Stage 1: Background literature review	Understand • Background theory and current state-of-the-art • Industry practices • Barriers and facilitators for consumer behaviours • Identify household food waste 'journey' • Outline the food waste behavioural ecosystem
Stage 2: Develop baseline measure of food waste in Australia	Understand • Nature and scope of changes needed in the supply chain • Identify leverage points to address in the research
Stage 3: Journey mapping	See • Consultation with consumers • Observation of behaviours within the behavioural ecosystem • Packaging and system innovation identification based on stages of the food waste journey • Collation of packaging alternatives and consumers' perceptions for Stage 4
Stage 4: Consumer perceptions of packaging	See • Consultation with consumers Assessment and validation of issues identified within Stages 2 and 3 • Ascertain locus and focus of action required • Create a summary of information for industry stakeholders from Stages 1, 2, 3, and 4
Stage 5: Think tank online survey	Understand • Consultation with affected stakeholders, i.e. industry and government • Assess responses to consumer research • Validate action loci and determine focus in conjunction with actors in the system

(continued)

approach to the 'wicked problem' of food waste requires multi-stakeholder engagement, as those who are affected may react unfavourably to posited solutions unless they have been included in the process of design. The research protocol for Project 1.2.2 was established to this end. Table 2 is a summary of the steps and a detailed explanation of each of the stages is outlined below.

Table 2 (continued)

Protocol stage	Design thinking stage
Stage 6: Publications and reports	***Ideation*** Ideate • Dissemination of reports to stakeholders and the public • Participation in industry forums and assimilation of feedback from affected participants • Assess barriers and facilitators to packaging innovation • Develop ideas based on feedback • Ideate preliminary prototypes of packaging
Stage 7: Packaging design consumer interviews	Ideate • Create packaging design stimulus material (idea generation) • Collate photos provided by journey mapping participants for good and bad ideas • Conduct in-home discussions about packaging design ideas, especially those in relation to reducing food waste and labelling
Stage 8: Design and develop packaging alternatives	Prototype and iterate • Design and create 25 two-dimensional packaging alternatives that aim to reduce food waste with a multi-disciplinary team of industrial design, consumer behaviour, and communication design researchers • Develop five ideas for each category of food (see the section on food waste in Australia)
Stage 9: Consumer perceptions of packaging alternatives	Evaluate • Online evaluation of packaging alternatives • Iteration of designs based on feedback from consumers
Stage 10: Think tank sprint series	***Implementation*** Narrative development • Present designs to industry • Iterate based on feedback, especially in regard to feasibility of adoption • Create narrative that can facilitate adoption for each food category
Stage 11: Design and develop	• Iterate based on feedback • Iterate if required • Create three-dimensional alternatives if required • Discuss barriers and facilitate to adoption of innovation
Stage 12: Concept testing with consumers	Proof of concept • Final testing of packaging alternatives with consumers in group format to facilitate idea generation • May use three-dimensional packaging alternatives for stimulus material

(continued)

Table 2 (continued)

Protocol stage	Design thinking stage
Stage 13: Report to industry and academic publications	Disseminate, practice, and build • Industry reports • Industry conference presentations • Academic publications, including book chapters and journal articles • Academic conference presentations

3.4.1 Stage 1: Systematised Literature Review

Stage 1 involved a systematised literature review (SLR) on consumer perceptions and understanding of packaging, packaging designs of on-pack communication, and guidance material for brand owners [13]. The review covered both academic and 'grey' literature, as there is a significant practical and theoretical contribution from organisations such as Worldwide Responsible Accredited Production (WRAP) and Industry Council for Packaging and the Environment (INCPEN) [56]. The aim of this stage was to understand consumer perceptions of the role of packaging and reducing food waste. The literature review was required to find what is already known about consumer behaviours in relation to packaged and non-packaged food consumption. It was important to understand the scope and nature of the academic knowledge base prior to positing design solutions for the use of packaging to reduce food waste [60], especially with a view to not 'reinvent the wheel' in any designs [99]. The focus of this project is food consumption and waste in the home; however, the scope of the review included any kind of packaging designed to reduce food waste to ensure a wide range of views were captured [75]. This review, combined with other stages of the project, enabled the team to posit packaging solutions that can reduce food waste. The review has been published if further detail is required [13].

3.4.2 Stage 2: Baseline Measure of Food Waste

One of the key aspects of the FFWCRC is that it is intended to produce real reductions in food waste over the coming decade (see https://fightfoodwastecrc.com.au/about-us/). Consequently, the first task was to establish a measure for assessing food waste in Australia in order to ensure that all the FFWCRC projects, including Project 1.2.2, are working towards a common and specific goal [22]. Using literature and publicly available secondary data, as well as data from various FFWCRC projects, including Project 122, the baseline measure (BLM) stage developed a benchmark against which we could assess the success or failure of food waste reduction initiatives across Australia. The outcome of the BLM stage was then used to inform other stages of Project 1.2.2 in terms of setting targets for reduction, as well as to inform other FFWCRC programs.

The baseline we established draws on a combination of sources: (1) a literature review, (2) publicly announced bin audit reports, and (3) analysis of primary survey data. Firstly, to establish the problem of FLW in Australia, food waste estimates and life cycle analysis (LCA) contributions for each food category were drawn from the baseline literature review [62]. Secondly, figures taken from a national food waste baseline [1] were contrasted against a local council bin composition audit, conducted for the Boroondara Council by EC Sustainable in 2017 [34], to present a sample of variances found in household food waste in Australia. Finally, primary data was then sourced from online survey results from two other FFWCRC projects (Projects FFWCRC1.2.1 and FFWCRC 1.2.2). The first study surveyed 95 food industry experts on perceptions and practices responding to food waste and their current adoption rate of save food packaging (SFP) design strategies (FFWCRC1.2.1 Stakeholder Review Survey). A second study of 1,015 participants sampled the Australian population, framing FLW in households and consumers' perceptions of food waste and food packaging. Estimates were made on weekly average food waste in cups (FFWCRC1.2.2 Existing Perceptions of Packaging Survey). Full reports and datasets can be downloaded from the FFWCRC website.

3.4.3 Stage 3: Journey Mapping (Rapid Ethnography)

The third stage of the research used a rapid ethnography [11] technique of journey mapping (JM) to follow food and possible wastage from the beginning of the consumer purchasing journey to the end. This stage of the project aimed to understand the role of packaging in the food waste journey, how packaging is used, how the consumers respond to packaging (changes, designs, styles, and types), and how the consumers waste (or not) their food. The JM process uses procedures of underlying methods associated with rapid ethnography [81, 90]. A journey map is built from the user's point of view [118]. The food waste journey starts with the perspectives of the user (i.e. their awareness of the aspects of food) [70]. It then moves to understand the sorts of things the user considers in the food consumption and food purchase journey, including emotional, social [125], and technical [110] issues in understanding food and food consumption and their relationship with food waste. JM asks, for example, does meal planning take place? If so, what are the barriers and benefits associated with meal planning? Are there more people in the household to be considered, and how are these people included in the decision-making process? The list is potentially endless, and it depends on the consumer and the behavioural ecosystem as to what things are considered. Furthermore, each consumer is likely to take a different journey [70] and it is the mapping of the multiple journeys that makes a food waste landscape visible to organisations that aim to make decisions based on the journey maps [115]. A journey map could easily take four or five hours of participant time. In this stage of the project, it was, therefore, not feasible to go through all components of the journey with each of the participants. However, if the participant was feeling generous with their time, we occasionally conducted a 'deep dive' into their food waste journey.

Figure 4 provides an example of a consumer and food waste journey. It shows the possible stages that appear as food waste is on the pathway from planning and purchase through to disposal. Each of the five categories of food—meat and seafood, bakery, packaged and processed foods, dairy and eggs, and fresh fruit and vegetables—was examined, producing insights that can be applied at the industry level as well as at the consumer level. Table 3 contains the details of the categories and the items included in the research.

Recruitment was managed and 37 journey mapping interviews were conducted in-home around Melbourne, Victoria, by a research supplier. The research supplier utilised one qualitative field researcher to conduct all the field works. Using one field researcher assisted with the consistency of approach, as well as the ability to recruit respondents from geographic quadrants/blocks in Melbourne—East, West, South, North, and Inner Melbourne—to ensure, both, strategic demographic coverage and expedition of researcher mobility.

The participant selection criteria sought and found respondents that were as follows:

- Primary or joint food purchasers and preparers in a household.
- A mix of male and female (approximately 1/3 male and 2/3 female).
- Aged 18 and over.
- A cross-section of the Victorian population by age, education levels, household income, and household structure (natural fall out).
- Regular users (purchasers and consumers) of at least one, but manifestly a variety or, most often, all of the five food categories wasted. The interviews segmented and focussed on one of the five food categories—that is on, either, meat and seafood, bakery, packaged and processed foods, dairy and eggs, or and fresh fruit and vegetables. Though discussion, in the journey mapping sessions, invariably covered far more than the one food group.
- The intention was to conduct 10 interviews each for the five food categories (50 total). Scheduling commitment conflicts of the researcher, however, lead to an assessment being made that there was sufficient coverage of categories and data to answer the core research question with confidence overall. And that the project would not proceed with further data collections after the 37th interview. Category representation ranged from 6 to 10 of each food category.

Each journey mapping session was audio recorded, while the interviewer created a handwritten map during the session as well as taking photos, after the session, of the respondent's food and the places it was stored and disposed of in. The RMIT team then transcribed the interviews and analysed the transcriptions along with the digitised journey maps and food images, with the images providing a useful context and even contradictory contrast to what was said during the journey mapping session.

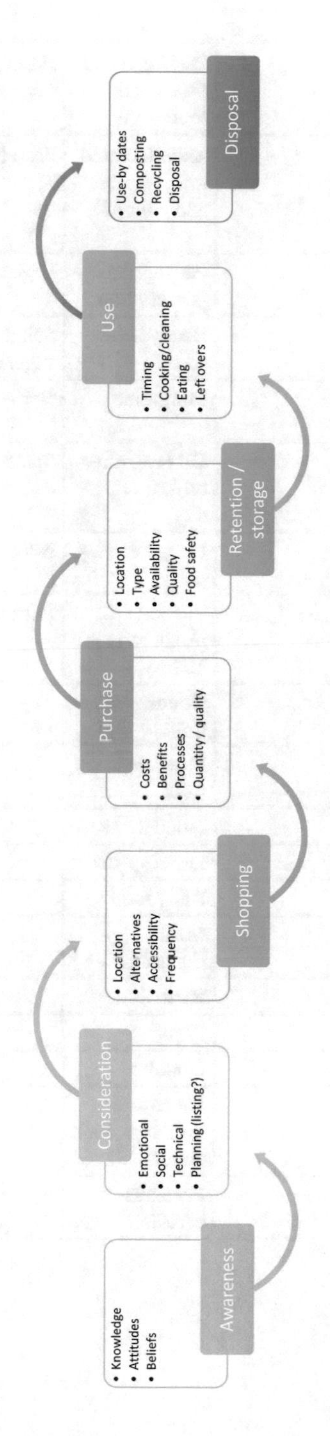

Fig. 4 Example of the dimensions of a consumer food waste journey map

Table 3 Detailed food categories

Meat and seafood	Bakery	Packaged and processed foods	Fresh fruit and vegetables	Dairy and eggs
Fish (all types)	Baked goods	Manufactured milk (e.g. soy and almond)	Fruit (all types)	Milk (cow, sheep, goat, i.e. from mammals)
Fresh meat (all types)	Biscuits and cookies	Packaged grocery items	Vegetables	Cheese
Frozen meat (all types)	Bread	Ready meals	Salad vegetables	Cream/sour cream
Seafood (all types)	Cakes	Margarine	Herbs	Eggs
	Flat bread	Oil (e.g. olive oil, canola, etc.)	Spices	Yoghurt
	Rolls	Preserves (e.g. jam)	Nuts	Butter/ghee
	Wraps	Pickled products (e.g. olives)	Pulses (fresh)	Kefir/Yakult
		Condensed milk		Butterfat
		Powdered milk		Ice cream
		Dried herbs		
		Dried spices		
		Baby food		
		Meal replacements		
		Rice		
		Pasta		
		Noodles		
		Sauces, dressings, and condiments		
		Breakfast cereals		

(continued)

Table 3 (continued)

Meat and seafood	Bakery	Packaged and processed foods	Fresh fruit and vegetables	Dairy and eggs
		Chocolate and confectionary		
		Canned beans and pulses		
		Canned vegetables		
		Sugar and sweeteners		

Derived from: https://sustainabletable.org.au/all-things-ethical-eating/seasonal-produce-guide/; https://www.foodsafety.edu.au/so-what-is-the-most-popular-vegetable-in-australia; https://www.euromonitor.com/packaged-food

3.4.4 Stage 4: Existing Perceptions of Packaging and Food Waste

Stage Four of the project sought to establish measurable benchmarks against which proposed solutions could be evaluated through an existing packaging perceptions survey (EPPS). This stage evaluated Australian consumers' perceptions of the role of packaging in reducing food waste via an online survey, utilising the Qualtrics XM platform (an institutional tool used for all project surveys). The survey was designed based on the insights gained from the food waste journey mapping stage.

The methodology drew on a mixed-methods approach, also known as multi-method [54]. Questions of both a quantitative and qualitative nature were included in the survey design. The key research question underpinning this component of the research was as follows:

Could packaging play a role in decreasing food waste, and if yes, how much and what sort?

By assessing consumer perceptions of food packaging and food waste generated in the household, the purpose of the EPPS was to quantify and validate for the purposes of decision-making drawing on the insights taken from the journey mapping stage of the project.

A sample size of 1015 participants was surveyed. Recruitment of participants was achieved through an online market research agency, which managed the survey distribution in-field, recruiting participants from their existing Australian panellists.

The participant selection criteria sought and found representatives that were as follows:

- Primary or joint food purchasers and preparers in a household.
- A mix of male and female (approximately 1/3 male and 2/3 female).
- Aged 18 and over (targeting a normal distribution of ages).
- A cross-section of the nation's population within each State and Territory, based on reported ratios from the Australian Demographic Statistics (ABS 2019).

- A cross-section of education levels, household income, and household structure (natural fall out).
- Regular users of one or more of the food categories wasted. The survey filtered participants (approximately 200 users per group—20% of sample) into one of the five food categories for specific enquires relating to meat and seafood, bakery, packaged and processed foods, dairy and eggs, and fresh fruit and vegetables.

Participants were screened out at the start of the survey if they did not provide consent, and/or they did not meet the selection criteria stipulated above. Participants were omitted from data analysis if they returned incomplete or automatically completed surveys.

3.4.5 Stage 5: Think Tank

Stage 5 was designed to consult with the industry in three discrete parts. Firstly, online surveys with industry participants where insights from the previous stages (SLR, JM, EPPS, and BLM) were disseminated to participants as part of the survey design. Participants were asked specific questions in relation to their assessment of the findings of the previous stages, especially in relation to their capacity to act on the issues raised by the participants. Participation was voluntary and was sourced via email invitation distributed by RMIT researchers to leading industry stakeholders within the five food categories. A total of 155 participants completed the industry survey. Seventy-five of these responses were omitted due to partial completion and refusal of consent, resulting in the analysis of 80 responses.

Large businesses (identified as employing 200 or more employees) represented over half of the respondents. Demonstrating a broader selection of persons, respondents served in roles at the CEO/MD executive level (17), marketing (15), and sustainability managers (10). All food categories were represented, including meat, seafood, fresh produce, dairy and eggs, bakery, food and/or beverages, ready meals, and processed foods, packaging supplier/designer/consultant. However, participants in the food and/or beverage roles and packaging supplier/designer/consultants were the most highly represented, with 28% and 34% of all surveyed responses.

Participants who currently were not or were unsure about whether they considered consumer food waste when designing packaging represented over a fifth of the sample collected, highlighting the need for further implementation of save food strategies that impact consumer food waste within the supply chain.

The second part of Stage Five consisted of an online workshop that discussed the results of previous stages of the project, in order to listen, learn, and capture in-depth insights [64] into the main themes gathered from the previous survey [63]. The principal purpose of this workshop was to ideate designs that would be feasible from an industry standpoint prior to being worked up into designs for the consumer research component [71].

The insights gathered from these two industry sources and the consumer research then informed the final part of Stage Five, several smaller focussed sessions (named

'Think Tank Sprint Series' [64]). The purpose of these focussed sessions was to 'sense check' the ideas and existing designs with specific industry representatives before conducting the final stages of the consumer research. The sessions established an extensive understanding of the barriers and facilitators to the implementation of the designs [90].

3.4.6 Stage 6: Publication Milestones

Throughout the research process, there was the potential to publish data and insights from the different modules of work as reports and articles as well as present at conferences and seminars. There has been no specific time frame established for the delivery of these additional publications. The purpose of the publication stage (PUB) is to ensure that reporting on the project is undertaken throughout the duration of the project (3 years) rather than waiting until the final phases of the research are complete. This enables the research outcomes to be applied, tailored, and implemented within the various contexts of use as knowledge is generated [42]. The rapid dissemination of knowledge through publication allows for iteration and further ideation and expands the knowledge network, as feedback is received throughout the conduct of the research. It also provides external validation of the methodology and ensures rigour in the approach [106].

3.4.7 Stage 7: Labelling and Packaging Design Impact on Consumer Decision-Making

Stage 7 of the project was packaging impact interviews (PII). The aim of this stage of the research was to understand the impact of packaging and labelling on consumer decision-making about food. Food labelling plays a role in helping consumers reduce their household food waste [126]. However, Chu and colleagues [23] have identified the need for further investigation of the role of date and storage labelling on foods, and their impact on consumer food waste on a systemic level. This stage explored the ways in which consumers use (or do not use) date, storage, and other labellings to make decisions about purchasing and using food, and how these decisions might contribute to or reduce food waste.

This study forms part of a multi-disciplinary, multi-phased project. As such, a mixed-method approach [132] was adopted to allow for cross-validation between qualitative and quantitative investigations across the project. In order to develop an approach for semi-structured interviews, the research team drew on existing literature from the systemic review, qualitative data from the journey mapping stage, and quantitative data from the existing perceptions survey stage.

This study used interviews about packaging that the consumers interact with, as well as examples of packaging that the research team had identified that may address issues related to household food waste. The in-home interviews were conducted virtually, via a secure online meeting platform provided by the research supplier.

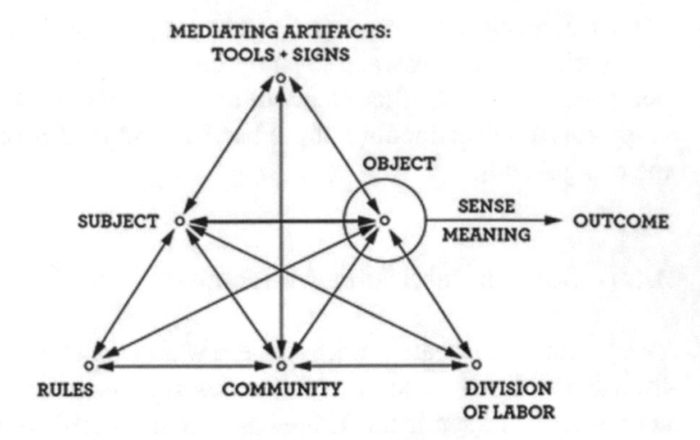

Fig. 5 Theoretical framework for conduct and analysis of interviews—Activity theory and its constructs ([36] p. 30)

Participants were interviewed about on-pack information for foods across the five food categories identified in the earlier stages of the broader project.

Consistent with prior research, activity theory was chosen as the theoretical and analytical framework for this stage of the project [23]. Activity theory conceptualises human interactions with artifacts within the context of real-life circumstances as a system of interrelationships between six different elements: (1) 'object' or central issues of the activity being considered, (2) 'subject' or the particular activities considered, (3) 'rules' or guidelines and limitations managing the system, (4) 'community' or stakeholders involved in the system, (5) 'division of labour' or working groups, disciplines, and roles in which the actors perform, and (6) 'mediating artifacts: tools and signs', or instruments used by actors within the system that changes with experience and knowledge [36]. Collectively, these function to produce a 'sense' or 'meaning' that results in a mutual 'outcome', as shown in Fig. 5.

This method allows a direct understanding of how consumers think about and engage with labelling and on-pack information when they are making decisions about purchasing, storing, consuming, and discarding food items. Activity theory allows an assessment of the interactions and the interrelationships that exist between individual consumers, the packaging, and other elements that impact their decisions.

A research supplier was asked to recruit for and conduct 60-min interviews with 50 grocery buyers/preparers (10 in each of the five food categories) using stimulus material provided by the RMIT research team. These materials were developed based on real-life on-pack information examples from each of the five food categories. The real-life examples were chosen based on findings from the earlier journey mapping stage of the research. For instance, storage was identified as a potential issue contributing to household food waste, as was packaging functionality such as resealability. As such, real-life labels that included storage instructions or pointed to the resealability of the packaging were used as prompts in the interviews.

Participants for this stage were sampled under the same conditions as Stage 3. Ten users in each of the five main food categories under consideration participated (50 people in total). Participants were also asked to identify and provide pictures of on-pack information shown by investigators that they felt was helpful or otherwise useful for them, especially in using packaging to help prevent food waste in their household.

The key research question for this stage of research was as follows:

What are the labelling and on-pack impacts on consumer decision-making? Does on-pack information help/hinder? What/sort/how/why? What is the role of date labelling?

The discussion guide explored the following topics:

- How often and in what situations consumers look at and gain understanding from on-pack information and labelling.
- What sort of on-pack information is used by consumers, how it is used, and why.
- Whether on-pack labelling helps or hinders consumer decision-making.
- The role of date labelling and the level of consumer understanding of different kinds of date labelling (i.e. 'use by' and 'best by' dates).
- The role of storage advice on-pack (e.g. messaging that instructs consumers to store food in the refrigerator or freezer.)
- How consumers think on-pack information and labelling could be changed or improved to help them reduce their household food waste.

Interviews with consumers were recorded and transcribed, and this data, along with images supplied by consumers, were analysed by the RMIT University team. The insights revealed complex relationships between consumers' existing knowledge, regular buying habits, in-store messaging, understanding of date labels, design layout of on-pack information and labelling, and the use of on-pack information. The insights gathered in this stage and insights from the journey mapping stage were then fed into Stage 8: Design and develop alternatives to create alternative packaging concepts.

3.4.8 Stage 8: Design and Develop Alternatives Based on Previous Stages of Research

Alongside the design thinking research process that provided the overarching framework for this research, the UK Design Council's 'double diamond' design method [114] was used as a guiding conceptual framework when designing packaging with environmental considerations. The comprised Stage 5 of the project is the design and development project (DDP).

The double diamond method has previously been used with life cycle assessment (LCA) as a process to address sustainability issues [24, 67]. Figure 6 illustrates that, when used with LCA, the double design method breaks strategy/design development into four key stages: (1) problem exploration, where assessment is used to develop an environmental benchmark of a set issue (in this case packaging designs); (2) problem definition, to determine the major practices that contribute to environmental impacts of the issue; (3) designing alternatives, where problems with practices (e.g. packages) are analysed and changes proposed by key stakeholders (e.g. consumers, designers, and industry participants); and (4) validation, where assessment is used to quantify

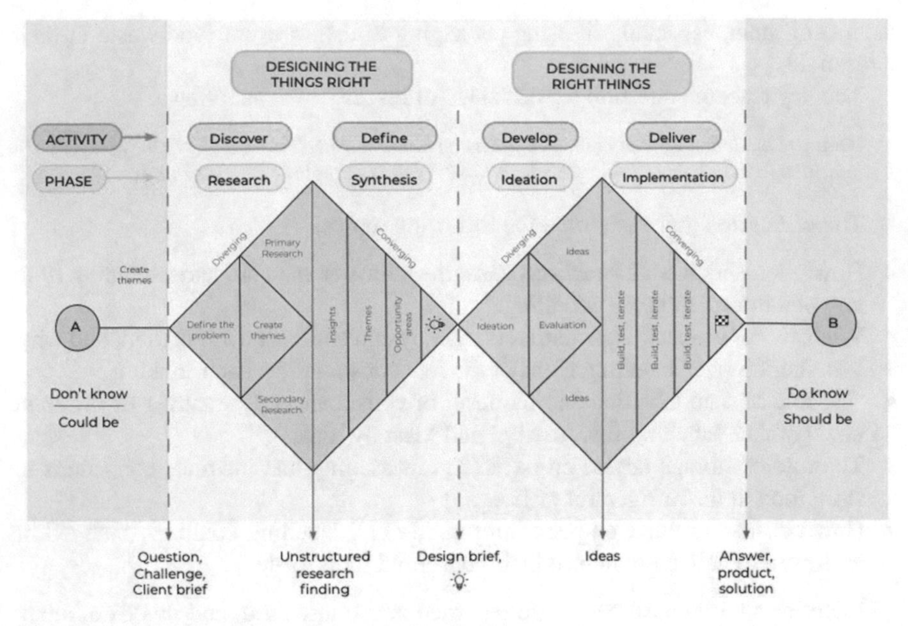

Fig. 6 *Double diamond/LCA process (based on* [67])

what alternatives or modified practices will make a difference to reduce impacts (i.e. packaging solutions that are environmentally sustainable and consumer friendly).

In this stage of the research, LCA benchmarking was substituted with the results of the earlier stages of the research program. The benchmark of the problem was derived from the results of:

- Stage 1 Baseline literature review.
- Stage 4 Existing perceptions of packaging survey.

The journey mapping stage identified contextual information contributing towards understanding home practices, including human behaviours, while consumers dealt with food, packaging, and food waste. It was then possible to identify what practices contribute to food waste occurring or being reduced, and hence arrive at a problem definition.

The insights provided by industry stakeholders in Stage 5 of the survey and workshop and the packaging impact interviews in Stage 7, themselves based on insights from previous stages, provided guidance on packaging alternatives that could address the problems identified and which are both feasible and practical.

In this stage of the research, design scenarios were developed as the precursor to the validation stage of the double diamond. These designs were developed based on data from the previous stages, through a series of conceptualising procedures consisting of mind mapping, categorising/coding design ideas with a matrix, and 2D visual prototypes, with a view to developing 3D prototypes at a later stage, depending on elapsed time and budget availability [58]. The design team collated these ideas in a 10 to 15-page report for the industry, and a series of visual representations were taken forward into other stages of the research (for validation and proof of concept research).

To decide on what packaging designs to develop for industry and consumer feedback, a design matrix was created. This method was separated into four distinct phases in line with design practice (with a fifth phase linked to Stage 9 and Stage 10):

Phase 1: Empathise—Collection of Insights

Phase 1 consisted of the collection of consumers' packaging-specific perceptions, behaviours, and/or needs from insights gathered from the previous stages of the research. Here, the insights were also allocated to specific clusters describing the type of packaging element (e.g. resealability, easy to open, storage, education, etc.).

Phase 2: Define—Categorisation of Insights

In Phase 2, the team separated all insights into the five food categories (bakery; dairy and eggs; packaged and processed; fresh fruit and vegetables; and meat and seafood), with all other insights categorised under a miscellaneous 'all food categories' sector.

Phase 3: Ideate—Populating Packaging Design Matrix

The design team collectively transferred the clustered and category-specific insights into a design matrix in Phase 3 to create the formula for each packaging design. The layout of this matrix allowed five packaging designs for each food category to be developed. To ensure each of the designs represented the breadth of the category, a different product variation was allocated to each and was then plotted against all suitable packaging clusters developed in Phase 1.

Phase 4: Prototype—Design Selection

All packaging design formulas were then analysed by the design team to collectively select the most appropriate packaging clusters per category, ensuring a range of elements were spread over the five designs per category. This phase also included some basic thumbnail sketching of ideas as a visualisation of possible final designs. Once the conceptualised packaging elements that were plotted on the matrix were found to be considerate of consumers, industry, and environmental needs, they were then visualised into 25 packaging formats. Each packaging idea was presented from a front-view 'hero' image, which is the prominent feature of the visual imagery. The design team created the two-dimensional visualisations of the agreed designs using Adobe Photoshop software, ready for review in the next stage of the project. The pack formats were founded on the 3D functional attributes required for each food category. Finally, the on-pack communication, callouts, branding, and attention-grabbing details were overlaid to communicate the functionality and value of the product and to call for behavioural change.

Phase 5: Test—Think Tank Appraisal of Packaging Visual Prototypes (Set for Stage 9 and Stage 10)

An online survey was designed to test consumer responses to the 25 packaging formats. Also a 'sprint series' think tank was set up to engage industry professionals from the food categories represented and capture their appraisals of the 25 packaging formats. This stage provided a collective review of the packaging concepts to check the relayed insights are meaningful and valuable to industry and consumers alike.

3.4.9 Stage 9: Labelling and Packaging Design's Impact on Consumer Decision-Making (CCS)

Stage 9 consisted of a consumer choice survey. Designs developed in Stage 8 were rendered into two-dimensional images with up to 30 words of descriptive text. The scoring matrix developed by [40] was adapted for the survey, as were dimensions of food waste from [130], in order to ensure that direct comparisons could be made between the European and Australian contexts. Consumer choice criteria about eco-friendliness were based on prior work by the research team [85]. An additional set of questions about the influence of packaging design on purchase intent is also posed to participants [44]. Questions about the impact of labelling on consumer choice were also included for each category [41, 103, 134].

The key research question posed in the survey was as follows:

What is labelling and packaging design's impact on consumer decision-making?

As with Stage 4, the survey was distributed via an online market research agency and the participant profile was consistent with that and other stages. That is, participants for Stage 9 were a

> Reasonable cross-section of the Australian population by gender, age, education level, household income, and household structure, for those food purchasers/preparers aged 18 years and over (with natural fallout).

Fifty users in each of the five main food waste categories under consideration were targeted and filtered into relevant food category packaging design review segments (250 people in total).

3.4.10 Stage 10: Think Tank Sprint Series

Stage Ten of this protocol was a series of category-specific online group discussions about the issues raised by the research. The purpose of the 'sprint series' think tanks was to ascertain within industry and stakeholder responses to both the research and the potential packaging designs. Stage 10 provided a form of sense-checking, which was deemed necessary because the feasibility of design implementation is premised on the specific barriers and opportunities that each industry manufacturing context presents [33]. The workshops conducted in Stage 5 were used to identify general industry-level principles of design and issues faced at a supply chain level [113]. The sprint series think tanks provided a deep dive into the specific questions facing manufacturers and suppliers in each category of food [63].

Furthermore, each category of food faces its own complexities when it comes to the potential to implement design ideas [82]. Consequently, prior to posing design solutions to industry, the research aimed to assess feasibility with affected stakeholders [50].

Questions asked in this stage covered two main topics: (1) the perceived barriers and opportunities for change within the industry sectors, and (2) the feasibility and acceptability of the packaging designs.

This stage of the research adopted a problem-solving approach to the issue of implementing packaging design innovation within organisations and industries [59]. Issues raised here were considered in the final stages of the project described below.

3.4.11 Stage 11: Design and Develop Solutions from Previous Research Stages

Using the outcomes from previous stages, the design team developed solutions to a 2D concept testing level (i.e. not full 3D prototypes). Additional techniques such as personal development [108], user journey mapping [70, 102], and systems mapping [105] were used to propose how both consumer and industry navigate 'pain' points [26], opportunities for change [30, 92], user interactions [81, 90], and supply chain issues [139]. These dimensions were derived from the Stage 10 think tank sprint series.

This stage was conducted concurrently within the sprint series whereby issues raised in each of the think tanks sessions were included in the design of the packaging as an iterative process of improvement [20]. Iterating throughout the process ensures that design is optimised for scalability and use immediately following the design process [69]. The sprint series took place over six months during 2021, so there was time to include industry perspectives of the designs during the generative design period [20].

Prior to inclusion into the final stage, i.e. Stage 12, the designs were evaluated by an expert panel of designers to ensure that environmental criteria were met [87]. This panel was made up of two academic experts in design and issues related to the environment, five industrial designers with expertise in packaging design, as well as two senior academics with expertise in consumer research.

3.4.12 Stage 12 Concept Testing Solutions

The final stage of the project involved taking the final versions of the package designs into consumer focus groups for each category in order to assess their acceptability within a social setting [71]. This final stage ensured that 'group think' and social pressures could be assessed [37], bearing in mind that the earlier stages of consumer research in the project were with individuals. Furthermore, this stage was an important way to decrease industry risk in decision-making based on the findings of the research [133].

It is also important to understand barriers and opportunities for design in consumers' own words [63]. It is especially necessary to ensure that language used by those implementing packaging is consistent with any education campaigns that may be necessary [14]. Consumer education may be required when what the consumer wants (e.g. no plastic) is not feasible if the goal is to reduce food waste [10].

This stage of the research also identified both met and unmet needs and wants within the consumer groups [8]. While the packaging designers were appraised of the extant consumer research as part of the 'understand and see' design thinking phase, Stage 12 ensured there was a cross-section of views available [52], and that open discussions about the proposed designs could take place [11]. This stage was also needed to uncover any issues with the packaging design that may not have been considered as part of the earlier phases of the research [100]. It also provided a final opportunity to iterate designs before proposing final design solutions to the industry [63].

Final concept testing of ideas was undertaken using four online focus group interviews (segmented by household structure) [111]. The key research question posed by this stage was as follows:

What are the potential solutions?

Participants for this stage were sampled under the same conditions as the other consumer research stages (particularly the other qualitative consumer stages—Stages 3 and 7). Four focus groups of 8–10 people each represented a cross-section of regular consumers of all/each of the five main food waste categories under consideration (approx. 40 people in total).

4 Discussion—Key Learnings from Project 1.2.2

4.1 Inspiration and Insight Development

The inspiration phase of the design thinking process involves observing, understanding, and building deep empathy for the end-user [55]. This enables the research team to define the end-users' needs, contexts, and history. The research phase typically involves framing the design challenge or problem; building a team; and undertaking background research, immersion, and consultation with the end-user.

4.1.1 Frame the Design Challenge and Build a Team

The initial phase of the design thinking process involves 'framing' the design challenge or problem [55]. To ensure a variety of perspectives and insights into the design challenge, it is advised that the research team is multi-disciplinary [55]. The Project 1.2.2 academic team consisted of industrial designers, engineers, communication specialists, consumer behaviour experts, and communication designers. The industry partners comprised a large-scale food retailer, a government environmental agency, and a packaging industry peak body. Several market research agencies were also involved in the project's scoping to ensure access to appropriate consumer groups at various stages.

One of the key learnings in this phase was the need to ensure there was a shared language in the project's stages [18], with regular stakeholder updates throughout the process. This was particularly necessary within the interdisciplinary teams; for example, a seemingly simple term like 2D (i.e. two-dimensional) was not understood by industry partners and caused confusion until it could be discussed in a meeting. It must also be noted that meetings of all partners were often difficult to arrange, due to the number of people and organisations involved.

Another of the key challenges faced was the nature of casual and contract staffing in Australia [47]. In a long-term project such as this, this caused problems when there were changes to the teams throughout the project's lifetime. Casual staffs are also likely to go with a regular source income instead of continuing with an hourly rate position. Furthermore, the nature of 'special projects' in the industry may mean that industry partners are needing to be re-briefed every six months, as the project officer may have moved on in the interim. Thus, maintaining momentum can be difficult when the project team changes across the course of the project.

An additional challenge of a project of this nature is the necessity to establish a diffused leadership model early on and to ensure that the various moving parts were intermeshed appropriately. The use of project management software actually caused some issues for team members due to unnecessary complications with planning, as people were not comfortable with 'being ordered around by a piece of software' and said software did not ameliorate the complexities of working on a multi-level multi-method project [112]. Those planning future projects of this type could consider appointing a full-time (non-academic) project manager with expertise in administering complex projects in design settings.

When it came to ensuring that the skills required were available in the research team, the project-specific skills were exemplary. However, the nature of the team meant that a significant proportion of time was spent learning new skills from each other 'on the job', so that the team members could understand each other, and collaborative tasks could be completed. Also, the natural curiosity that attends an academic research environment meant that time was spent learning things that could be done more efficiently by others. Consequently, deadlines were rarely met, as waiting for someone to 'catch up' was common. Another attendant issue was that people needed to feel comfortable with delegating to others, collaborating, and sharing tasks with others. This was not always the case, thus building collaboration skills is essential in long-term multi-method situations, especially where people are used to working autonomously or alone [138].

4.1.2 Undertake Background Research

Consideration of the broader context of the design challenges is pertinent in the design thinking process [55]. The background research for this project consisted of a literature review, industry seminars in 2019, an industry needs assessment, and consultations with the industry. The setting up of the FFWCRC took many years of extensive industry negotiation prior to inception. All of this information is fed into the project plan. However, the consumer research component was somewhat of an afterthought, and some catching up was required.

Given that there is a significant amount of research conducted into the 'green consumer' (more than 20,000 publications listed on Scholar Google as of March 2021), it was somewhat problematic to assimilate this large volume of research. However, the systematised literature review (Langley et al., 2021) focussed on packaging and consumers, and while it produced an overview of the state of the knowledge base, it was potentially too narrow in that it focussed mostly on packaging design. In order to close this gap, the primary research (survey, interviews, groups) was very wide ranging. However, the primary research was possibly too broad in scope to be readily implemented by industry partners. Consequently, the research translation task has taken longer than originally envisaged.

4.1.3 Consultation with End-Users

A priority of the design thinking process is end-user engagement [55]. Consumer consultation was a critical aspect of Project 1.2.2, as was industry consultation. When it comes to packaging design, there are various types of 'user'. There are those that have to implement designs within their organisations (industry), as well as buyers and consumers who use and dispose of the packaging in their daily lives. The various stages of research undertaken were principally designed to heighten consultation, but were also expanded to include multiple stakeholders and multiple stages (see Table 2).

A challenge arose from this high-intensity data collection and consultation: how to translate this information into something practical and usable by partners. Early on in the process, it became evident that there was simply too much information to be meaningfully incorporated into partners' organisational decision-making processes. To this end, the project partners requested short form 'highlights' reports, infographics, and videos supported by downloadable datasets, rather than lengthy reports. This was also supported by a series of presentations and conference Q&A sessions to foster broader industry engagement with the research results.

Another interesting 'challenge' that had beneficial outcomes once it was resolved was the learning that accompanied attempts to communicate across the different 'planes' of the system (i.e. between and across academic and industry partners). People tend to develop shorthand language as a means of communicating within the group [28]. This works very effectively when 'insiders' are only speaking to each other. However, the nature of this project ensured that everyone was an 'outsider' and needed to learn about 'others' as the project progressed [80]. The learning across the project was therefore deeper and more effective as a consequence.

4.1.4 Consultation with Experts

To ensure a variety of perspectives, it is recommended that 'experts' in the field are engaged to provide their insights into the design challenge [55]. Project 1.2.2 had an interdisciplinary team consisting of industry experts (partners and think tank participants), engineers, designers (including industrial designers), and an expert panel who assessed proposed packaging designs.

Learnings from this stage of the research were focussed on the timely and committed engagement of expertise in the various project stages. The project had tight timelines with a number of contingent stages. However, while the project required expertise at specific times, given that the experts were volunteers, it meant that the experts would (not surprisingly) prioritise their main role over participation in the project. As a result, timelines were often not met. Future projects could consider paying experts for their input. However, given the scale of Project 1.2.2, this would probably result in a project that was financially unworkable. Additionally, industry group participation was an essential part of this project and getting competitors into the same 'room' was often contentious. Future projects might consider how to manage this somewhat fraught process to enable open system-wide participation towards a common societal-level goal.

Another learning from this project was the reliance on consumers to solve the packaging problem, or what Brennan and colleagues (2016) call 'consumer myopia'. The assumption that consumers can, and do, actively choose between types of packaging was evident throughout the various stages of consumer research. However, our research also shows that consumers choose what is presented to them in the retail environment. This dilemma will need to be better understood in future research of this nature. Inferring that consumers have power in the system, merely because they buy the end product, is problematic.

4.1.5 Collaborative Consultation with Experts and End-Users

The literature review indicated that a majority of studies collect data from experts and end-users independently. The iterative nature of the research was unique (in our experience). As described previously, this can have both positive and negative impacts on the research. Aligning industry expectations of academic research with market research was interesting; some of the activities included in Project 1.2.2 are not standard practice in the industry, which is much more familiar with the conduct of market research. Acting in an unconventional manner is necessary for innovation [112]. However, not following conventions provides a myriad opportunity for misunderstanding, as people do not know what to expect. Another (somewhat amusing) issue arose as a result of sharing the consumer research outcomes with the industry before reports had been published, leading the industry representatives to assert that the consumers were 'wrong' in their perspectives. Explaining that consumers perspectives are neither right nor wrong, but are what they are, will be a challenge for future research.

4.1.6 Immersion and Observation

Immersion into the lives or situations of end-users can provide great insight into their needs [55]. To this end, Project 1.2.2 enlisted ethnographic techniques such as journey mapping, in-home research, and visits to markets and food shopping settings, as well as utilising design thinking, packaging design, and confirmatory research. Information about food waste was also ascertained by analysing bin audit data from secondary sources. COVID-19 lockdowns meant that there were difficulties in accessing behavioural infrastructure and specific environments. While COVID-19 and the pandemic permitted a much broader range of people to participate—because there was no need for face-to-face contact—it also meant that home environments were not entirely visible to the team. This may require follow up in future research.

This research also shows the importance of observation and immersion as approaches that allow deep understandings of complex issues. Self-reported data from these studies indicated that consumers, in their own minds, don't waste food. They do, however, redefine food so that the word 'waste' does not apply. They also said they did not waste food while sending in photographs of very full refrigerators with more food than can possibly be eaten before it expires. This research illustrates that triangulation is essential, especially where consumers say one thing and do something else entirely.

4.2 Ideation

The ideation phase is an iterative process where researchers identify opportunities and ideas to address the design challenge [55]. It involves a divergence and convergence of ideas to identify a feasible, end-user-driven approach to tackle the design challenge.

To make decisions on how to develop packaging concepts, all packaging-related insights from previous research stages were collated. These insights were then grouped according to food categories and allocated to clusters describing the essence of the insight (e.g. resealability, easy to open, education, etc.). From here, the insights were plotted onto a design matrix to build a formula for five separate packaging concepts per food category. This stage focussed on representing the breadth of the category through product variations and described a wide range of packaging cluster selections.

4.2.1 Team Consideration and Brainstorming

Team consideration and brainstorming involves the group-based discussion of findings and insights from the inspiration phase. This usually involves an initial divergence and converging of ideas and priorities to pursue further [55]. One of the essentials for this project was the capturing of the interdisciplinary insights necessary for breakthrough innovation [21] while ensuring feasibility [46]. Stage 8 began the process of developing 'blue sky' ideas that could then be filtered for achievability by the subsequent stages of the project. This ensured that ideas were able to be put forward without having to be practical at the outset. The process of deciding which ideas went forward into later stages provided as much insight into the barriers and opportunities for change as any of the formal research stages. This avoided eliminating ideas before they had a chance to be thought through in terms of a variety of interdisciplinary perspectives.

4.2.2 Design Efficiency Within Constraints

The matrix method used in Stage 8 to converge ideas [55] further helped to streamline the process of synthesising insights from previous stages of the research. As such, no further research was really needed in developing the concepts, as is usually the case when designers design. However, once the matrix was discussed and design directions agreed, there was a point of difference amongst the team in terms of how the concept visualisations might be realised. Some wanted to develop a series of thumbnail sketches before moving to Adobe Photoshop to visualise the final concepts. Although this was a good idea to settle the visualisations through 'drawing' first, it was pointed out that the time allotted for visualising 25 concepts was not going to allow for this to occur for every category and concept. The team agreed that the thumbnailing approach would therefore be used for one category to help develop five concepts, so that the team could map out the most efficient way to work on the other four categories. This collective approach to determining the right process to design with set resources available was a key learning from the project. It showed that adaptiveness, collaboration, and efficiency can be achieved concurrently when teams work constructively within a constrained context. Thus, the concepts that resulted still captured the breadth and depth of the previous research stages, while the team made sure the resources provided to complete the task were respected.

4.2.3 Identify Key Principles and Ideas

Identifying key principles involves a convergence of ideas to identify the elements of the solution [55]. However, as mentioned previously, employing so many diverse perspectives represented a challenge to convergence. Bringing the ideas back to the practical possibility of being implemented meant that some otherwise excellent ideas were not able to be put forward. For example, one of the designs presented earlier in the project was to re-cycle food waste into packaging. Unfortunately, that idea did not make it to the final feasibility stages because it was not about reducing food waste per se but about eliminating plastic. Sometimes being practical and sticking to the brief mean that good ideas get wasted [79].

4.2.4 Co-Creation, Rapid Prototyping, and Refining Prototypes

Rapid prototyping involves the development of low fidelity prototypes [55] for the purposes of ideation and testing. Project 1.2.2. involved a complex interplay between stages of co-creation and prototyping as well as refining prototypes. One of the learnings from this process was the necessity to be flexible and adaptable as the environment and the participants change. Design research is characterised by being able to pivot quickly as user desires change as a result of actions and interactions between actors in the system. Importantly, when consumers don't see themselves as the problem when it comes to food waste, co-creation also involves shifting people's perspectives to ensure they understand that things have to change and they have to be integral to that change.

4.3 Implementation

The implementation phase of the design thinking process involves implementing a higher fidelity version of the prototype. The implementation phase continues to be iterative and evolves according to end-user feedback [55]. For Project 1.2.2, the design thinking approach also allowed for the rapid innovation of methods along the way, especially in relation to adapting to the exigencies of the 2020–2021 global pandemic. This chapter has been written around the implementation of the project;

however, the implementation of the project's outcomes has yet to occur. The implementation phase will continue beyond the end of the project because once the designs are delivered to the industry partners, there will be an additional phase for industry implementation of outcomes.

4.3.1 Practical Considerations for Large-Scale Dissemination and Long-Term Engagement

Design studies should consider the viability of the long-term implementation and sustainability of the final prototype [55]. Pragmatism is one of the foundational principles of design research. It has to work in order to be successful. The Project 1.2.2 protocol ensures that when packaging designs are implemented, they will have an excellent chance of reducing consumer food waste.

A final key learning is that complex projects such as this one, while having the capacity to create world-changing outcomes, are expensive in terms of time, effort, and money. The commitment of industry, government, and academic institutions is essential. A vast amount of non-monetary contribution has made this project an extraordinary one.

Acknowledgements The authors would like to acknowledge the funding of the Fight Food Waste Cooperative Research Centre for Project 1.2.2 and our partners Woolworths and Sustainability Victoria. The Cooperative Research Centre Program supports industry-led collaborations between industry, researchers and the community. The Fight food Waste Cooperative Research Centre gratefully acknowledges the Australian Government Cooperative Research Centre Program financial contribution through the Cooperative Research Centres program as well as the participants of this project. We would also acknowledge the editorial support of Dr Eloise Florence.

References

1. ARCADIS (2019) National food waste baseline: final assessment report. Retrieved from https://www.environment.gov.au/system/files/pages/25e36a8c-3a9c-487c-a9cb-66ec15ba61d0/files/national-food-waste-baseline-final-assessment.pdf
2. Aday MS, Yener U (2015) Assessing consumers' adoption of active and intelligent packaging. Br Food J 117(1):157–177. https://doi.org/10.1108/BFJ-07-2013-0191
3. Aitsidou V, Michailidis A, Partalidou M, Iakovidou O (2019) Household food waste management: socio-ecological dimensions. Br Food J 121(9):2163–2178
4. Aktas E, Sahin H, Topaloglu Z, Oledinma A, Huda AKS, Irani Z, Sharif AM, van't Wout T, Kamrava M (2018) A consumer behavioural approach to food waste. J Enterp Inf Manage 31(5):658–673. https://doi.org/10.1108/jeim-03-2018-0051
5. Amato M, Fasanelli R, Riverso R (2019) Emotional profiling for segmenting consumers: the case of household food waste. Calitatea 20(S2):27–32
6. Aschemann-Witzel J, Jensen JH, Jensen MH, Kulikovskaja V (2017) Consumer behaviour towards price-reduced suboptimal foods in the supermarket and the relation to food waste in households. Appetite 116:246–258. https://doi.org/10.1016/j.appet.2017.05.013

7. Barska A, Wyrwa J (2016) Consumer perception of active and intelligent food packaging. Zagadnienia Ekonomiki Rolnej 4(349):134–155. https://doi.org/10.5604/00441600.1225668
8. Belk RW (2007) Handbook of qualitative research methods in marketing. Edward Elgar Publishing
9. Bhattacharya A, Nand A, Prajogo D (2021) Taxonomy of antecedents of food waste - a literature review. J Clean Prod 291:1259102
10. Blanc S, Massaglia S, Brun F, Peano C, Mosso A, Giuggioli NR (2019) Use of bio-based plastics in the fruit supply chain: an integrated approach to assess environmental, economic, and social sustainability. Sustainability 11(9):2475. Retrieved from https://www.mdpi.com/2071-1050/11/9/2475
11. Brennan L, Fry M-L, Previte J (2015) Strengthening social marketing research: Harnessing "insight" through ethnography. Australas Mark J 23(4):286–293
12. Brennan L, Previte J, Fry M-L (2016) Social marketing's consumer myopia. J Soc Mark 6(3):219–239. https://doi.org/10.1108/JSOCM-12-2015-0079
13. Brennan L, Langley S, Verghese K, Lockrey S, Ryder M, Francis C, Phan-Le NT, Hill A (2021). The role of packaging in fighting food waste: a systematised review of consumer perceptions of packaging. J Clean Prod 281:125276
14. Brooks V (2003) Exploitation to engagement: the role of market research in getting close to niche targets. Int J Mark Res 45(3):337–354
15. Brown T (2008) Design thinking. Harvard Bus Rev 86(6):84
16. Bucci M, Calefato C, Colombetti S, Milani M, Montanari R (2010) Fridge fridge on the wall: what can I cook for us all?: an HMI study for an intelligent fridge. In: Paper presented at the international conference on advanced visual interfaces, AVI 2010, Roma, Italy
17. Buchanan R (1992) Wicked problems in design thinking. Des Issues 8(2):5–21
18. Buchanan R (2019) Systems thinking and design thinking: the search for principles in the world we are making. She Ji: J Des Econ Innovation 5(2):85–104
19. Cammarelle A, Lombardi M, Viscecchia R (2021) Packaging innovations to reduce food loss and waste: are Italian manufacturers willing to invest? Sustainability 13(4):1963
20. Chen S (2019) The design imperative: the art and science of design management. Springer.
21. Cheng CC, Chen JS (2013) Breakthrough innovation: the roles of dynamic innovation capabilities and open innovation activities. J Bus Ind Mark 28(5):444–454
22. Chinen K, Endo H, Matsumoto M, Han YS (2021) Embedding a sustainability focus in packaging development processes. In: Kishita Y, Matsumoto M, Inoue M, Fukushige S (eds) EcoDesign and sustainability II. Springer, pp 49–60
23. Chu W, Williams H, Verghese K, Wever R, Glad W (2020) Tensions and opportunities: an activity theory perspective on date and storage label design through a literature review and co-creation sessions. Sustainability 12(3):1162
24. Clune SJ, Lockrey S (2014) Developing environmental sustainability strategies, the double diamond method of LCA and design thinking: a case study from aged care. J Clean Prod 85:67–82
25. Corradini MG (2018) Shelf life of food products: from open labeling to real-time measurements. Ann Rev Food Sci Technol 9:251–269
26. Davis SM, Gourdji J (2018) Making the healthcare shift: the transformation to consumer-centricity. Morgan James Publishing
27. de Almeida Oroski F (2020) Exploring food waste reducing apps—a business model lens. In: Food waste management. Springer, pp 367–387
28. Deutsch CP (1981) The behavioral scientist: insider and outsider. J Soc Issues 37(2):172–191
29. Devin B, Richards C (2018) Food waste, power, and corporate social responsibility in the Australian food supply chain. J Bus Ethics 150(1):199–210
30. Diaz-Ruiz R, Costa-Font M, Gil JM (2018) Moving ahead from food-related behaviours: an alternative approach to understand household food waste generation. J Clean Prod 172:1140–1151. https://doi.org/10.1016/j.jclepro.2017.10.148
31. do Canto NR, Grunert KG, De Barcellos MD (2021) Circular food behaviors: a literature review. Sustainability 13(4):1872

32. Druckman J, Bayes R, Bolsen TA (2020) Research agenda for climate change communication and public opinion: the role of consensus messaging and beyond. Environmental Communication 1–19
33. Durose C, Perry B, Richardson L (2021) Co-producing research with users and communities. In: Loeffler E, Bovaird T (eds) The Palgrave handbook of co-production of public services and outcomes. Springer, pp 669–691
34. EC Sustainable (2020) Boroondara council food wastre bin audit and analysis partnering with EC sustainable. Retrieved from https://www.ecsustainable.com/audit-and-analysis/
35. Edwards J, Othman M, Crossin E, Burn S (2018) Life cycle assessment to compare the environmental impact of seven contemporary food waste management systems. Bioresour Technol 248:156–173
36. Engeström Y (2005) Developmental work research: Expanding activity theory in practice, vol 12, Lehmanns Media
37. Esteban-Bravo M, Vidal-Sanz JM (2021) Marketing research methods: quantitative and qualitative approaches. Cambridge University Press.
38. Farr-Wharton G, Foth M, Choi JH-J (2014) Identifying factors that promote consumer behaviours causing expired domestic food waste. J Consum Behav 13(6):393–402. https://doi.org/10.1002/cb.1488
39. Faulkner GP, Livingstone MBE, Pourshahidi LK, Spence M, Dean M, O'Brien S, Kerr MA (2017) An evaluation of portion size estimation aids: consumer perspectives on their effectiveness. Appetite 114:200–208
40. Fausset CB, Baranak A, Farmer SK, Mann EL, Harrington CN, Price CE, Ray JB (2014) Developing a scoring matrix to evaluate the usability of consumer packaging: a pilot study. In: Paper presented at the proceedings of the human factors and ergonomics society annual meeting.
41. Fernández-Serrano P, Tarancón P, Besada C (2021) Consumer information needs and sensory label design for fresh fruit packaging: an exploratory study in Spain. Foods 10(1):72
42. Field B, Booth A, Ilott I, Gerrish K (2014) Using the knowledge to action framework in practice: a citation analysis and systematic review. Implementation Sci 9(1):1–14
43. Food and Agriculture Organisation (2019) The state of food and agriculture: moving forward on food loss and waste reduction. Retrieved from http://www.fao.org/3/ca6030en/ca6030en.pdf
44. Fraser A (2018) The influence of package design on consumer purchase intent. In: Pascall MA, Han JH (eds) Packaging for nonthermal processing of food. John Wiley & Sons, pp 225–249
45. Ganglbauer E, Fitzpatrick G, Comber R (2013) Negotiating food waste: using a practice lens to inform design. ACM Trans Comput Hum Interact 20(2):1–25. https://doi.org/10.1145/2463579.2463582
46. Gardien P, Rincker M, Deckers E (2015) Innovating Innovation: Introducing the rapid cocreation approach to facilitate breakthrough innovation. In: Paper presented at the 11th European Academy of Design Conference
47. Gilfillan G (2018) Characteristics and use of casual employees in Australia. Statistics and mapping, parliamentary services, Canberra, Australia.
48. Hall-Phillips A, Shah P (2017) Unclarity confusion and expiration date labels in the United States: a consumer perspective. J Retailing Consum Serv 35:118–126. https://doi.org/10.1016/j.jretconser.2016.12.007
49. Hamby A, Pierce M, Daniloski K (2019) Co-creating social change using human-centred design. In: Macro-social marketing insights: systems thinking for wicked problems, 110
50. Hanssen OJ, Møller H, Svanes E, Schakenda V (2012) Life cycle assessment as a tool in food waste reduction and packaging optimization: packaging innovation and optimization in a life cycle perspective. In: Curran MA (ed) Life cycle assessment handbook: a guide for environmentally sustainable products. John Wiley & Sons, pp 345–367
51. Hebrok M, Boks C (2017) Household food waste: drivers and potential intervention points for design – an extensive review. Journal of Cleaner Production 151(380–392). https://doi.org/10.1016/j.jclepro.2017.03.069

52. Hennink M, Hutter I, Bailey A (2020) Qualitative research methods. Sage
53. Hoyer WD, Chandy R, Dorotic M, Krafft M, Singh SS (2010) Consumer cocreation in new product development. J Serv Res 13(3):283–296
54. Hunter A, Brewer JD (2015) Designing multimethod research. In Hesse-Biber SN, Burke Johnson R (eds) The Oxford handbook of multimethod and mixed methods research inquiry. Oxford Publishing, pp 185–205
55. IDEO (2015) The field guide to human centered design
56. INCPEN, WRAP (2019) Key findings report: UK survey 2019 on citizens' attitudes & behaviours relating to food waste, packaging and plastic packaging. Retrieved from https://www.wrap.org.uk/sites/files/wrap/Citizen-attitudes-survey-food-waste-and-packaging.pdf
57. Keller A, Wyles KJ (2021) Straws, seals, and supermarkets: topics in the newspaper coverage of marine plastic pollution. Mar Pollut Bull 166:112211
58. Kokotovich V (2008) Problem analysis and thinking tools: an empirical study of non-hierarchical mind mapping. Des Stud 29(1):49–69
59. Lake D, McFarland A, Vogelzang J (2020) Creating resilient interventions to food waste: aligning and leveraging systems and design thinking. In: Närvänen E, Mesiranta N, Mattila M, Heikkinen A (eds) Food waste management: solving the wicked problem. Springer, pp 193–221
60. Lame G (2019) Systematic literature reviews: an introduction. In: Paper presented at the proceedings of the design society: international conference on engineering design
61. Langley J (2016) Participatory design: co-creation|co-production|co-design combining imaging and knowledge. In: Knowledge utilisation colloquium, Llandudno, Wales, June 2016. (Unpublished). Retrieved from http://shura.shu.ac.uk/14631/.
62. Langley S, Francis C, Ryder M, Brennan L, Verghese K, Lockrey S (2020) Consumer perceptions of the role of packaging in reducing food waste: baseline industry report. Retrieved from https://fightfoodwastecrc.com.au/project/consumer-perceptions-of-the-role-of-packaging-in-reducing-food-waste/
63. Laurel B (2003) Design research: methods and perspectives. MIT press.
64. Lewrick M, Link P, Leifer L (2018) The design thinking playbook: mindful digital transformation of teams, products, services, businesses and ecosystems. John Wiley & Sons
65. Licciardello F (2017) Packaging, blessing in disguise: review on its diverse contribution to food sustainability. Trends Food Sci Technol 65:32–39. https://doi.org/10.1016/j.tifs.2017.05.003
66. Lindh H, Williams H, Olsson A, Wikstrom F (2016) Elucidating the indirect contributions of packaging to sustainable development: a terminology of packaging functions and features. Packag Technol Sci 29:225–246. https://doi.org/10.1002/pts
67. Lockrey S, Verghese K, Crossin E, Young G (2020) Development of an environmental impact reduction strategy for Australia's Antarctic infrastructure. J Ind Ecol 24(4):804–814
68. Lockrey S, Verghese K, Danaher J, Newman L, Barichello V (2019) The role of packaging for Australian fresh produce. Australian Fresh Produce Alliance
69. Lu S, Liu A (2016) Innovative design thinking for breakthrough product development. Procedia CIRP 53:50–55
70. Ludwig T, Wang X, Kotthaus C, Harhues S, Pipek V (2017) User narratives in experience design for a B2B customer journey mapping. In: Burghardt M, Wimmer R, Wolff C, Womser-Hacker C, (eds) Mensch und Computer 2017 - Tagungsband. Regensburg: Gesellschaft für Informatik e.V.., pp 193–202. https://doi.org/10.18420/muc2017-mci-0108
71. Lune H, Berg BL (2017) Qualitative research methods for the social sciences. Pearson
72. Lusk JL, Roosen J, Bieberstein A (2014) Consumer acceptance of new food technologies: causes and roots of controversies. Ann Rev Resour Econ 6:381–405. https://doi.org/10.1146/annurev-resource-100913-012735
73. Lydekaityte J (2019) Smart interactive packaging as a cyber-physical agent in the interaction design theory: a novel user interface. In: Paper presented at the IFIP conference on human-computer interaction

74. Males J, Van Aelst P (2021) Did the blue planet set the agenda for plastic pollution? An explorative study on the influence of a documentary on the public, media and political agendas. Environ Commun 15(1):40–54
75. Malone L (2019). Desire lines: a guide to community participation in designing places. Routledge
76. Masson M, Delarue J, Blumenthal D (2017) An observational study of refrigerator food storage by consumers in controlled conditions. Food Qual Prefer 56:294–300. https://doi.org/10.1016/j.foodqual.2016.06.010
77. Mazzucchelli A, Gurioli M, Graziano D, Quacquarelli B, Aouina-Mejri C (2021) How to fight against food waste in the digital era: key factors for a successful food sharing platform. J Bus Res 124:47–58
78. McCarthy B, Liu H-B (2017) Waste not, want not. Br Food J 119(12):2519–2531. https://doi.org/10.1108/bfj-03-2017-0163
79. McKeown M (2008) The truth about innovation. Pearson Education India
80. McNess E, Arthur L, Crossley M (2015) 'Ethnographic dazzle' and the construction of the 'Other': revisiting dimensions of insider and outsider research for international and comparative education. Compare: J Comp Int Educ 45(2):295–316
81. Millen DR (2000) Rapid ethnography: time deepening strategies for HCI field research. In: Paper presented at the 3rd conference on designing interactive systems: processes, practices, methods, and techniques, New York City, New York, USA. https://doi.org/10.1145/347642.347763
82. Møller H, Hagtvedt T, Lødrup N, Andersen JK, Madsen PL, Werge M, Aare AK, Reinikainen A, Rosengren Å, Kjellén J (2016) Food waste and date labelling: issues affecting the durability. Nordic Council of Ministers
83. National Academies of Sciences, Engineering, and Medicine (2020) A national strategy to reduce food waste at the consumer level. The National Academies Press. https://doi.org/10.17226/25876
84. Nemat B, Razzaghi M, Bolton K, Rousta K (2019) The role of food packaging design in consumer recycling behavior—a literature review. Sustainability 11(16):4350
85. Nguyen AT, Parker L, Brennan L, Lockrey S (2020) A consumer definition of eco-friendly packaging. J Clean Prod 252. https://doi.org/10.1016/j.jclepro.2019.119792
86. Otten JJ, Diedrich S, Getts K, Benson C (2018) Commercial and anti-hunger sector views on local government strategies for helping to manage food waste. J Agric Food Syst Commun Dev 8(B):55–72
87. Parashar S, Sood G, Agrawal N (2020) Modelling the enablers of food supply chain for reduction in carbon footprint. J Clean Prod 275:122932
88. Pellegrini G, Sillani S, Gregori M, Spada A (2019) Household food waste reduction: Italian consumers' analysis for improving food management. Br Food J 121(6):1382–1397. https://doi.org/10.1108/BFJ-07-2018-0425
89. Pereira de Abreu DA, Cruz JM, Paseiro Losada P (2012) Active and intelligent packaging for the food industry. Food Rev Int 28(2):146–187. https://doi.org/10.1080/87559129.2011.595022
90. Pink S, Morgan J (2013) Short-term ethnography: intense routes to knowing. Symb Interact 36(3):351–361
91. Porat R, Lichter A, Terry LA, Harker R, Buzby J (2018) Postharvest losses of fruit and vegetables during retail and in consumers' homes: quantifications, causes, and means of prevention. Postharvest Biol Technol 139(C):135–149. https://doi.org/10.1016/j.postharvbio.2017.11.019
92. Porpino G, Parente J, Wansink B (2015) Food waste paradox: antecedents of food disposal in low income households. Int J Consum Stud 39(6):619–629. https://doi.org/10.1111/ijcs.12207
93. Quested T, Parry A (2017) WRAP Household food waste in the UK, 2015. Retrieved from https://wrap.org.uk/resources/report/household-food-waste-uk-2015

94. Redmond EC, Griffith CJ (2005) Consumer perceptions of food safety education sources. Br Food J 107(7):467–483. https://doi.org/10.1108/00070700510606882
95. Reeves S, Peller J, Goldman J, Kitto S (2013) Ethnography in qualitative educational research: AMEE Guide No. 80. Med Teach 35(8), e1365–e1379. https://doi.org/10.3109/0142159X.2013.804977
96. Rhein S, Schmid M (2020) Consumers' awareness of plastic packaging: more than just environmental concerns. Resourc Conserv Recycl 162:105063
97. Richter B (2017) Knowledge and perception of food waste among German consumers. J Clean Prod 166:641–648
98. Rittel WJ, Webber MM (1973) Dilemmas in a general theory of planning. Policy Sci 4(2):155–169
99. Rodgers PA, Galdon F, Bremner C (2020) Design research-in-the-moment: eliciting evolutive traces during the Covid-19 crisis. Strategic Des Res J 13(3):312–326
100. Rodgers PA, Milton A (2013) Research methods for product design. Laurence King
101. Romani S, Grappi S, Bagozzi RP, Barone AM (2018) Domestic food practices: a study of food management behaviors and the role of food preparation planning in reducing waste. Appetite 121:215–227. https://doi.org/10.1016/j.appet.2017.11.093
102. Rosenbaum MS, Otalora ML, Ramírez GC (2017) How to create a realistic customer journey map. Bus Horiz 60(1):143–150. https://doi.org/10.1016/j.bushor.2016.09.010
103. Rønnow HN (2020) The effect of front-of-pack nutritional labels and back-of-pack tables on dietary quality. Nutrients 12(6):1704
104. Searchinger T, Waite R, Hanson C, Ranganathan J (2019) Creating a sustainable food future: a menu of solutions to feed nearly 10 billion people by 2050. Retrieved from https://www.wri.org/publication/creating-sustainable-food-future.
105. Sedlacko M, Martinuzzi A, Røpke I, Videira N, Antunes P (2014) Participatory systems mapping for sustainable consumption: discussion of a method promoting systemic insights. Ecol Econ 106:33–43
106. Sibley K, Li L, Abbott JH (2016) Increasing the impact of peer-reviewed publications through tailored dissemination strategies: perspectives for practice feature in JOSPT. J Orthop Sports Phys Ther 46(7):500–501. https://doi.org/10.2519/jospt.2016.0110
107. Simms C, Trott P, van den Hende E, Hultink EJ (2020) Barriers to the adoption of waste-reducing eco-innovations in the packaged food sector: a study in the UK and the Netherlands. J Clean Prod 244:118792
108. Siricharoen WV (2021) Using empathy mapping in design thinking process for personas discovering. In Vinh PC, Rakib A (eds) Context-aware systems and applications, and nature of computation and communication. ICCASA 2020, ICTCC 2020. Lecture notes of the institute for computer sciences, social informatics and telecommunications engineering, vol 343. Springer. https://doi.org/10.1007/978-3-030-67101-3_15
109. Spang ES, Achmon Y, Donis-Gonzalez I, Gosliner WA, Jablonski-Sheffield MP, Momin MA, Tomich TP (2019) Food loss and waste: measurement, drivers, and solutions. Annu Rev Environ Resour 44:117–156
110. Stewart B (2012) Packaging design and development. In: Emblem A (ed) Packaging technology: fundamentals, materials and processes. Elsevier, pp 411–440
111. Stewart DW, Shamdasani P (2017) Online focus groups. J Adv 46(1):48–60
112. Swink M (2006) Building collaborative innovation capability. Res Technol Manage 49(2):37–47
113. Szaky T (2019) The future of packaging: from linear to circular. Berrett-Koehler Publishers
114. Technopolis Group (2015) Innovation by design: how design enables science and technology research to achieve greater impact. Retrieved from https://www.designcouncil.org.uk/resources/report/innovation-design
115. Temkin BD, McInnes A, Zinser R (2010) Mapping the customer journey: best practices for using an important customer experience tool. Retrieved from http://crowdsynergy.wdfiles.com/local--files/customer-journey-mapping/mapping_customer_journey.pdf

116. Thompson B, Toma L, Barnes AP, Revoredo-Giha C (2018) The effect of date labels on willingness to consume dairy products: implications for food waste reduction. Waste Manage 78:124–134. https://doi.org/10.1016/j.wasman.2018.05.021
117. Trivedi RH, Patel JD, Acharya N (2018) Causality analysis of media influence on environmental attitude, intention and behaviors leading to green purchasing. J Clean Prod 196:11–22
118. Tueanrat Y, Papagiannidis S, Alamanos E (2021) Going on a journey: a review of the customer journey literature. J Bus Res 125:336–353
119. van Geffen L, van Herpen E, van Trijp H (2020) Household food waste—how to avoid it? an integrative review. In: Närvänen E, Mesiranta N, Mattila M, Heikkinen A (eds) Food waste management: solving the wicked problem. Springer, pp 27–55
120. van de Grift TC, Kroeze R (2016) Design thinking as a tool for interdisciplinary education in health care. Acad Med 91(9):1234–1238
121. van Holsteijn F, Kemna R (2018) Minimizing food waste by improving storage conditions in household refrigeration. Resour Conserv Recycl 128:25–31
122. Velasco C, Spence C (2019) Multisensory product packaging: an introduction. In: Velasco C, Spence C (eds) Multisensory packaging: designing new product experiences. Springer, pp 1–18
123. Verghese K, Lewis H, Lockrey S, Williams H (2015) Packaging's role in minimizing food loss and waste across the supply chain. Packag Technol Sci 28:603–620. https://doi.org/10.1002/pts.2127
124. Verghese K, Lockrey S (2019) National food waste baseline-Final assessment report. Retreived from https://www.environment.gov.au/system/files/pages/25e36a8c-3a9c-487c-a9cb-66ec15ba61d0/files/national-food-waste-baseline-final-assessment.pdf
125. Verhulst N, Vermeir I, Slabbinck H, Lariviere B, Mauri M, Russo V (2020) A neurophysiological exploration of the dynamic nature of emotions during the customer experience. J Retail Consum Serv 57:102217
126. WRAP (2017) Labelling guidance: best practice on food date labelling and storage advice. Retrieved from https://wrap.org.uk/sites/default/files/2020-08/Development%20of%20best%20practice%20on%20food%20date%20labelling%20and%20storage%20advice.pdf
127. Wikström F, Verghese K, Auras R, Olsson A, Williams H, Wever R, Soukka R (2018) Packaging strategies that save food: a research agenda for 2030. J Ind Ecol 23(3):532–540. https://doi.org/10.1111/jiec.12769
128. Wikström F, Williams H, Trischler J, Rowe Z (2019) The importance of packaging functions for food waste of different products in households. Sustainability 11(9). https://doi.org/10.3390/su11092641
129. Williams H, Wikström F, Otterbring T, Lfgren M, Gustafsson A (2012) Reasons for household food waste with special attention to packaging. J Clean Prod 24:141–148. https://doi.org/10.1016/j.jclepro.2011.11.044
130. Williams H, Lindström A, Trischler J, Wikström F, Rowe Z (2020) Avoiding food becoming waste in households–the role of packaging in consumers' practices across different food categories. J Clean Prod 121775
131. Wilson NLW, Rickard BJ, Saputo R, Ho ST (2017) Food waste: the role of date labels, package size, and product category. Food Qual Prefer 55:35–44. https://doi.org/10.1016/j.foodqual.2016.08.004
132. Wisdom J, Creswell JW (2013) Mixed methods: integrating quantitative and qualitative data collection and analysis while studying patient-centered medical home models. Agency for Healthcare Research and Quality, Rockville. Retrieved from https://pcmh.ahrq.gov/page/mixed-methods-integrating-quantitative-and-qualitative-data-collection-and-analysis-while.
133. Witell L, Kristensson P, Gustafsson A, Löfgren M (2011) Idea generation: customer co-creation versus traditional market research techniques. J Serv Manage 22(2):140–159
134. Woolley K, Risen JL (2021) Hiding from the truth: when and how cover enables information avoidance. J Consum Res 47(5):675–697

135. Wurster S, Schulze R (2020) Consumers' acceptance of a bio-circular automotive economy: explanatory model and influence factors. Sustainability 12(6):2186
136. Young W, Russell SV, Robinson CA, Barkemeyer R (2017) Can social media be a tool for reducing consumers' food waste? A behaviour change experiment by a UK retailer. Resour Conserv Recycl 117:195–203. https://doi.org/10.1016/j.resconrec.2016.10.016
137. Young CW, Russell SV, Robinson CA, Chintakayala PK (2018) Sustainable retailing – Influencing consumer behaviour on food waste. Bus Strat Environ 27(1):1–15. https://doi.org/10.1002/bse.1966
138. Yuille J, Varadarajan S, Vaughan L, Brennan L (2014) Leading through design: developing skills for affinity and ambiguity. Des Manage J 9(1):113–123
139. Yun B, Bisquert P, Buche P, Croitoru M, Guillard V, Thomopoulos R (2018) Choice of environment-friendly food packagings through argumentation systems and preferences. Ecol Inf 48(March):24–36. https://doi.org/10.1016/j.ecoinf.2018.07.006

UV-Shielding Biopolymer@Nanocomposites for Sustainable Packaging Applications

Akshay S. Patil, Omkar S. Nille, Govind B. Kolekar, Daewon Sohn, and Anil H. Gore

Abstract Biopolymers include chitosan, cellulose, starch, proteins, sodium alginate, agar, gelatin, polyvinyl alcohol, etc., and their nanocomposites have opened new doors to researchers in sustainable packaging applications due to overwhelming intrinsic properties. Biopolymer@nanocomposites based packaging materials along with UV-shielding capability have great importance in sustainable packaging applications. The incorporation of chemically derived UV-shielding agents in the bio-based polymeric matrix is not too safe for the long-term packaging of food and other related materials. Designing biopolymer-based nanocomposite films with the aid of biocompatible UV-shielding nanomaterials (ZnO, TiO_2, SiO_2, CeO_2, carbon-based nanomaterials, etc.) will be more beneficial for sustainable packaging applications and exploration of possible alternative economical materials. Designing biopolymer-based UV-shielding films by using ecofriendly, biocompatible, and sustainable nanomaterials of different kinds has great importance in science and technology. Various protocols for the synthesis of UV-shielding nanomaterials are vital for covering the entire UV region. Especially, nanomaterials having the ability to absorb in UV region and re-emission in a visible region have utmost importance for sustaining mechanical properties of UV-shielding films. Optimizing the appropriate concentration of nanomaterials into a polymeric matrix to maintain physical properties, like transparency, tensile strength (strain–stress), crystallinity, flexibility, etc., is also more significant. Along with UV-shielding ability, antibacterial properties can also be imparted by adding chitosan-like biopolymer and antibacterial nanomaterials (Ag_2O, ZnO, etc.) into the biopolymers matrix. The utilization of waste-based biomass and metal resources to design sustainable, low-cost, and biodegradable composites will help in the more pronounced commercialization of packaging materials by the economical

A. S. Patil · D. Sohn
Department of Chemistry and Research Institute for Convergence of Basic Science, Hanyang University, Seoul 04763, South Korea

A. S. Patil · O. S. Nille · G. B. Kolekar · A. H. Gore (✉)
Fluorescence Spectroscopy Research Laboratory, Department of Chemistry, Shivaji University, Kolhapur 416004, Maharashtra, India
e-mail: anil.gore@utu.ac.in

A. H. Gore
Tarsadia Institute of Chemical Science, Uka Tarsadia University, Bardoli 394350, Gujarat, India

S. S. Muthu (eds.), *Sustainable Packaging*, Environmental Footprints and Eco-design of Products and Processes, https://doi.org/10.1007/978-981-16-4609-6_7

ways on large. Various UV-shielding biopolymer@nanocomposite and their possible sustainable packaging applications are systematically summarized in this chapter.

Keywords Nanomaterials · Biopolymer · Nanocomposite · UV-shielding · Sustainable packaging

1 Introduction

Sustainable packaging currently plays a significant role in maintaining the quality of food products by protecting them from environmental, chemical, and physical challenges and improving their shelf-life. For long-time preservation of food from UV rays, microorganisms, pathogens, bacteria, and prolonged durability of any kinds of materials, it is necessary to store it safely in a specific place with suitable packaging materials. The anthropogenic activities, industrialization, air pollution, and greenhouse gases have created holes in the ozone layer; hence the UV rays coming from sunlight easily pass through the ozone layer, which is very harmful to living and non-living things. These UV rays create many adverse effects such as degradation of organic and inorganic dyes and loss of mechanical strength of coloring pigments and breaking and cracking of plastic and dullness of plastics, yellowing papers, etc. [1]. Consequently, it reduces the efficiency and lifetime of materials. Another critical environmental problem is raised due to global warming, greenhouse gases, chlorofluorocarbon gas (CFC), etc. The ozone layer acts as a protector to the earth from hazardous radiation emitted from the sun. The stratosphere of the earth's atmosphere is covered with an ozone layer that absorbs UV rays and prevents the earth's surface from dangerous UV rays. However, recently, due to rapid growing industrialization, increased automobiles, deforestation, greenhouse gases, CFC gases, etc., the ozone layer is primarily damaged, and a large hole is generated in the ozone layer. The UV rays are highly energetic electromagnetic radiations having the potential to harm human health. These hazardous UV rays hit on the earth's surface and also on humans and animals.

Food directly exposed to sunlight for a long time may be dangerous because, UV light can affect the safety and shelf-life of food, and food kept in a wet or shady place can instantaneously capture the moisture and humidity and starts to decompose and spoils rapidly. So, to avoid or overcome these problems, we need to pack the food and edible materials with smart and active packaging materials. Modern trends and changing lifestyles are the enticements for the evolution of innovative and novel food packaging techniques without compromising safety and quality characteristics [2]. Continuous exposure to UV radiation affects the eyes, dangerous skin diseases like skin burning, skin cancer, and it also affects plants, fruits, vegetables, etc.

Also, synthetically derived polymers such as polystyrene (PS), polymethylacrylate (PMA), polyvinylidene fluoride (PVF), etc. are mostly used in food packaging applications due to adorable properties like high tensile strain and thermal stability. To enhance the mechanical proprieties of plastic material, various synthetic monomers, polymers, synthetic additives, and stabilizers are used as precursor materials. These synthetically derived materials at one side achieved all the desired properties and were largely utilized for small to large scale applications. Unfortunately, uncontrolled use, no recycling, and the non-degradable properties of these blessed materials have accumulated every corner of our green earth. This is ultimately causing one of the biggest problems of pollution. After its utilization, millions of tons of plastic and related materials become non-degradable garbage and polluting soil and water sources every year. Plastic pollution is potentially causing significant damage to marine life, human health, food, wildlife, etc. Recent eye-opening studies have revealed that these plastic materials do not degrade easily, but after several years, they break down into the smallest micro-plastic particles and monomers like bisphenol-A, etc. These micro-plastic and monomer materials are contaminated into water bodies and affect human and aquatic life and cause several genetic mutations, cancer, and ultimately death. There is a lack of techniques that could help to degrade plastic materials. Hence in search of alternatives to synthetic plastic materials, researchers have greatly focused their attention toward biopolymers due to their biocompatible, cost-effective, and biodegradable properties.

Hence, there is an urgent need to resolve the above mentioned non-degradable plastic pollution and UV radiation-related problems. Scientists and researchers around the world have given attention to the resolution of alternatives to plastic and the designing of UV-blocking materials. The sustainable, readily available, cheap, biocompatible, and biomass-derived materials can be used as a easily biodegradable UV-blocking plastic materials. Biopolymers obtained from various natural resources like cellulose, starch, chitosan, alginate, agar, etc. can be used for the designing sustainable UV-blocking food packaging materials.

After taking a lot of effort on the above mentioned problems, researchers are finding a solution that will be very impactful or revolutionary in the packaging field. It is in the view of ecofriendly and sustainable packaging, which we can use in every sector. It is very promising that nanotechnology can be used in food packaging because, this technique could enhance food protection and superiority while decreasing the use of precious raw materials and waste generation. The biopolymer films can be used along with the nanomaterials in smart and active packaging. Biopolymers are ecofriendly as well as biodegradable materials, which can be easily degraded after its use without any adverse effects on the environment. Biopolymer@nanocomposite films can be prepared by various methods which includes electrospinning, spin coating, film casting, dip-dry coating, spray coating, etc. Also, the nanomaterials (NMs) can be synthesized by using sustainable biomass sources (biomass, bio-wastes, and agricultural wastes) and some other metal precursors having fascinating UV-blocking properties. These potent UV-blocking NMs can be easily embedded into polymer matrix with improved mechanical and UV-blocking properties without disturbing their intrinsic properties.

This book chapter aims to show innovative and alternative approaches toward the designing of UV-blocking as well as food packaging biopolymer@nanocomposite materials. Also, the chapter promotes and generates awareness regarding innovative, cheap, and biodegradable ways to develop active UV-blocking packaging materials.

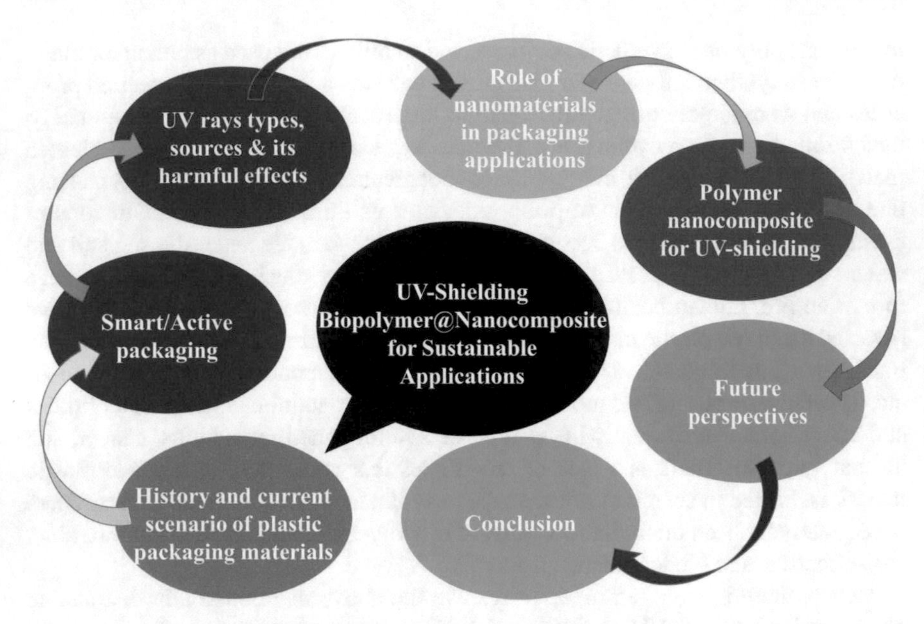

Fig. 1 The road map of biopolymer@nanocomposite for sustainable UV-shielding applications

Herein, it describes the history, first time use of plastic as packaging material, and current scenario of plastic packaging materials, as well as the current market valuation of plastic packaging material at the global level. This book chapter also increases our knowledge about the different types and ranges of UV rays and their harmful effects on human health, food, and materials. Moreover, this book chapter, not only reveals what is smart/active packaging but also provides in detail information about the physical and chemical properties of nanomaterials and the principal role of nanomaterials in sustainable UV-shielding applications. Similarly, this book chapter explains systematically about the biopolymers and synthetic polymers and their nanocomposites for the UV-blocking and smart/active packaging materials to improve the shelf-life of food and their real-life applications. The systematic way/road map of the book chapter is shown in Fig. 1.

2 History and Current Scenario of Plastic Packaging Materials

Nowadays, we have plenty of options for packaging and storage, but materials that have been traditionally used in food packaging include tin-free boxes, cardboards, aluminum foils, cans, glasswares, papers, laminates, plastic wrappers, or boxes. The correct selection of plastic packaging materials retains product quality and freshness during distribution and storage for a long duration. Moreover, globally plastic-based packaging materials are widely used because of flexibility, high stretching capacity, lightweight, and more transparent materials. New plastic packaging materials are helping to decrease food waste. It helps to preserve the veggies and culmination sparkling and keep their nutrition properties for an extended time. These bags were the best food storage bags, and they were used for the packaging of sandwiches or rolls (CSL STYLE ERROR: reference with no printed form).

In recent revolutionary decades, plastic materials are used in various fields due to their broad applicability. In the 1800s, plastics were discovered, including cellulose nitrate, styrene, and vinyl chloride, but until the twentieth century, they were not used in any packaging. Some of the first uses were during World War II with commercialization for food packaging occurring after the war [3]. According to Mordor Intelligence report, the plastic packaging market was valued at US$318.92 billion in 2017, and it will be estimated to reach a value of US $399.44 billion by 2023 at a CAGR of 3.44% over the forecast period of 2018–2023 (CSL STYLE ERROR: reference with no printed form). Dozens of different synthetic plastics, totaling about 200 million tons per year, are produced worldwide [4]. Plastic packaging industries are growing due to the increasing use of packaging materials in various applications, including food and beverages, personal care, household care, consumer electronics, and construction.

In food packaging, polyethylene was one of the first polymers used as packaging material. There is a varied choice of plastic films made from different types of petroleum-based plastic polymers, which are hazardous as well as toxic to human life and the environment, and mainly it is non-biodegradable, which exists in the environment for a long time. Each polymer has its range of mechanical tensile strain and stress, optical, thermal, and moisture/gas barrier properties. Accordingly, because of their unique properties and the physical or chemical behavior of materials, polymers are used in the field of packaging.

3 What is Smart/Active Packaging?

Generally, packaging material interacts with the packed food and plays a crucial role in extending the shelf-life of food and preventing microorganisms and bacterial growth as well as it works like gas and UV light barriers. This material we can call active packaging material [5]. Smart packaging is purposeful to monitor food quality and provide information about the worth of the packaged food or its surrounding environment to determine the safe shelf-life [6]. In the United States, Japan, and Australia, smart packaging is already successfully applied to extend the shelf-life of food quality and safety [5].

Recently, Swarup Roy et al. have reported gelatin/carrageenan-based color-indicator film integrated with shikonin and propolis for smart food packaging applications [7]. The prepared film showed excellent pH-responsive color change over a wide pH range from 2 to 12; also, the film showed antimicrobial and antioxidant activity. Such a kind of film was effectively used in the monitoring freshness of packaged milk. To take an advantage of each material's functional or aesthetic properties, today's food packages often combine numerous materials. Advances in the field of food packaging research can influence the environmental impact of packaging.

For a long time, banana leaves, bark, etc. have been used as packaging material for wrapping and storing to improve the shelf-life of food. As time passes, food packaging material has also been changed. The space of banana leaves and bark is taken by plastic, cardboard, aluminum, glass, wood boxes, and papers. Glass and metals are also fully resistant to gases and water vapors, thus providing an effective barrier to the exchange of materials between the air inside the kit and the external atmosphere [8]. These materials has also some drawbacks, like glass material is fragile, it can break easily, and such a type of material is too heavy and much costlier. Plastic packaging material also affects human life. They are non-biodegradable, toxic, and their environmental impacts are high. So, it currently needs to find out alternative options for the above mentioned materials.

4 Types of UV Rays, Sources, and Their Harmful Effects on Human Health, Foods, and Materials

In the electromagnetic spectrum, seven types of rays are there, out of which, the region from 100 to 400 nm is recognized as the ultraviolet region, and 400–800 nm is a visible region (UV–visible region). UV radiation has a shorter wavelength and higher energies, situated between the infrared and X-ray regions. UV radiation is conventionally classified into three different regions, mainly ultraviolet-A (UV-A 320–400 nm), ultraviolet-B (UV-B 280–320 nm), and ultraviolet-C (UV-C 100–280 nm). This division was given by the Commission International de l'Eclairage (CIE) and corresponds largely to the effects of UV-rays on biological tissue [9].

Sun is the primary source of natural ultraviolet radiation. The UV rays which come from sunlight easily pass through the ozone layer, which is very harmful to living and non-living things. In the stratospheric ozone layer, UV-B radiation is strongly absorbed (nearly 95%), but some part of UV-B and most of UV-A reaches earth surfaces and are responsible for health-related problems, like skin burning, and it promotes skin aging, skin cancer, cellular DNA damages, etc. [10]. UV–visible light has been reported to accelerate the oxidation process of food and subsequently cause its deterioration and nutrient loss [11]. Artificially, the sources of UV light are produced in the field of medical, recreational, and occupational. Arc lamps and incandescent lights are the most common sources of artificial UV light. As well as electric welding arcs, which are widely used in industries including aerospace, mechanical, construction, and automobile, are the significant sources of UV light.

UV rays create many detrimental effects along with degradation of natural and inorganic dyes and coloring pigments [1], lack of mechanical energy, i.e., breaking and cracking of plastic, and dullness of plastics, yellowing papers. All of the UV-C region and maximum of the UV-B radiation is absorbed with the aid of the earth's ozone layer, so almost all the ultraviolet radiation acquired on the earth is UV-A. Majorly, both UV-A and UV-B radiations can affect health when exposed for a long time to UV light, thus increasing the risk of skin burning, cataracts, blindness, and in severe cases, many more health hazards may cause.

Organic materials damaged by absorbing UV light is well known to all of us. The main reason behind the degradation of wood is also ultraviolet light. The UV light reacts with the surface of wood, which first results in the color change and roughness of the wood, then slowly it mechanically becomes weak, and finally cracks. Fine art in galleries and monuments in museums suffers a significant photodegradation, as art pieces are exposed for long periods to natural and artificial illumination. The dyes in canvases and photos undergo a progressive vanishing and finally losing their coloration [9].

UV light affects the DNA system; also can cause critical damage to human tissues. UV-C will be able to damage DNA and other molecules and is often used as a germicidal agent. Exposure to both types of rays UV-B and UV-A for a long time prompts tanning and burn from the sun (erythema) and can affect the immune system. UV-A has a longer wavelength, so it can easily penetrate deeper into the skin and plays a role in skin photo-aging.

5 Role of Nanomaterials in UV-Shielding and Sustainable Packaging

Nanotechnology plays a crucial role in the field of sustainable packagings, such as active and smart packaging to block UV light as well as control the growth of bacteria in food. The physicochemical properties of bulk materials drastically change at the nano level, they improve properties like high surface to volume ratio, less toxic, biocompatible, and the optical and electrical properties can improve the UV-shielding capacity. Also, nanomaterials give enhanced mechanical and thermal stability to the composite, and is having good applications in the sustainable packaging. It also works as a gas barrier between the internal food material and the outer atmosphere. Some nanomaterials have the ability to show antimicrobial activity against both gram-positive and gram-negative types of bacteria. Nanotechnology in antimicrobial packaging creates interest and is evolving with the application of nanotechnology due to its critical role in improving microbial safety and extending the shelf-life of food products [12]. The chemically derived UV-shielding agents embedded in the bio-based polymer composite are not much safe for a long time as a concern to health issues and safety of food. The different kinds of nanomaterials are used in sustainable packaging applications, like organic and inorganic nanomaterials based on their physical and chemical properties. Numerous inorganic UV absorbing materials based on particles or thin film coatings have been developed. Inorganic oxides,

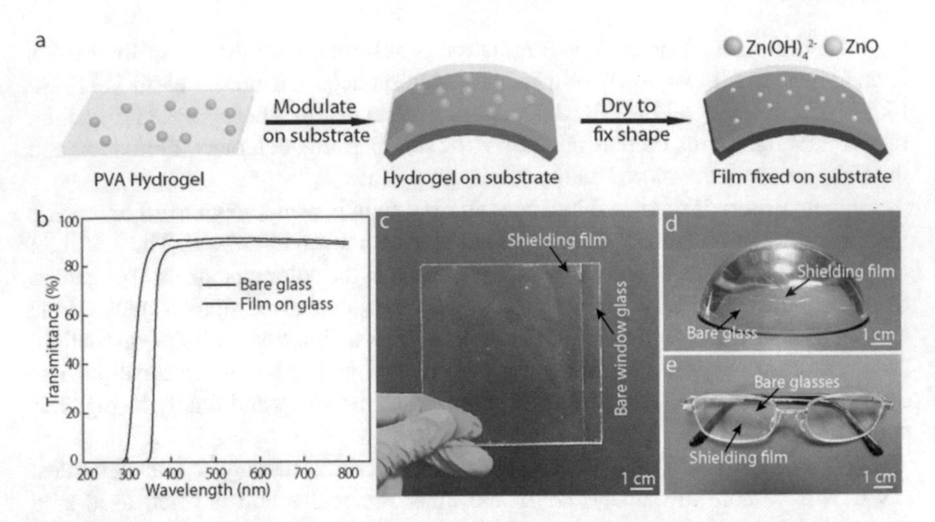

Fig. 2 **a** The schematic image shows the process of modulating PVA/ZnO film on a substrate. The obtained hybrid film was able to modulate on a window glass (**c**), semi-sphere (**d**), and glasses (**e**), respectively. The UV–vis spectra of bare and film-coated window glass were shown in (**b**)

for example, TiO_2, CeO_2, and ZnO are the most widely used materials for protection against UV radiations [13–15].

Zhao et al. developed ZnO nanomaterial through in situ method with polyvinyl alcohol (PVA) polymeric film for the UV-shielding application [16] (Fig. 2). The synthesized PVA/ZnO film absorption capacity is over 99% of UV light between 200 and 350 nm wavelength region; likewise, its optical transparency in the visible region was similar to that of neat PVA film (87–90%). Figure 2 shows a schematic image of PVA/hydrogel modulate on the substrate like window glass and bare glass as a UV-shielding application.

The main role of nanomaterials in packaging is to enhance the UV-blocking capacity of packaging materials by controlling the thickness as well as transparency. Other main benefits of nanomaterials are to absorb the UV light maximum, and it transmits in the visible region, so packaging materials are not affected by the heat, and the shelf-life of packaging materials is automatically increased [17]. Nanomaterials play a significant role in emerging several assets of biodegradable food packaging materials such as mechanical properties, water or gas barriers, and antimicrobial activity, thus leading to the increased shelf-life of stored food by the prevention of spoilage to some extent.

6 Biopolymer and Synthetic Polymer Nanocomposites in UV-Shielding and Sustainable Packaging Applications

Nowadays, most of the packaging materials are synthetically derived petroleum-based polymers presently used for food packaging. Synthetically derived polymers have overwhelming properties to attract everyone; they have high mechanical stress–strain capacity, high thermal stability, ease of handling, cheap rate, and so on, but if we think of environmental concern, we cannot use such synthetically derived materials because they are non-degradable, toxic, and creates ecological issues. Biopolymers are the polymers acquired from natural sources and utilized for different organic and mechanical applications. A biopolymer should be non-poisonous, non-antigenic, non-aggravation, non-cancer-causing, serializable, and adequately accessible for their far-reaching applications. Biopolymers made from different regular assets, like starch, cellulose, chitosan, and different proteins from plant and creature roots [4], have been considered as attractive options for non-biodegradable petroleum-based plastic packaging materials since they are plentiful, inexhaustible, modest, harmless to the ecosystem, biodegradable, and biocompatible. A graphical image shows some examples of biopolymers and synthetically derived polymers in Fig. 3.

A researcher Jong-Whan Rhim and Perry K.W. Ng, in their review well mentioned about how biopolymer-based packaging materials are useful. According to them, biopolymer-based packaging materials are biodegradable as well as edible, also it shows antimicrobial and antioxidant activity. Additionally, biopolymer@nanocomposite provides improved strength and water resistance capacity to the packaging materials with gas barrier properties. Biopolymer@nanocomposite based packaging materials with their special characteristics and various applications are mentioned in Table 1. Biopolymer films additionally may fill in as gas and solute boundaries and supplement different kinds of

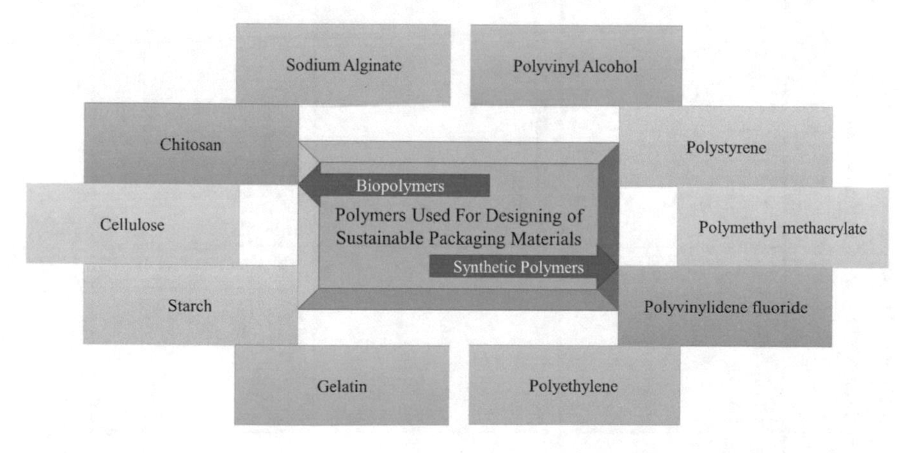

Fig. 3 Naturally derived biopolymer and synthetically derived polymers for sustainable packaging applications

Table 1 Biopolymer and synthetic polymer embedded with nanomaterial films and its applications in sustainable active food packaging and UV-blocking

S. No.	Biopolymer	Synthetic polymer	Embedded nanomaterials	Special characteristics/features of polymer composites	Applications	References
1	Polylactic acid (PLA)	–	Cu-doped ZnO powder functionalized with Ag nanoparticles	Bio-nanocomposite offering suitable mechanical and thermal properties	The designed bio-nanocomposite showed good barrier properties toward UV light, water vapors, oxygen, and carbon dioxide along with good antimicrobial activity	Vasile et al. [23]
2	–	2, 3-Dialdehyde starch (DAS)	Magnetic nanoparticles (MNPs)	MNPs/DAS composite film having hydrophobic nature as well as good mechanical properties	Composite film for food packaging application and can be used for oxygen-sensitive foods like chilled meat because the film contains oxygen scavenging MNPs (Fe_3O_4)	Keshk et al. [27]
3	Chitosan (CS)	Polyvinyl alcohol/chitosan (PVA/CS)	Silicon dioxide (SiO_2)	Enhanced mechanical and tensile properties for food packages to extend the preservation time	The silica embedded PVA/CS biodegradable film improved the preservation time of cherries three times by reducing the permeability of oxygen and moisture	Yu et al. [28]

(continued)

Table 1 (continued)

S. No.	Biopolymer	Synthetic polymer	Embedded nanomaterials	Special characteristics/features of polymer composites	Applications	References
4	Polylactic acid (PLA)	–	Titanium dioxide (TiO_2)	PLA/TiO_2 composites have good mechanical, thermal, photocatalytic, and antimicrobial properties	The PLA/TiO_2 composites can be used as an antimicrobial agent as well as can be used in food packaging to improve the shelf-life of fruits and vegetables	Kaseem et al. [13]
5	–	PVA-montmorillonite K10 clay nanocomposite	Silver nanoparticles	Antimicrobial activity against common foodborne pathogens like *S. typhimurium* and *S. aureus*, also having greater mechanical properties, high water resistivity, and light barrier capacity	PVA-MMT nanocomposite film used for packaging of chicken sausage sample	Mathew et al. [26]
6	Polylactic acid (PLA)	–	Titanium dioxide (TiO_2)	TiO_2 embedded PLA composite having potential antibacterial properties	The TiO_2/PLA composite can be used as a food packaging material with enhanced antimicrobial activity	González et al. [19]

(continued)

Table 1 (continued)

S. No.	Biopolymer	Synthetic polymer	Embedded nanomaterials	Special characteristics/features of polymer composites	Applications	References
7	Tapioca starch/decolorized hsian-tsao leaf gum (dHG)	–	0.1–0.3% cinnamon oil or 0.3–0.6% GSE (grape seed extract)	PVA/Dpa-h films exhibit stronger UV-shielding capabilities and can almost block the complete UV region (200–400 nm)	Applications of edible coatings on minimally processed vegetable	Lai et al. [29]
8	Starch	Starch/polyvinyl alcohol (PVA)	Citric acid	Strong antimicrobial activity against Listeria monocytogenes and *Escherichia coli*, highly transparent and good tensile strength	Antimicrobial food packaging application	Wu et al. [30]
9	–	Polyvinyl alcohol (PVA)	Dopamine-melanin solid nanoparticles (Dpa-s NPs) and hollow nanoparticles (Dpa-h NPs)	Dpa-h NPs enhances the UV light absorption efficiency and gives improved UV-shielding property to nanocomposite film	The PVA/Dpa-h nanocomposite film blocks the UV region (200–400 nm) completely. The transparent nanocomposite film can be used for UV-shielding applications	Wang et al. [31]

(continued)

Table 1 (continued)

S. No.	Biopolymer	Synthetic polymer	Embedded nanomaterials	Special characteristics/features of polymer composites	Applications	References
10	α-cellulose	–	Cerium dioxide (CeO_2)	The nanocomposite film containing 2.95% CeO_2 nanoparticles by weight had >75% light transmittance (550 nm), high UV-shielding properties, and a certain degree of hydrophobicity	CeO_2 shows higher transparency to visible light, which offers an attractive way for other conventional applications such as coating, personal care products, etc.	Singh et al. [32]
11	–	Polystyrene	Cerium dioxide (CeO_2)	The CeO_2 nanoparticles were incorporated into polystyrene (PS) matrix and dispersed homogeneously, with an average crystallite size of 3–5 nm and a bandgap at 3.01 eV	UV-shielding applications	Liu et al. [14]
12	–	Polystyrene	Titanium dioxide (TiO_2)	TiO_2 content 0.5 wt.% extensive UV-blocking effect in the region below 355 nm was observed	UV-shielding applications	Acharya et al. [18]

(continued)

Table 1 (continued)

S. No.	Biopolymer	Synthetic polymer	Embedded nanomaterials	Special characteristics/features of polymer composites	Applications	References
13	Chitosan	–	Curcumin grafted 2, 2, 6, 6-tetramethylpiperidine-1-oxyl (TEMPO) cellulose nanofiber	The composite film showed good mechanical strength along with water insolubility, antioxidant capacity, and antibacterial activity against *E. coli*	Biodegradable, antibacterial properties with strong UV-blocking ability of composite film used in food packaging applications to maintain the food quality and enhance the shelf-life	Zhang et al. [33]
14	Chitosan (CH)	–	Ellagic acid (EA)	A thin film composed of CH and EA has high mechanical properties (Young's modulus = 3.21–3.57 GPa) and thermal stability up to 215–220 °C, as well as these thin films totally inhibited the growth of both gram-positive (*S. aureus*) and gram-negative (*P. aeruginosa*) food pathogenic bacteria	The CH/EA films inhibit the UV light and protect from foodborne pathogens and avoid oxidative degradation. The CH/EA film used for UV-blocking ecofriendly food packaging application	Vilela et al. [34]

(continued)

Table 1 (continued)

S. No.	Biopolymer	Synthetic polymer	Embedded nanomaterials	Special characteristics/features of polymer composites	Applications	References
15	–	Polymethyl-methacrylate	Zinc oxide (ZnO)	ZnO/PMMA nanocomposite film having good UV-blocking capacity even after long exposure to UV light	Application in the museums and art galleries for absorbing ultraviolet radiations falling on the objects which are vulnerable to it	Rawat et al. [15]
16	–	Polyvinyl alcohol (PVA)	Lignin nano-micelle	High tensile strength 91.3 MPa when adding only 2% of lignin Nano-micelle, good thermal stability, and excellent UV-blocking efficiency	The biodegradable PVA/LNM nanocomposite film having excellent water vapor barrier and UV-shielding properties, which can be used in food or medical packaging	Zhang et al. [35]
17	Chitosan and gelatin	–	Boric acid (BA), polyethylene glycol (PEG) as a plasticizer	The film with flexible and transparent in nature having UV light barrier properties	Biodegradable chitosan and gelatin films are used as packaging films	Ahmed and Ikram [36]

(continued)

Table 1 (continued)

S. No.	Biopolymer	Synthetic polymer	Embedded nanomaterials	Special characteristics/features of polymer composites	Applications	References
18	–	Polyvinylidene fluoride (PVDF)	Silica-coated nickel oxide nanoparticles (NiO@SiO_2)	The nanocomposite films (PSNO15-15 wt.% loading of the SNO nanoparticles) block UV radiations up to 99%	NiO@SiO_2/PVDF nanocomposite film for UV protection and electromagnetic interference (EMI) shielding application	Dutta et al. [37]
19	–	Polyvinyl alcohol (PVA)	Metal–organic frameworks ZnO/C and ZnO/C-W	The ZnO/C-W/P film with 6 wt.% of nanocomposite can block nearly 97% UV-C (100–280 nm), more than 95% of UV-B (280–315 nm), and more than 96% of UV-A (315–400 nm)	Ultraviolet blocking films	Tarasi and Morsali [38]
20	Sodium alginate	–	Thymol	A film having high tensile strength with UV-blocking capability also shows antioxidant and antibacterial activities	Application in fresh-cut apple packaging	Chen et al. [39]

(continued)

Table 1 (continued)

S. No.	Biopolymer	Synthetic polymer	Embedded nanomaterials	Special characteristics/features of polymer composites	Applications	References
21	Nano-fibrillated cellulose	–	Mango leaf extract (MLE)	thermal stability up to 250 °C, good mechanical performance (Young's modulus >4.7 GPa), UV light barrier properties, antioxidant capacity antimicrobial activity against *Staphylococcus aureus* and *Escherichia coli*	Food packaging application	Bastante et al. [40]
22	Polyvinyl alcohol/liquefied ball-milled chitin (PVA/LBMC)	–	Silica	A composite film having good tensile strength and thermal stability	The biodegradable PVA/LBMC film is used for food packaging	Zhang et al. [41]
23	Carboxymethyl chitosan	–	Nano MgO	CMCS/MgO composites exhibited excellent antimicrobial activity against Listeria monocytogenes and Shewanella Baltica	The CMCS/MgO film used for food packaging with improved waterproof and antibacterial performance	Wang et al. [42]

(continued)

Table 1 (continued)

S. No.	Biopolymer	Synthetic polymer	Embedded nanomaterials	Special characteristics/features of polymer composites	Applications	References
24	Agar	–	Copper sulfide nanoparticles (CuS NP)	In vitro analysis showed excellent biocompatibility of CuS NP and Agar/CuS NP nanocomposite films on skin fibroblast L929 cell lines with cell viability above 90%. Also, they exhibited antibacterial activity against foodborne pathogenic bacteria, *E. coli*, and some activity against L. monocytogenes	Agar/CuS NP composite films used in food packaging as well as biomedical applications	Roy et al. [43]
25	Gelatin	–	Curcumin	1.5% of curcumin improved the UV-blocking effect by more than 99% at a loss of 5.7% of transparency	Active food packaging application	Roy and Rhim [44]

(continued)

Table 1 (continued)

S. No.	Biopolymer	Synthetic polymer	Embedded nanomaterials	Special characteristics/features of polymer composites	Applications	References
26	–	Polyvinyl alcohol (PVA)	Waste tea residue carbon dots (WTR-CDs)	Composite thin film has highly transparent with high mechanical tensile stress and strain. The composite, PVA@WTR-CDs-3 films were succeeded to block full (100%) of UV-C (230–280 nm) and UV-B (280–315 nm) region, while 20–60% of UV-A (315–400 nm)	A next-generation food packaging material for UV-blocking applications	Patil et al. [17]

bundling by improving the quality and expanding the shelf-life of usability of food sources.

Moreover, biopolymer-based wraping materials have some valuable properties as hustling materials in improving food quality and increase the period of usability through limiting microbial development in the material. Packaging materials prepared from biopolymers such as chitosan, starch, cellulose, agarose, gelatin, sodium alginate, polylactic acid due to non-toxicity can also be used in edible packaging. As well known that every material has pros and cons, similarly, biopolymers have some restrictions due to their poor mechanical properties as there are benefits along with some drawbacks also. The utilization of biopolymers has been limited due to their typically poor mechanical and hindrance properties, which might be improved by incorporating nanomaterials. Nanotechnology attracted all research fields to involve in better sustainable development because of its admirable properties. Nanomaterials have characteristic and distinguishable properties like high mechanical strengths, high surface area, sensing ability, fluorescence, etc.; therefore, nanomaterials have attracted all researchers to do work in this area. Table 1 gives detailed arrangements of different kinds of biopolymer, and synthetic polymer embedded with UV-shielding materials (nanomaterials, extract) for sustainable active food packaging and UV-blocking applications are discussed.

Organic and inorganic UV-shielding nanomaterials such as carbon quantum dots (CQDs), ZnO, silver nanoparticles, TiO_2, SiO_2, CeO_2, etc. are mainly used along with biopolymers, viz., chitosan, polylactic acid, starch, cellulose, gelatin, sodium alginate, etc. for UV-shielding and sustainable food packaging applications [9, 13, 14, 16–24]. Embedded nanomaterial in the biopolymer enhanced the physical as well as chemical properties of biopolymer@nanocomposite and provides higher mechanical stress and strain, thermal stability, heat resistance to the materials as compared to the conventional and neat polymer.

The research group of Periayya Uthirakumar proposed the carbon quantum dots/N-doped ZnO nanoparticle (CQDs/N-ZnO) embedded with polymer polymethyl methacrylate (PMMA) film for UV-shielding application. The prepared polymer nanocomposite film showed a maximum ~92% UV light-shielded (direct daylight) containing 1.0 wt.% of CQD/N-ZnO with a maintained thickness which was 250 μm [22].

Samuel C. Hess et al. found a great solution on UV-blocking using carbon quantum dots in an aqueous solution. Synthesis of carbon quantum dots composed of polymer is an easy and cheap method for the UV-blocking application. In this study, they prepared a one-pot CQDs reaction in an aqueous (PVA) polyvinyl alcohol solution. They have shown a commercial application on the glass vial, as is shown in Fig. 4 [25].

Nowadays, customers favor seeing products through packaging materials, which is an additional requirement in the packaging field; there is a high demand in the market for highly transparent UV-shielding materials. Recently on that problem, Patil et al. has found a satisfactory alternative to it [17]. They designed highly transparent, thin polymer nanocomposite films and showed the successful commercialization of UV-blocking and sustainable food packaging application on grapes fruits. They

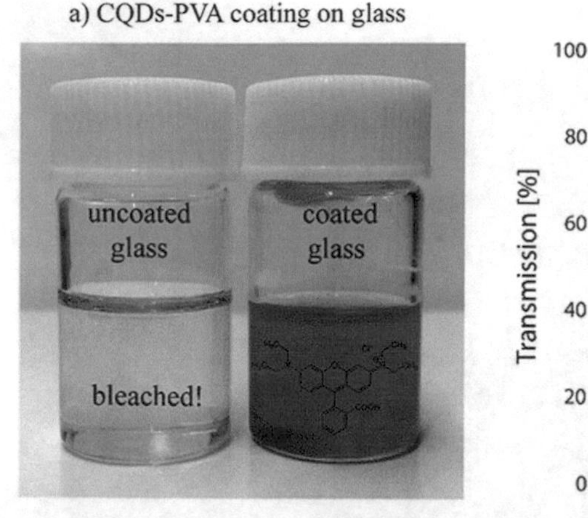

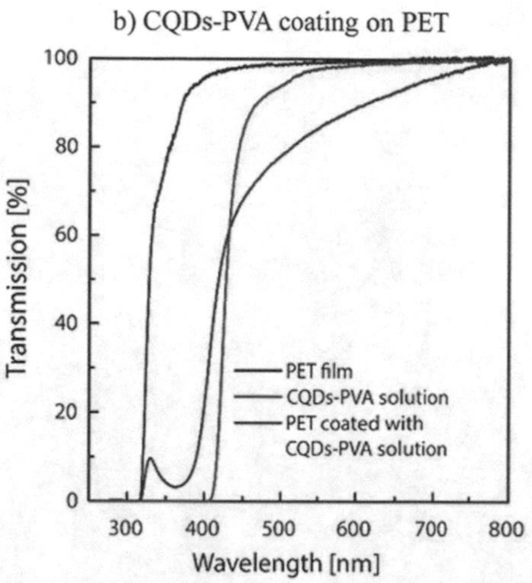

Fig. 4 **a** Bleaching experiment using rhodamine B in polyethersulfone/N-methyl-2-pyrrolidone solution in an uncoated (left) and a CQDs–PVA solution-coated (right) glass vial. Both vials were exposed to UV light (365 nm, 60 Hz) for 2.5 h. **b** UV–Vis transmission curves of pure PET (from PET bottle, black curve), a CQDs–PVA solution synthesized from an aqueous precursor solution, and a PET film coated with this CQDs–PVA solution and dried at 120 °C for 45 min (blue curve)

fabricated waste tea residue carbon dots (WTR-CDs) @PVA nanocomposite thin film by simple solvent casting method for UV-blocking application. Fabricated film PVA@WTR-CDs-3 successfully blocks complete 100% of UV-C (230–280 nm) and UV-B (280–315 nm) while 20–60% block UV-A (315–400 nm).

For UV-blocking and sustainable food packaging, they have shown real commercialized application on grapes as a model fruit (Fig. 5). The two sets of fruits were exposed to the UV lamp for a period of 30 h. Cup A was taken without wrapping, while cups B and C were wrapped with the pristine PVA film and PVA@WTR-CDs composite thin films, respectively. There are massive eye observable changes that were visible in the color and form of grapes best after the 15 h of exposer to the UV light. The grapes in cup A and cup B have been barely brownish in the shade as compared to the grapes in cup C. Then UV irradiation was continued up to 30 h, and changes have been pictorially proven in Fig. 5.

Shiji Mathew and co-worker's examined the improvement of biodegradable PVA-montmorillonite K10 clay nanocomposite blended films with in situ generated ginger extract-mediated silver nanoparticles for chicken sausages application [26]. The prepared PVA/$AgNO_3$/ginger/MMT (PAGM) extract film shown antibacterial activity against common food pathogens *S. typhimurium* (MTCC 1251) and *S. aureus* (MTCC 96). The prepared PAGM nanocomposite film as an active packaging material, the chicken sausage was packed in pouches made from PAGM films for application. As a control pack, normal polythene pouches were used (Fig. 6). PAGM nanocomposite film pouches strongly inhibited bacterial growth as compared to the

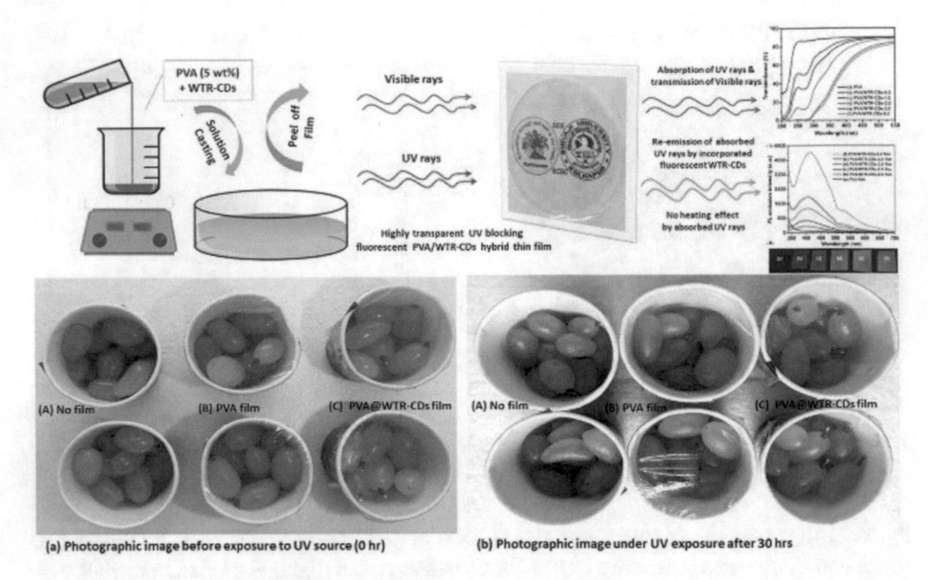

Fig. 5 Schematic and mechanism for fabrication of pristine PVA and PVA@WTR-CDs composite thin films and photographic images for fabricated composite films in food packaging applications under before (**a**) and after (**b**) exposure to the UV light (30 h) Cup **a** without films, **b** wrapped with pristine PVA films, and **c** wrapped with PVA@WTR-CDs composite films

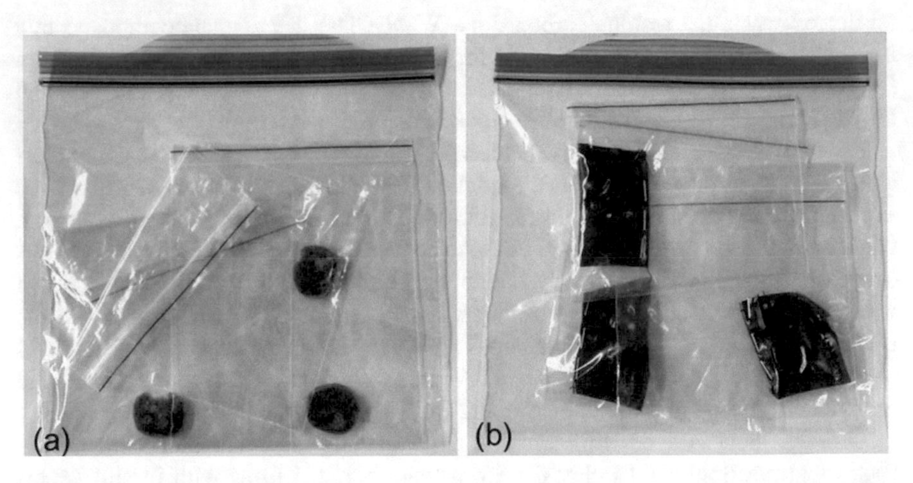

Fig. 6 Digital image of chicken sausage sample packed in **a** control polythene pouches and **b** PAGM pouches (to be reproduced in color on the web and in black-and-white in print)

control, which means polythene pouches are not efficient to retard the growth of bacteria on packed food as compared to PAGM pouches.

7 Future Perspectives

Nowadays, packaging materials are continuously improving with technology. In the forthcoming years, packaging materials will become technology-based due to changing lifestyles, people are more conscious about their health, and increasing demand for healthy and high-quality foods. The packaging materials will be sustainable, bio-based, edible, smart and intelligent, sensitive as well as extends the shelf-life and safety of the food.

Biopolymer@nanocomposite-based packaging materials will provide sustainability to balance/maintain/stability of the ecological system. The research and development department will generate more improvements in the area of biopolymer@nanomaterials-based sustainable packaging, and they will provide high-quality packaging materials which have high mechanical strength, full range of UV-blocking, high thermal stability, antimicrobial, and gas barrier properties, as well as monitor the food and extend the shelf-life of the food.

A great challenge in this area is to develop flexible, water-insoluble, thinner, and more transparent with high mechanical and thermal stability with easily biodegradable properties, which may be the next area for the development in sustainable packaging applications.

8 Conclusion

In this chapter, detailed information about the biopolymers with their nanocomposite for sustainable packaging application is discussed. Also, in this chapter, the history of traditional food packaging and its current developments are well discussed. The advancement in nanotechnology and its increased applicability in sustainable and UV-shielding packaging is revolutionary due to its distinguishable and characteristic properties in UV barrier as well as a biodegradable alternative to non-decomposable plastic. The various biopolymers and their nanocomposites in the development of real-life applications as active and smart packaging are tabulated in Table 1. The current chapter well describes the role of nanomaterials as a UV-blocking, antimicrobial, water, and gas barrier. From this book chapter, we can conclude that the biopolymer and their nanocomposites are having real-life UV-shielding properties, which lead to improving the quality of food materials and extending the shelf-life. This biopolymer-based nanomaterial composite food packaging can be the future of real-life active, smart, reusable, easily biodegradable, and smart packaging materials.

References

1. Hattori H, Ide Y, Sano T (2014) Microporous titanate nanofibers for highly efficient UV-protective transparent coating. J Mater Chem A 2:16381–16388. https://doi.org/10.1039/c7ta00397h
2. Majid I, Ahmad Nayik G, Mohammad Dar S, Nanda V (2018) Novel food packaging technologies: innovations and future prospective. J Saudi Soc Agric Sci 17:454–462. https://doi.org/10.1016/j.jssas.2016.11.003
3. Risch SJ (2009) Food packaging history and innovations. J Agric Food Chem 57:8089–8092. https://doi.org/10.1021/jf900040r
4. Rhim JW, Ng PKW (2007) Natural biopolymer-based nanocomposite films for packaging applications. Crit Rev Food Sci Nutr 47:411–433. https://doi.org/10.1080/10408390600846366
5. Ardnt G, Armstrong W, Cooksey K et al (2010) The Wiley encyclopedia of packaging technology
6. Murugesan GS, Sathishkumar M, Swaminathan K (2006) Arsenic removal from groundwater by pretreated waste tea fungal biomass. Bioresour Technol 97:483–487. https://doi.org/10.1016/j.biortech.2005.03.008
7. Roy S, Rhim JW (2021) Preparation of gelatin/carrageenan-based color-indicator film integrated with shikonin and propolis for smart food packaging applications. ACS Appl Bio Mater 4:770–779. https://doi.org/10.1021/acsabm.0c01353
8. Chaudhary P, Fatima F, Kumar A (2020) Relevance of nanomaterials in food packaging and its advanced future prospects. J Inorg Organomet Polym Mater 30:5180–5192. https://doi.org/10.1007/s10904-020-01674-8
9. Zayat M, Garcia-Parejo P, Levy D (2007) Preventing UV-light damage of light sensitive materials using a highly protective UV-absorbing coating. Chem Soc Rev 36:1270–1281. https://doi.org/10.1039/b608888k
10. Lucas RM, Yazar S, Young AR et al (2019) Human health in relation to exposure to solar ultraviolet radiation under changing stratospheric ozone and climate. Photochem Photobiol Sci 18:641–680. https://doi.org/10.1039/C8PP90060D
11. Nerín C, Tovar L, Djenane D et al (2006) Stabilization of beef meat by a new active packaging containing natural antioxidants. J Agric Food Chem 54:7840–7846. https://doi.org/10.1021/jf060775c
12. Duncan TV (2011) Applications of nanotechnology in food packaging and food safety: barrier materials, antimicrobials and sensors. J Colloid Interface Sci 363:1–24. https://doi.org/10.1016/j.jcis.2011.07.017
13. Kaseem M, Hamad K, Rehman ZU (2019) Review of recent advances in polylactic acid/TiO2 composites. Materials (Basel) 12. https://doi.org/10.3390/ma12223659
14. Liu KQ, Kuang CX, Zhong MQ et al (2013) Synthesis, characterization and UV-shielding property of polystyrene-embedded CeO_2 nanoparticles. Opt Mater (Amst) 35:2710–2715. https://doi.org/10.1016/j.optmat.2013.08.012
15. Rawat A, Soni U, Malik RS, Pandey SC (2018) Facile synthesis of UV blocking nano-sized zinc oxide and polymethyl-methacrylate polymer nanocomposite coating material. Nano-Struct Nano-Objects 16:371–380. https://doi.org/10.1016/j.nanoso.2018.09.002
16. Zhao Z, Mao A, Gao W, Bai H (2018) A facile in situ method to fabricate transparent, flexible polyvinyl alcohol/ZnO film for UV-shielding. Compos Commun 10:157–162. https://doi.org/10.1016/j.coco.2018.09.009
17. Patil AS, Waghmare RD, Pawar SP et al (2020) Photophysical insights of highly transparent, flexible and re-emissive PVA @ WTR-CDs composite thin films: a next generation food packaging material for UV blocking applications. J Photochem Photobiol A Chem 400. https://doi.org/10.1016/j.jphotochem.2020.112647
18. Acharya AD, Sarwan B, Sharma R et al (2017) UV-shielding efficiency of TiO2-polystyrene thin films prepared by solution cast method. J Phys Conf Ser 836:3–7. https://doi.org/10.1088/1742-6596/836/1/012048

19. González EAS, Olmos D, Lorente M, ángel, et al (2018) Preparation and characterization of polymer composite materials based on PLA/TiO2 for antibacterial packaging. Polymers (Basel) 10:1–14. https://doi.org/10.3390/polym10121365
20. Jiang Y, Song Y, Miao M et al (2015) Transparent nanocellulose hybrid films functionalized with ZnO nanostructures for UV-blocking. J Mater Chem C 3:6717–6724. https://doi.org/10.1039/c5tc00812c
21. Tang XZ, Kumar P, Alavi S, Sandeep KP (2012) Recent advances in biopolymers and biopolymer-based nanocomposites for food packaging materials. Crit Rev Food Sci Nutr 52:426–442. https://doi.org/10.1080/10408398.2010.500508
22. Uthirakumar P, Devendiran M, Yun JH et al (2018) Role of carbon quantum dots and film thickness on enhanced UV shielding capability of flexible polymer film containing carbon quantum dots/N-doped ZnO nanoparticles. Opt Mater (Amst) 84:771–777. https://doi.org/10.1016/j.optmat.2018.08.016
23. Vasile C, Râpă M, Ştefan M et al (2017) New PLA/ZnO:Cu/Ag bionanocomposites for food packaging. Express Polym Lett 11:531–544. https://doi.org/10.3144/expresspolymlett.2017.51
24. Wang Y, Su J, Li T et al (2017) Bioactive chitosan/ellagic acid films with UV-light protection for active food packaging. Food Hydrocoll 73:120–128. https://doi.org/10.1021/acsami.7b08763
25. Hess SC, Permatasari FA, Fukazawa H et al (2017) Direct synthesis of carbon quantum dots in aqueous polymer solution: one-pot reaction and preparation of transparent UV-blocking films. J Mater Chem A 5:5187–5194. https://doi.org/10.1039/c7ta00397h
26. Mathew S, S S, Mathew J, Radhakrishnan EK, (2019) Biodegradable and active nanocomposite pouches reinforced with silver nanoparticles for improved packaging of chicken sausages. Food Packag Shelf Life 19:155–166. https://doi.org/10.1016/j.fpsl.2018.12.009
27. Keshk SMAS, El-Zahhar AA, Haija MA, Bondock S (2019) Synthesis of a magnetic nanoparticles/dialdehyde starch-based composite film for food packaging. Starch/Staerke 71:1–7. https://doi.org/10.1002/star.201800035
28. Yu Z, Li B, Chu J, Zhang P (2018) Silica in situ enhanced PVA/chitosan biodegradable films for food packages. Carbohydr Polym 184:214–220. https://doi.org/10.1016/j.carbpol.2017.12.043
29. Lai TY, Chen CH, Lai LS (2013) Effects of tapioca starch/decolorized hsian-tsao leaf gum-based active coatings on the quality of minimally processed carrots. Food Bioprocess Technol 6:249–258. https://doi.org/10.1007/s11947-011-0707-3
30. Wu Z, Wu J, Peng T et al (2017) Preparation and application of starch/polyvinyl alcohol/citric acid ternary blend antimicrobial functional food packaging films. Polymers (Basel) 9:1–19. https://doi.org/10.3390/polym9030102
31. Wang Y, Su J, Li T et al (2017) A novel UV-shielding and transparent polymer film: when bioinspired dopamine-melanin hollow nanoparticles join polymers. ACS Appl Mater Interfaces 9:36281–36289. https://doi.org/10.1021/acsami.7b08763
32. Singh HK, Kumar S, Bamne J et al (2020) Analyzing the synthesis of various inorganic nanoparticles and their role in UV-shielding. Mater Today Proc. https://doi.org/10.1016/j.matpr.2020.09.499
33. Zhang X, Li Y, Guo M et al (2021) Antimicrobial and UV blocking properties of composite chitosan films with curcumin grafted cellulose nanofiber. Food Hydrocoll 112. https://doi.org/10.1016/j.foodhyd.2020.106337
34. Vilela C, Pinto RJB, Coelho J et al (2017) Bioactive chitosan/ellagic acid films with UV-light protection for active food packaging. Food Hydrocoll 73:120–128. https://doi.org/10.1016/j.foodhyd.2017.06.037
35. Zhang X, Liu W, Liu W, Qiu X (2020) High performance PVA/lignin nanocomposite films with excellent water vapor barrier and UV-shielding properties. Int J Biol Macromol 142:551–558. https://doi.org/10.1016/j.ijbiomac.2019.09.129
36. Ahmed S, Ikram S (2016) Chitosan and gelatin based biodegradable packaging films with UV-light protection. J Photochem Photobiol B Biol 163:115–124. https://doi.org/10.1016/j.jphotobiol.2016.08.023

37. Dutta B, Kar E, Sen G et al (2020) Lightweight, flexible NiO@SiO2/PVDF nanocomposite film for UV protection and EMI shielding application. Mater Res Bull 124. https://doi.org/10.1016/j.materresbull.2019.110746
38. Tarasi S, Morsali A (2021) Fabrication of transparent ultraviolet blocking films using nanocomposites derived from metal-organic frameworks. J Alloys Compd 868. https://doi.org/10.1016/j.jallcom.2021.158996
39. Chen J, Wu A, Yang M et al (2021) Characterization of sodium alginate-based films incorporated with thymol for fresh-cut apple packaging. Food Control 126. https://doi.org/10.1016/j.foodcont.2021.108063
40. Bastante CC, Silva NHCS, Cardoso LC et al (2021) Biobased films of nanocellulose and mango leaf extract for active food packaging: supercritical impregnation versus solvent casting. Food Hydrocoll 117. https://doi.org/10.1016/j.foodhyd.2021.106709
41. Zhang J, Xu WR, Zhang YC et al (2020) In situ generated silica reinforced polyvinyl alcohol/liquefied chitin biodegradable films for food packaging. Carbohydr Polym 238. https://doi.org/10.1016/j.carbpol.2020.116182
42. Wang Y, Cen C, Chen J, Fu L (2020) MgO/carboxymethyl chitosan nanocomposite improves thermal stability, waterproof and antibacterial performance for food packaging. Carbohydr Polym 236. https://doi.org/10.1016/j.carbpol.2020.116078
43. Roy S, Rhim JW, Jaiswal L (2019) Bioactive agar-based functional composite film incorporated with copper sulfide nanoparticles. Food Hydrocoll 93:156–166. https://doi.org/10.1016/j.foodhyd.2019.02.034
44. Roy S, Rhim JW (2020) Preparation of antimicrobial and antioxidant gelatin/curcumin composite films for active food packaging application. Colloids Surf B Biointerf 188. https://doi.org/10.1016/j.colsurfb.2019.110761

Design and Development of Robust Optimization Model for Sustainable Cross-Docking Systems: A Case Study in Electrical Devices Manufacturing Company

Hamdi G. Resat, Pelin Berten, Zeliha Kilek, and M. Batuhan Kalay

Abstract This study presents a novel bi-objective solution methodology designed for sustainable material handling operations in cross-docking areas of companies. After successfully applied linearization techniques over a predefined non-linear model, a proposed mixed-integer linear programming model is designed by considering decrease of reception process lead-time; elimination of redundant material handling movements; as well as reduction in waiting time of pallets under sorting and transfer processes in packaging operations. We present a bi-objective modeling approach and outline linearization of the non-linear mathematical programming problem to minimize material handling cost and total sustainability score in terms of carbon dioxide emissions by using epsilon constraint method to find out the Pareto frontier. An illustrative data set collected from one of the leading electrical devices manufacturing companies is used to illustrate the proposed solution approach. The solution approach is shown as a comparative study after specifying some pre-processing and symmetry breaking measures, valid inequalities, and logic cuts.

Keywords Bi-objective optimization · Discrete optimization · Sustainability · Cross-docking · Emission optimization · Warehouse management · Linearization · Mixed integer linear programming

1 Introduction

Efficiency assessments get more critical positions under warehouse management systems while global market conditions become more competitive for real-life applications. Many companies use warehouse management systems (WMS) in order to convert their material handling operations from workforce-dependent systems to automated ones. More on-time deliveries, better quality conditions, and reduced

---

H. G. Resat (✉) · P. Berten · Z. Kilek · M. B. Kalay
Department of Industrial Engineering, Izmir University of Economics, Izmir 35330, Turkey
e-mail: giray.resat@ieu.edu.tr

S. S. Muthu (eds.), *Sustainable Packaging*, Environmental Footprints and Eco-design of Products and Processes, https://doi.org/10.1007/978-981-16-4609-6_8

operational costs in material handling processes (receiving, sorting, storing, value-added services, etc.) are expected after the companies satisfy such flexibility transformation in their inventory management systems. Although cost, time, and space management conditions seem to be primal assessment criteria of warehouse management systems, one of the main pillars of the sustainability concept (environmental conditions) recently gets significant importance by companies after the European Green Deal regulations are announced [30]. According to the International Energy Agency (IEA) report in 2019, global energy-related CO_2 emissions was around 33 giga-tons and 21% of that CO_2 emissions was by manufacturing industries [18]. Although manufacturing systems have such significant effects over the environment, 15% of global carbon emissions result from transport and logistics operations [36].

The dynamic market conditions require more flexible and customized systems to satisfy customer demands. Management of the robust inventory systems is one of the actions in order to satisfy this fluctuated environment. However, efficient time and cost management in warehouses are one of the critical issues, and there are many studies to deal with time and cost consuming activities in inventory management systems. At that point, cross-docking becomes a good option for short-period (no more than 24 h) storage of products to eliminate unnecessary inventory levels [13]. The activities in cross-docking operations can be classified into three main classes that are loading, sorting (including integration and consolidation), and unloading [1].

Ladier [20] states an integer programming model for truck-scheduling problems by considering inbound and outbound trucks and gets a solution with the help of three different heuristics methodologies. The employee timetable problem is addressed with three mixed-integer problems. Reactive scheduling which is called "online scheduling" in the cross-docking literature is used for re-optimizing the existing schedule when an unforeseeable event takes place. Golshahi-Roudbaneh et al. [14] present a truck-scheduling problem in cross-docking systems and design four meta-heuristics with the objective function of minimizing make-span and determining an order for both receiving and shipping trucks. Furthermore, two heuristics are proposed to solve the truck-scheduling problem. Molavi et al. [27] propose a truck-scheduling model in a cross-docking area with consideration of fixed due dates via minimizing the delivery cost of retained products and delayed shipments. A mixed-integer linear programming model is developed to minimize total logistics cost including delivery and penalty cost of delayed loads at the end of the planning period. Arabani et al. [4] apply five meta-heuristics to schedule trucks in cross-docking system with temporary storage which are Ant Colony Optimization (ACO), Particle Swarm Optimization (PSO), Genetic Algorithm (GA), Tabu Search (TS), and Differential Evaluation (DE). The meta-heuristics algorithms are affected by their parameters and factors that take different values. Taguchi plans and ANOVA tables are designed in order to use these parameters ideally. Wisittipanich and Hengmeechai [35] propose a mixed-integer programming for truck sequencing and door assigning in a multi-door cross-docking system to minimize total time spent on operational activities. A modified particle swarm optimization that is called GLNPSO is used for solving the problem. GLNPSO and original PSO results are compared, and the GLNPSO is justified due to its capability of finding high quality results with quick convergence. In the study of [33], a truck-scheduling problem at a cross-docking is established where an inbound truck with partially loaded/unloaded can be used as an outgoing truck. The mathematical model is formed as mixed-integer programming in order to find the best dock door and assign destinations with the scheduling of trucks to minimize make-span which can be solved with a hybrid heuristics-simulated annealing. Dulebenets [11] studies in scheduling of inbound and outbound trucks at a cross-docking facility via formulating a mixed-integer linear programming model that aims to minimize the total service cost of trucks. The proposed algorithm is also compared with the other five meta-heuristic algorithms which are commonly used in cross-docking such as tournament selection, first come first served policy with inbound truck precedence constraints (FCFS-ITPC), custom mutation operator, ranking selection, and migration. Guemri et al. [15] use a probabilistic tabu search method as a heuristic algorithm that tries to solve the cross-docking assignment problem (CDAP). At the end of this study, two different heuristic algorithms with a probabilistic approach are introduced. These algorithms include the objective of CDAP which is to minimize the material handling costs while finding the optimal truck assignment within the cross-docking area. In addition, when a comparison is made, the heuristic performs better than the other algorithms in terms of central processing unit (CPU) time consumption. Different from other studies, Rijal et al. [32] try to solve the dock door assignment and truck scheduling problem at the same time rather than solving them consecutively. Also, this study takes both inbound and outbound activities into consideration rather than traditional truck scheduling problems that account for either inbound or outbound activities solely. A flexible approach that assumes dock doors and can serve for both inbound and outbound activities is applied for dock door assignment. Adaptive Large Neighborhood Search (ALNS) is applied for solving the mathematical model because this algorithm considers assignment and scheduling decisions simultaneously. The U-shaped and I-shaped cross-dock terminals are investigated to see the impacts. Fard and Vahdani [12] try to avoid the waiting queue during the assignment of outbound trucks to outbound doors. M/M/1 stochastic model is exercised for the objective of minimizing the time spent waiting in the queue. The model has two objective functions. First objective function is designed for minimizing the cost for holding inventory, time for customer delivery, and waiting time in the queue. The other objective is about minimization of the energy consumption by forklifts during these operations. Two different meta-heuristic algorithms are presented for proposed models with more than one objective function. Khalili-Damghani et al. [19] study the cross-dock truck scheduling with multi periods and aim to decrease costs of operations, prevent accumulation of products, reduce the storage time, and minimize the delays in shipment of products by using genetic algorithms (GA). The comparison between these algorithms is made based on the results that are attained by CPLEX solver in GAMS. Liao et al. [21] developed multiple algorithms to minimize the total weighted tardiness while considering berth allocation and a cross-dock sequencing problem with multiple doors. The considered sequencing problem assumes leaving schedules for outgoing trucks are fixed and there are multiple doors available for incoming trucks. At the end of this study, six different algorithms are presented, and differential evolution (DE) generates the best results when confronted with the other alternatives represented in the article.

Cross-docking assignment problem (CDAP) is established as a mixed-integer programming that will be moved between the point of departure and arrival destinations via incoming and outgoing doors. In the study of [13], CPLEX is applied to find an optimum solution, and the results are compared with the ones obtained from the heuristics methodologies. Nikolopoulou et al. [29] include cross-docking stations to a delivery routing problem. The costs that occur during the movement of goods between the points of departure and arrival destinations are examined. A local search approach is used for building a meta-heuristic algorithm that minimizes the total transportation costs. The results show that when the departure and arrival destinations are closer, there will be the lowest costs if direct shipment is used. Birim [6] studies on minimizing the total costs that incur during transport operations while identifying the best routes. A simulated annealing heuristics method is used to find an optimum route. In their study, Mohtashami et al. [26] generate a generic model that tries to minimize multiple outcomes: transportation costs, number of travels made by trucks, and total time spent. They propose two population-based meta-heuristics to solve their multi-objective optimization model that are non-dominated sorting genetic algorithm (NSGA-II) and the multi-objective particle swarm optimization (MOPSO). They find that NSGA-II algorithm shows better performance over MOPSO algorithm according to numerical results. Zhang et al. [37] generate two heuristic algorithms that identify the fastest option for where to store the incoming goods generating efficient lists that include amount of goods, customer-specific marks, and location of storage. The model is developed as a mixed-integer program. Alamri and Syntetos [3] highlight the weaknesses of LIFO and FIFO methods and develop a new method for inventory management strategy plan named Allocation-In-Fraction-Out (AIFO). AIFO represents a strategy for rare situations and offers to either implement FIFO or LIFO. A general mathematical formulation of LIFO and FIFO models are developed with the motivation of investigation and numerical comparison. Agustina et al. [2] study for just-in-time deliveries of the food industry with cross-docking operations considering the minimum cost of holding, and the model also considers the arrival time of deliveries and punishment for the ones that are not on time. In order to solve this, the authors use mixed-integer programming and integrate vehicle scheduling and routing into an innovative and all-inclusive model that was previously modeled separately. They also introduce the notion of delivery time windows and customer zones to reduce the sets of values of the choice variables. The efficient use of land enables multi-floor cross-dock warehouses to generate more functional working capital. This opportunity gives an advantage to the responsiveness of supply chain operations. However, it gains importance to being cost-effective on cross-docking operations because of the high volume of product flow in loading and unpacking operations. In the study of [34], three strategies are chosen to solve a multi-floor, cross-dock door assignment problem (MCDAP) which minimize the costs that incur during material handling activities. Zuluaga et al. [38] study the reverse cross-docking linear programming model to optimize the total costs via assignment decision in which the model optimizes the system by assigning the boxes to traditional warehouses or applying cross-docking. Luo et al. [22] implement a meta-heuristic algorithm HGA-OLS on the mathematical model of MTO-CD (Make-to-order)—(cross-docking) synchronization in order to maximize the overall efficiency. This model is formulated to find synchronized decisions of production scheduling and warehousing activities. The algorithm combines genetic algorithms with local search and opposition-based learning to increase and improve offspring individuals. Two experiments are performed to make the decisions of coordination of manufacturing plant and warehousing. On the other hand, different scenarios are coded in MATLAB. Nassief et al. [28] mention that the objective function is related to weighted travel time between doors in the cross-docking facility. Moeller [25] states line sequence optimization (LSO) method with possible strict time-windows on planning and controlling the order-picking processes by considering the productivity and integration of functionality. Boysen et al. [7] propose a mathematical model to minimize the order spread (the number of conveyor sections from the first occurrence of the order to the last) in an Automated Storage and Retrieval System (ASRS) and solve the problem by developing simulation-based dynamic programming model with some heuristics algorithms in order to evaluate the impact of release sequence on the usage of packaging centers and the working programs of packaging labors. Boysen et al. [8] refer to an automated sorting system (ASS) from three perspectives which are inbound stations, the main conveyor, and outbound stations. If there are multiple inbound stations to satisfy the large variety of customer demands, variable destinations should be assigned to minimize the cumulative operation time of a cross-docking station. The authors implemented a mixed-integer programming (MIP) and branch and bound (B&B) models for the described problem. Manzini et al. [23] present the improvement and implementation of extended models, resolutions, and technologies in warehousing and material handling systems. Heragu et al. [16] try to decide the optimal size of areas in a warehouse by considering the amount of inventory while minimizing the material handling costs. As a result of the study, a well-structured mathematical model is developed that locates the goods to three zones of the warehouse. In addition, Mital et al. [24] present a modeling framework and an efficient algorithm for designing a material handling system and a warehouse by identifying system configurations and measuring the trade-off between costs, performance, and risk. The implementation of material handling and storage systems is focused on the evaluation of a Pareto-optimal graph. Furthermore, Huang et al. [17] present an integrated model to choose the location and to determine the space for warehouses in each network. The first stage is shipping the cargoes from suppliers to warehouses, and the second stage is delivering from warehouses to assembly plants. In this model, time of storage is not certain in warehouses. Arrival time of items is random because of the distances and the other factors. Unknown stock time and random arrival time cause an inefficient space in the warehouses. The objective of the problem is to minimize the total cost including transportation cost and warehouse operation costs. Behnamian et al. [5] perform the imperialist competitive algorithm (ICA) and simulated annealing (SA) which are the two heuristics proposed near optimal solution sets. Model of cross-docking networks with time-varying uncertain demands is not able to solve with exact solution algorithms, therefore, SA gives a faster, ICA gives a better solution in this study.

The main novelties of this study are to:

- Design a bi-objective optimization model for cross-docking operations to provide alternative solution sets under objectives of minimization of the total cost, processing time, and CO_2 emissions.
- Consider sustainability concept by integrating environmental effects of transportation activities that occurred during material handling operations at cross-docking areas.
- Generate Pareto solutions for sustainable cross-docking operations to decision makers and validate the proposed model by using real-life data sets.

The rest of the chapter is organized as follows: After this detailed literature review and problem description section, Sect. 2 will outline the proposed methodology for the bi-objective model. Section 3 contains the computational detail and data set of proposed model. Details of obtained results by using developed bi-objective MILP model are shared in Sect. 4. Section 5 will present some potential work and an evaluation of the results after obtaining study outputs for decision makers.

2 Methodology

The primary aim of this chapter is to develop a bi-objective MILP model that minimizes the waiting number of pallets and total pallet waiting times on the cross-docking areas by creating balance between them. This model will provide a deep comparison analysis between the current situation and proposed scenarios by considering many parameters such as sorting times, pallet controlling times, conveyor capacities, etc.

Sets

i Inbound pallets ($i = 1, \ldots, I$).
j Outbound pallets ($j = 1, \ldots, J$).
k Product type ($k = 1, \ldots, K$).
t Time period ($t = 1, \ldots, T$).

Scalar

A Handling setup time for each product type.
C_p Cost of waiting per pallet on reception area.
C_w Cost of waiting per pallet on conveyor.
M a large positive value.
$\in$ a small epsilon value.
Γ, Θ, Ψ Constants for carbon-footprint calculations.

Parameters

Cap_k	Capacity of conveyor dedicated for product k.
GE_k	Controlling time of pallet including product k.
KE_i	Entering time of pallet i to reception area.
RE_t	Number of pallets came to reception area in period t.
U_k	Required time to sort product k from inbound pallets.
r_{ik}	Number of product k on inbound pallet i.
H_i	Completion time of handling of pallet i in cross-docking area.

Variables

$C_i \in R^+$	The entering time of inbound pallet i into handling area.
$F_i \in R^+$	The leaving time of inbound pallet i from handling area.
$L_{jk} \in R^+$	The leaving time of k type product from outbound pallet j.
$P_i \in R^+$	The waiting time of pallet i before handling process on reception area.
$ZT_{jk} \in R^+$	The leaving time from conveyor of product k type to outbound pallet j.
$\gamma_{ijk} \in R^+$	Minimum possible leaving time value of k type outbound pallet j (which is connected with inbound pallet i) from handling area.
$\delta_{jk} \in R^+$	The leaving time of fully filled k type outbound pallet j from handling area.
$ZK_{jk} \in R^+$	The starting quality control time of k type outbound pallet j.
$\varphi_{jk} \in R^+$	The leaving time from conveyor of fully filled k type outbound pallet j.
$B_{kt} \in Z^+$	Number of k type waiting pallets on conveyor area in period t.
$D_t \in Z^+$	Number of waiting pallet in reception area before handling in period t.
$s_{jk} \in Z^+$	Number of type k product loaded in outbound pallet j.
$SN_{kt} \in Z^+$	Number of leaving k type product outbound pallets in period t.
$x_{ijk} \in Z^+$	Number of product k transferred from inbound pallet i to outbound pallet j.
$GN_{kt} \in Z^+$	Number of controlled k type pallet in period t.
Ω	Slack variable

$$m_{jkt} \begin{cases} 1 & \textit{If k type outbound pallet j leaves the handling area in period t} \\ 0 & \textit{otherwise} \end{cases}$$

$$m_{1_{jkt}} \begin{cases} 1 & \textit{If k type outbound pallet j don't leaves the handling area after period t} \\ 0 & \textit{otherwise} \end{cases}$$

$$m_{2_{jkt}} \begin{cases} 1 & \textit{If k type outbound pallet j don't leaves the handling area after period t} \\ 0 & \textit{otherwise} \end{cases}$$

n_{it} $\begin{cases} 1 \; if \; inbound \; pallet \; i \; enter \; to \; handling \; area \; in \; period \; t \\ 0 \quad otherwise \end{cases}$

$n_{1_{it}}$ $\begin{cases} 1 \; if \; inbound \; pallet \; i \; enter \; to \; handling \; area \; after \; period \; t \\ 0 \quad otherwise \end{cases}$

$n_{2_{it}}$ $\begin{cases} 1 \; if \; inbound \; pallet \; i \; enter \; to \; handling \; area \; before \; period \; t \\ 0 \quad otherwise \end{cases}$

y_{jk} $\begin{cases} 1 \; if \; product \; k \; is \; transferred \; to \; outbound \; pallet \; j \\ 0 \quad otherwise \end{cases}$

v_{ijk} $\begin{cases} 1 \; if \; product \; k \; is \; transferred \; from \; inbound \; pallet \; i \; to \; outbound \; pallet \; j \\ 0 \quad otherwise \end{cases}$

g_{jt} $\begin{cases} 1 \; if \; k \; type \; outbound \; pallet \; j \; leaves \; conveyor \; in \; period \; t \\ 0 \quad otherwise \end{cases}$

g_{1jt} $\begin{cases} 1 \; if \; k \; type \; outbound \; pallet \; j \; leaves \; conveyor \; in \; period \; t \\ 0 \quad otherwise \end{cases}$

g_{2jt} $\begin{cases} 1 \; if \; k \; type \; outbound \; pallet \; j \; leaves \; conveyor \; in \; period \; t \\ 0 \quad otherwise \end{cases}$

π_{jkt} $\begin{cases} 1 \; if m_{jt} and y_{jk} both take value of \; 1 at the same time \\ 0 \quad otherwise \end{cases}$

θ_{jkt} $\begin{cases} 1 \; if g_{jt} and y_{jk} both take value of \; 1 at the same time \\ 0 \quad otherwise \end{cases}$

Model Formulation

$$Minf_1 = \sum\nolimits_{j \in J} \sum\nolimits_{k \in K} (\varphi_{jk} C_w) + \sum\nolimits_{i \in I} ((F_i - KE_i) C_p) - \Omega \in \tag{1}$$

$$\sum_{j \in J} \sum_{k \in K} (\varphi_{jk} - \delta_{jk}) \Gamma + \sum_{i \in I} \sum_{j \in J} \sum_{k \in K} (x_{ijk} \Theta)$$

$$+ \sum_{k \in K} \sum_{t \in T} ((B_{kt} + SN_{kt} + GN_{kt}) \Psi) + \Omega = f_2^{UP} \tag{2}$$

$$\sum\nolimits_{j \in J} |K| y_{jk} = \sum\nolimits_{i \in I} r_{ik}, \forall k \in K \tag{3}$$

$$\sum\nolimits_{j \in J} x_{ijk} \le r_{ik}, \forall i \in I, k \in K \tag{4}$$

$$\sum\nolimits_{i \in I} x_{ijk} \le |K| y_{jk}, \forall j \in J, k \in K \tag{5}$$

$$x_{ijk} \le |K| \forall i \in I, j \in J, k \in K \tag{6}$$

$$y_{jk} \geq y_{(j+1)k} \forall j \in J, k \in K \tag{7}$$

$$\sum\nolimits_{i \in I} \sum\nolimits_{j \in J} x_{ijk} = \sum\nolimits_{i \in I} r_{ik}, \forall k \in K \tag{8}$$

$$KE_i \leq C_i, \forall i \in I \tag{9}$$

$$F_i \geq C_i + H_i, \forall i \in I \tag{10}$$

$$F_i \leq C_{(i+1)}, \forall i \in I \tag{11}$$

$$H_i = \sum\nolimits_{k \in K} (r_{ik} U_k), \forall i \in I \tag{12}$$

$$x_{ijk} \leq M v_{ijk}, \forall i \in I, j \in J, k \in K \tag{13}$$

$$x_{ijk} \geq v_{ijk}, \forall i \in I, j \in J, k \in K \tag{14}$$

$$L_{jk} \geq F_i v_{ijk}, \forall i \in I, j \in J, k \in K \tag{15}$$

$$L_{jk} \leq L_{k(j+1)}, \forall j \in J, k \in K \tag{16}$$

$$\delta_{jk} \geq L_{jk} y_{jk}, \forall j \in J, k \in K \tag{17}$$

$$\delta_{jk} \leq ZK_{jk}, \forall j \in J, k \in K \tag{18}$$

$$ZT_{jk} \geq ZK_{jk} + GE_k y_{jk}, \forall j \in J, k \in K \tag{19}$$

$$ZT_{jk} \leq ZK_{jk}, \forall j \in J, k \in K \tag{20}$$

$$\varphi_{jk} \geq ZT_{jk} y_{jk}, \forall j \in J, k \in K \tag{21}$$

$$0 \leq C_i - t < n_{it} \forall i \in I, t \in T \tag{22}$$

$$\sum\nolimits_{t \in T} n_{it} = 1, \forall i \in I \tag{23}$$

$$0 \leq L_{jk} - t < m_{jkt}, \forall j \in J, k \in K, t \in T \tag{24}$$

$$0 \leq \varphi_{jk} - t < g_{jkt}, \forall j \in J, k \in K, t \in T \tag{25}$$

$$\varphi_{jk} \leq |T| \forall j \in J, k \in K \tag{26}$$

$$D_t + RE_t - \sum\nolimits_{i \in I} n_{it} = D_{(t+1)}, \forall t \in T \tag{27}$$

$$SN_{kt} = \sum\nolimits_{j \in J} y_{jk} m_{jkt}, \forall k \in K, t \in T \tag{28}$$

$$B_{kt} + SN_{kt} - GN_{kt} = B_{k(t+1)}, \forall k \in K, t \in T \tag{29}$$

$$GN_{kt} = \sum\nolimits_{j \in J} y_{jk} g_{jkt}, \forall k \in K, t \in T \tag{30}$$

Equation (1) indicates one of the objective functions of this model, that is, the sum of the total cost of pallet waiting times on the reception area and in the system. Equation (2) shows the second objective function of this model related with carbon footprint assessments. This function consists of three parts considering waiting times of pallets at final destinations, total handled products throughout the system, and total number of pallets used in the system, respectively. Equation (3) indicates that the total number of products on the inbound pallets must be equal to the total number of products that can be carried on outbound pallets. Equation (4) ensures that total number of k type products which are transported from inbound pallet i to any outbound pallet j, should be less than or equal to total number of k type product on that inbound pallet i. Equation (5) shows that the total number of k type products which are transported to outbound pallet j from any inbound pallet i should be less than or equal to total number of products that can be carried on outbound pallets. Equation (6) ensures that the total number of k type products which are transported from pallet i to j must be less than or equal to the total number of products that can be carried on outbound pallets. Equation (7) indicates that each k type pallet should wait until the previous pallet is fully filled. Equation (8) ensures that the total number of transferred products throughout the system is equal to the total number of products which are entered to system under inbound pallets i. Equation (9) shows that each inbound pallet firstly enters the reception area, then enters the handling area. Equation (10) indicates that leaving time from the handling area of pallet i is greater than or equal to the sum of entering time to handle the area of pallet i and handling time of pallet i. Equation (11) ensures that each inbound pallet i can enter the handling area after the previous pallet left the handling area. Equation (12) shows that handling time of pallet i is equal to the sum of each k type product handling time in that pallet i. Equation (13) indicates that if there is any k type product transported from pallet i to pallet j, then these two pallets are dependent on each other. Equation (14) ensures that if inbound pallet i and outbound pallet j are dependent with each other, there must be at least one product which is transported between them. Equation (15) shows that entering time of conveyor of k type outbound pallet j must be later than leaving time of inbound pallet i at handling area. Equation (16) shows that each k type outbound pallet j must wait until the previous pallet is fully filled. Equation (17) ensures that leaving time from the handling area of filled k type outbound pallet j cannot be greater than leaving time from handling area of filled or empty k type outbound pallet j. It also indicates that if the leaving time from handling area of filled k type outbound pallet j is positive, then the k type of outbound pallet j is filled with full capacity, and if k type outbound pallet j is not full, then the leaving time from handling area of filled k type outbound pallet j can take any value. Equation (18) shows that each outbound pallet firstly enters conveyors, then they start to be controlled. Equation (19) ensures that leaving time from the conveyor of k type outbound pallet j is equal to the sum of starting to control time of that pallet and controlling time of that pallet. Equation (20) shows that each k type outbound pallet j must be firstly controlled and then leaves to conveyor. Equation (21) satisfies that before k type outbound pallet j leaves from conveyor, it should be fully occupied with pallets. Equation (22) indicates the relationship between continuous time balance and discrete time balance of entering time and period to handling area of inbound pallet i. If entering time to the handling area of inbound pallet i is between t and $(t + 1)$, it means that inbound pallet i enters the handling area in period t. Equation (23) shows that each inbound pallet i can enter the handling area in one specific period. Equation (24) indicates the relationship between continuous time balance to discrete time balance of leaving time and period from handling area of k type outbound pallet j. If leaving time from the handling area of k type outbound pallet j is between t and $(t + 1)$, it means that k type outbound pallet j leaves from handling area in period t. Equation (25) indicates the relation between continuous time balance to discrete time balance of leaving time and period from conveyors of k type outbound pallet j. If leaving time from the conveyor of k type outbound pallet j is between t and $(t + 1)$, it means that k type outbound pallet j leaves from conveyor in period t. Equation (26) defines that boundary of continuous time balances matches with a maximum period number. Equation (27) shows that the number of pallets that comes in that period is equal to the difference between the number of waiting pallets in a period and the number of pallets handled in that period. Equation (28) shows that total number filled outbound pallet which enters to conveyor in period t is equal to sum of outbound pallets which leaves to handling in that period (with linearization equations). Equation (29) shows that if the number of pallets that comes to conveyor in that period is added to the amount of waiting pallets in that period in conveyor and subtract the total number of pallets controlled in that period, then the number of pallets at each conveyor in the next period is founded. Equation (30) shows that total number filled outbound pallet which leaves from conveyor in period t is equal to sum of outbound pallets which are controlled on that period.

Linearization of MINLP Model and Logical Sets

The non-linear equations are linearized in order to simplify the model and decrease the computational effort during solution processes. The following logical sets and equations are used for substituting the non-linear constraints. Therefore,

Equation (15) is removed and replaced with Eqs. (31)–(34).

$$L_{jk} \geq \gamma_{ijk}, \forall i \in I, j \in J, k \in K \quad (31)$$

$$\gamma_{ijk} \leq F_i, \forall i \in I, j \in J, k \in K \quad (32)$$

$$\gamma_{ijk} \leq M v_{ijk}, \forall i \in I, j \in J, k \in K \quad (33)$$

$$\gamma_{ijk} \geq F_i - M(1 - v_{ijk}), \forall i \in I, j \in J, k \in K \quad (34)$$

Equation (17) is removed and replaced with Eqs. (35)–(37).

$$\delta_{jk} \leq L_{jk}, \forall j \in J, k \in K \quad (35)$$

$$\delta_{jk} \leq My_{jk}, \forall j \in J, k \in K \quad (36)$$

$$\delta_{jk} \geq \delta_{jk} - M\left(1 - y_{jk}\right), \forall j \in J, k \in K \quad (37)$$

Equation (21) is removed and replaced with Eqs. (38)–(40).

$$\varphi_{jk} \leq ZT_{jk}, \forall j \in J, k \in K \quad (38)$$

$$\varphi_{jk} \leq My_{jk}, \forall j \in J, k \in K \quad (39)$$

$$\varphi_{jk} \geq ZT_{jk} - M\left(1 - y_{jk}\right), \forall j \in J, k \in K \quad (40)$$

Equation (22) is removed and replaced with Eqs. (41)–(45).

$$(C_i - t) - 1 - \in \leq M\left(1 - n_{1_{it}}\right), \forall i \in I, t \in T \quad (41)$$

$$(C_i - t) - 1 - \in \geq -Mn_{1_{it}}, \forall i \in I, t \in T \quad (42)$$

$$(C_i - t) + \in \geq -M\left(1 - n_{2_{it}}\right), \forall i \in I, t \in T \quad (43)$$

$$(C_i - t) + \in \leq Mn_{2_{it}}, \forall i \in I, t \in T \quad (44)$$

$$n_{it} = n_{1_{it}} + n_{2_{it}} - 1, \forall i \in I, t \in T \quad (45)$$

Equation (24) is removed and replaced with Eqs. (46)–(50).

$$\delta_{jk} - t - 1 - \in \leq M\left(1 - m_{1_{jkt}}\right), \forall j \in J, k \in K, t \in T \tag{46}$$

$$\delta_{jk} - t - 1 - \in \geq -Mm_{1_{jkt}}, \forall j \in J, k \in K, t \in T \tag{47}$$

$$\delta_{jk} - t + \in \geq -M\left(1 - m_{2_{jkt}}\right), \forall j \in J, k \in K, t \in T \tag{48}$$

$$\delta_{jk} - t + \in \leq Mm_{2_{jkt}}, \forall j \in J, k \in K, t \in T \tag{49}$$

$$m_{jkt} = m_{1_{jkt}} + m_{2_{jkt}} - 1, \forall j \in J, k \in K, t \in T \tag{50}$$

Equation (25) is removed and replaced with Eqs. (51)–(55).

$$\varphi_{jk} - t - 1 - \in \leq M(1 - g_{1_{jkt}}), \forall j \in J, k \in K, t \in T \tag{51}$$

$$\varphi_{jk} - t - 1 - \in \geq -Mg_{1_{jkt}}, \forall j \in J, k \in K, t \in T \tag{52}$$

$$\varphi_{jk} - t + \in \geq -M(1 - g_{2_{jkt}}), \forall j \in J, k \subset K, t \in T \tag{53}$$

$$\varphi_{jk} - t + \in \leq Mg_{2_{jkt}}, \forall j \in J, k \in K, t \in T \tag{54}$$

$$g_{jkt} = g_{1_{jkt}} + g_{2_{jkt}} - 1, \forall j \in J, k \in K, t \in T \tag{55}$$

Equation (28) is removed and replaced with Eqs. (56)–(59).

$$\pi_{jkt} \leq m_{jt}, \forall j \in J, k \in K, t \in T \tag{56}$$

$$\pi_{jkt} \leq My_{jk}, \forall j \in J, k \in K, t \in T \tag{57}$$

$$\pi_{jkt} \geq m_{jt} - M(1 - y_{jk}), \forall j \in J, k \in K, t \in T \tag{58}$$

$$SN_{kt} = \sum\nolimits_{j \in J} \pi_{jkt}, \forall j \in J, k \in K, t \in T \tag{59}$$

Equation (30) is removed and replaced with Eqs. (60)–(63).

$$\theta_{jkt} \leq g_{jt}, \forall j \in J, k \in K, t \in T \tag{60}$$

$$\theta_{jkt} \leq My_{jk}, \forall j \in J, k \in K, t \in T \tag{61}$$

$$\theta_{jkt} \geq g_{jt} - M\left(1 - y_{jk}\right), \forall j \in J, k \in K, t \in T \tag{62}$$

$$GN_{kt} = \sum\nolimits_{j \in J} \theta_{jkt}, \forall \in K, t \in T \tag{63}$$

As given in the model formulation section, the augmented epsilon method [31] is used to deal with bi-objective cases. Equation (1) is taken as the main objective function since cost assessments are considered primal activities under harsh competitive environments. Therefore, Eq. (2) is added as one of the constraints. The multiplication of slack variable (Ω) with some small epsilon value is subtracted from the cost function in this solution algorithm to deteriorate the objection function. However, the same slack variable is included into the Eq. (2) and the sum of the second objective function, and this slack variable should equal the upper limit of this objective function.

3 Computational Experiments

An illustrative example is carried out by using real-life data obtained from one of the largest electrical devices manufacturing companies in Turkey. The aim of this case study is to show details of a robust optimization approach and comprehensive calculations for sustainable cross-docking systems.

3.1 Data

The workflow in the warehouse and time balance of sorting operations in the warehouse are illustrated in Figs. 1 and 2, respectively. The goods arrive to warehouse $[KE_i]$ within pallets in a period $[RE_t]$, and they are sorted according to storage zones where they will be placed. The inbound pallets—unloaded from trucks and waits to be controlled and sorted—contain different types of items $[r_{ik}]$, while the outbound pallets only contain a single type of product, and they are sorted and controlled versions of the inbound pallets. When the trucks of suppliers arrive, the forklift operators unload the pallets in a shortest time and place them into the sortation area. After the inbound pallets are taken into the sorting area $[C_i]$, their barcodes are read, the items are controlled, and sorted at specific time $[U_k]$, according to their types before inbound pallet sortation is finalized $[F_i]$, according to total time they sorted $[H_i]$. The classified items are placed on the outbound pallets. While the operators organize outbound pallets, they do not consider using the maximum pallet capacity $[Cap_k]$. After the outbound pallets are organized $\left[L_{jk}\right]$, forklift operators place the

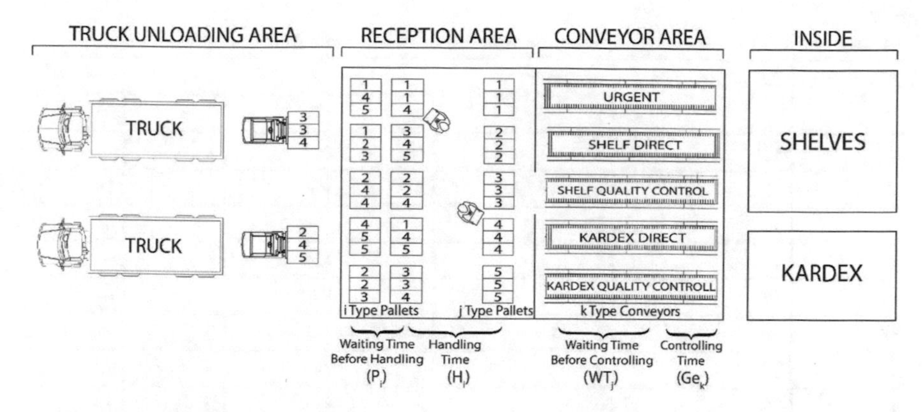

Fig. 1 Warehouse workflow scheme

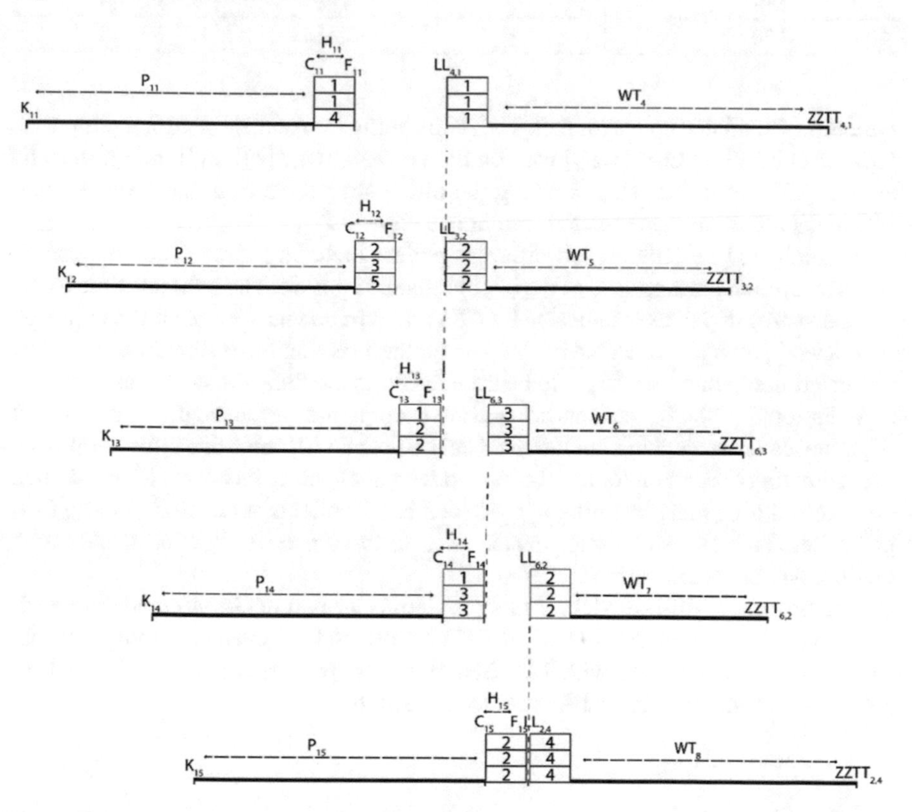

Fig. 2 Timeline of inbound and outbound pallets

Table 1 Product types (unit) and entering times to reception area (min) of inbound pallets

i	KE_i(min)	r_{ik} (units)				
		1	**2**	**3**	**4**	**5**
1	11	2	–	–	1	–
2	12	–	1	1	–	1
3	13	1	2	–	–	–
4	14	1	–	2	–	–
5	15	–	3	–	–	–
6	16	1	–	–	2	–
7	17	1	–	–	–	2
8	18	–	–	–	2	1
9	19	–	–	2	–	1
10	20	–	–	1	1	1

pallets on the related conveyor belt. The items going into quality control section with specific controlling time [GE_k] wait on the conveyor belt [P_i] until their turn to be started [ZK_{jk}]. The last step is taking the outbound pallets from the conveyor belt and replacing them to the related warehouse zone [ZT_{jk}].

As indicated in Table 1, ten different pallets including three different product types entered the reception area during the planned period. The entrance time to the reception area of pallets is indicated by KE_i, and type and number of products carried by relevant pallet are indicated by r_{ik}. The pallets entering the reception area will be separated firstly and then begin to be transferred to the final storage areas.

Additionally, some constants are used in order to assess the environmental effect of proposed cross-docking operations. Carrano et al. [10] use a prescribed approach for estimating the carbon footprint of a wooden pallet during its life cycle, and carbon emissions during manufacturing of a wooden block pallet is assumed as 3.47 kg CO_2 per pallet. This emission factor is also used in the objective function related with environmental considerations.

The proposed MILP model for this problem is written in GAMS modeling environment, solved with IBM ILOG CPLEX 12.1 [9], and executed on a computer with Intel Core I5 2520M CPU with 2.50 GHz dual core processor and with 4.00 GB of RAM. An optimality gap of 1% is set for the solutions.

4 Results

The key concepts and analysis of solutions on an illustrative case are discussed in this section, as well as specifics of the illustrative example.

An illustrative example of the mathematical model can be seen in Fig. 3. In this chart, the entire process and product flow are summarized and visualized with time

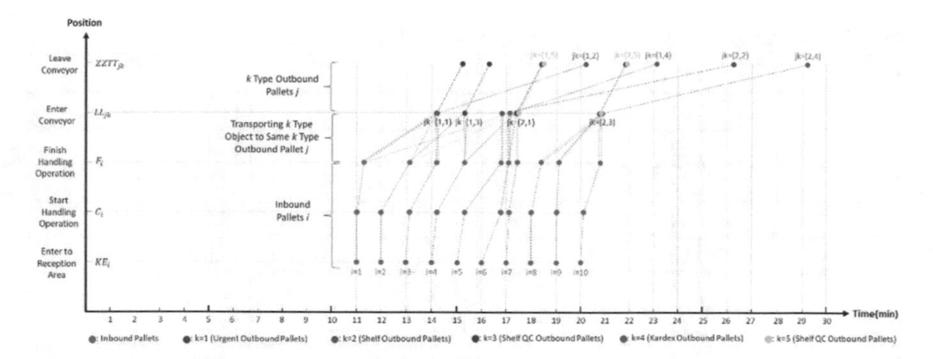

Fig. 3 Timeline of inbound and outbound pallets

indicators. Entering of inbound pallets to the reception area, from reception area to handling area, and leaving from handling area to final destinations are shown in Fig. 3.

The first inbound pallet enters the reception area at time KE_1 when $t = 11$, and it consists of two items with type $k = 1$ and one item with type $k = 4$. The inbound pallet waits in the reception area for a while, P_1 depending on the handling time of the predecessor inbound pallets. The inbound pallet enters the handling area at time C_1, when $t = 11$. The items with type $k = 1$ are put in the outbound pallet which consists of items with type $k = 1$, and items with type $k = 4$ are put in the outbound pallet which consists of items with type $k = 4$. The handling takes H_1=0.3 min. The same process is repeated for every pallet, considering every outbound pallet only contains single type items. An outbound pallet can leave the handling area when it reaches full capacity (3 items). Therefore, outbound pallet $j = 4$ leaves the handling area at time $\delta_{1,1} = 14.2$, after the 3rd inbound pallet is sorted, and another item with type $k = 1$ is placed on this outbound pallet. After the outbound pallet leaves the handling area, it is placed on the conveyor and waits to be placed on the related storage zone. The waiting time, $WT_{1,1}$, equals 1 min, and it depends on quality control time and the predecessor pallets in the conveyor. Finally, the outbound pallet leaves the conveyor at time $\varphi_{1,1} = 15.2$.

Entering time to reception area (C_i) and leaving time from reception area (F_i) for different inbound pallets are shown in Table 2.

Figure 4 shows how the products changed from inbound pallets to outbound ones on the material handling area and how they are classified. The arrows shown in bold indicate the final product that came to that outbound pallet as the last palette.

Tables 3 and 4 show when the outbound pallets entered the conveyors (δ_{jk}) and when they left from the conveyors (φ_{jk}).

The details of 50 different Pareto solutions of two objective functions obtained from proposed MILP model for decision makers by considering minimization of both total operational cost and carbon emission rates are given in Fig. 5.

Two solution sets are selected randomly (indicated with red markers in Fig. 5) and given in Table 5 in order to clarify how carbon footprint and cost values are

Table 2 Inbound pallets entering and leaving times of handling area (min)

i	KE_i	C_i	F_i
1	11	11	11.3
2	12	12	13.1
3	13	13.1	14.2
4	14	14.2	15.3
5	15	15.3	16.8
6	16	16.8	17.1
7	17	17.1	17.4
8	18	18	18.3
9	19	19	20.1
10	20	20.1	20.8

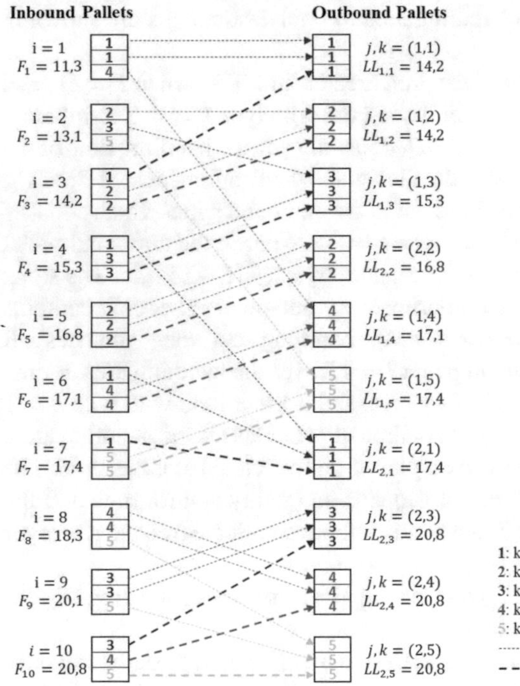

Fig. 4 Product transportation from inbound pallets to outbound pallets

Table 3 Conveyor entering times of *k* type outbound pallets *j* (min)

δ_{jk}	1	2	3	4	5
1	14.2	14.2	15.3	17.1	17.4
2	17.4	16.8	20.8	20.8	20.8

Table 4 Conveyor leaving times of k type outbound pallets j (min)

φ_{jk}	1	2	3	4	5
1	15.2	20.2	16.3	23.1	18.4
2	18.4	26.2	21.8	29.1	21.8

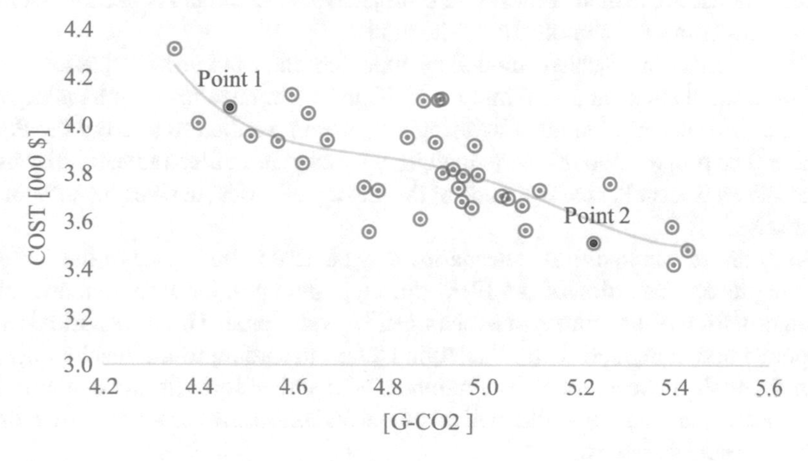

Fig. 5 Pareto solution set in terms of total cost and carbon emissions

Table 5 Alternative solutions of two objective functions

	$f_1(Cost)(000\ \$)$	$f_2(CarbonFootprint)$ (g-CO_2)
Single objective	4.1542	5.9
Bi-objective		
Point 1	4.0759	4.5
Point 2	3.5114	5.2

affected throughout the operations. Firstly, the model is run by including only a single objective function and the upper bounds of each objective function are found. Based on these values, the pareto frontier is found for the bi-objective model. As given in Table 5, while moving from "Point 1" to "Point 2," although the economic value of the system is reduced around 16%, carbon footprint is increased around 15%.

5 Conclusion

This study considers a decrease of material handling time in the reception area and minimizes total operational cost in warehouses as well as considering the sustainability perspective by including carbon footprint assessments. Extension of reception process lead-time (RPLT) due to bottleneck operations in material handling systems lead to accumulation of the received goods in reception area that affects production negatively and causes a delay in the entrance of products having priority into the warehouses. Therefore, a MINLP model for the bi-objective problem is introduced for cross-docking operations. After linearization of some constraints in the proposed MINLP model is satisfied, a proposed bi-objective MILP model is used to decrease computation time and complexity of the model.

We present a bi-objective modeling approach and outline linearization of the non-linear mathematical programming problem to minimize material handling cost and total carbon emission rates by using ε-constraint method to find out the Pareto frontier. The proposed solution methodology is tested in an illustrative case by using a real-life data set collected from one of the leading electrical devices manufacturing companies.

As future research, different scenarios may be tried in the layout of the sortation zone to prevent distortion of the FIFO during piling up in the reception area or an implementation of the conveyor system can be considered. These scenarios can be supported and compared with simulation studies according to the results obtained from the mathematical model. The numbers of electrical forklifts and operators can be considered as a part of mathematical model and illustrative cases can be extended with these new variables.

References

1. Abad HKE, Vahdani B, Sharifi M, Etebari F (2018) A bi-objective model for pickup and delivery pollution-routing problem with integration and consolidation shipments in the cross-docking system. J Clean Prod 193:784–801
2. Agustina D, Lee C, Piplani R (2014) Vehicle scheduling and routing at a cross docking center for food supply chains. Int J Prod Econ 152:29–41
3. Alamri AA, Syntetos AA (2018) Beyond LIFO and FIFO: exploring an allocation-in-fraction-out (AIFO) policy in a two-warehouse inventory model. Int J Prod Econ 206:33–45
4. Arabani AB, Ghomi SF, Zandieh M (2011) Meta-heuristics implementation for scheduling of trucks in a cross-docking system with temporary storage. Expert Syst Appl 38(3):1964–1979
5. Behnamian J, Fatemi Ghomi SMT, Jolai F, Telgerdi M (2018) Optimal design of cross docking supply chain networks with time-varying uncertain demands. J Ind Syst Eng 11(2):1–20
6. Birim Ş (2016) Vehicle routing problem with cross docking: a simulated annealing approach. Proc Soc Behav Sci 235:149–158
7. Boysen N, Fedtke S, Weidinger F (2018) Optimizing automated sorting in warehouses: The minimum order spread sequencing problem. Eur J Oper Res 270(1):386–400
8. Boysen N, Briskorn D, Fedtkea S, Schmickerathb M (2019) Automated sortation conveyors: a survey from an operational research perspective. Eur J Oper Res276(3):796–815
9. Cplex II (2009) 12.1. User's manual for CPLEX
10. Carrano AL, Thorn BK, Woltag H (2014) Characterizing the carbon footprint of wood pallet logistics. For Prod J 64(7–8):232–241. https://doi.org/10.13073/fpj-d-14-00011
11. Dulebenets MA (2019) A delayed start parallel evolutionary algorithm for just-in-time truck scheduling at a cross-docking facility. Int J Prod Econ 212:236–258
12. Fard SS, Vahdani B (2019) Assignment and scheduling trucks in cross-docking system with energy consumption consideration and trucks queuing. J Clean Prod 213:21–41

13. Gelareh S, Glover F, Guemri O, Hanafi S, Nduwayo P, Todosijević R (2020) A comparative study of formulations for a cross-dock door assignment problem. Omega 91:102015
14. Golshahi-Roudbaneh A, Hajiaghaei-Keshteli M, Paydar MM (2017) Developing a lower bound and strong heuristics for a truck scheduling problem in a cross-docking center. Knowl-Based Syst 129:17–38
15. Guemri O, Nduwayo P, Todosijević R, Hanafi S, Glover F (2019) Probabilistic Tabu search for the cross-docking assignment problem. Eur J Oper Res 277:875–885
16. Heragu SS, Du L, Mantel RJ, Schuur PC (2005) Mathematical model for warehouse design and product allocation. Int J Prod Res 43(2):327–338
17. Huang S, Wang Q, Batta R, Nagi R (2015) An integrated model for site selection and space determination of warehouses. Comput Oper Res 62:169–176
18. Iea. (n.d.). Global CO2 emissions in 2019—analysis. https://www.iea.org/articles/global-co2-emissions-in-2019. Accessed 19 May 2020
19. Khalili-Damghani K, Tavana M, Santos-Arteaga FJ, Ghanbarzad- Dashti M (2017) A customized genetic algorithm for solving multi-period cross-dock truck scheduling problems. Measurement 108:101–118
20. Ladier AL (2014) Scheduling cross-docking operations: integration of operational uncertainties and resource capacities (Doctoral dissertation), Université de Grenoble
21. Liao T, Egbelu P, Chang P (2013) Simultaneous dock assignment and sequencing of inbound trucks under a fixed outbound truck schedule in multi-door cross docking operations. Int J Prod Econ 141(1):212–229
22. Luo H, Yang X, Wang K (2019) Synchronized scheduling of make to order plant and cross-docking warehouse. Comput Ind Eng 138:106108
23. Manzini R, Bozer Y, Heragu S (2015) Decision models for the design, optimization and management of warehousing and material handling systems. Int J Prod Econ (170):711–716
24. Mital P, Goetschalckx M, Huang E (2015) Robust material handling system design with standard deviation, variance and downside risk as risk measures. Int J Prod Econ 170:815–824
25. Moeller K (2011) Increasing warehouse order picking performance by sequence optimization. Procedia-Social Behav Sci 20 (2011):177–185
26. Mohtashami A, Tavana M, Santos-Arteaga FJ, Fallahian-Najafabadi A (2015) A novel multi-objective meta-heuristic model for solving cross-docking scheduling problems. Appl Soft Comput 31:30–47
27. Molavi D, Shahmardan A, Sajadieh MS (2018) Truck scheduling in a cross docking systems with fixed due dates and shipment sorting. Comput Ind Eng 117:29–40
28. Nassief W, Contreras I, Jaumard B (2018) A comparison of formulations and relaxations for cross-dock door assignment problems. Comput Oper Res 94:76–88
29. Nikolopoulou AI, Repoussis PP, Tarantilis CD, Zachariadis EE (2017) Moving products between location pairs: cross-docking versus direct-shipping. Eur J Oper Res 256:803–819
30. Pianta M, Lucchese M (2020) Rethinking the European green deal: an industrial policy for a just transition in Europe. Rev Rad Pol Econ 52(4):633–641. https://doi.org/10.1177/0486613420938207
31. Resat HG, Turkay M (2015) Design and operation of intermodal transportation network in the Marmara region of Turkey. Transp Res Part E Logist Transp Rev 83:16–33
32. Rijal A, Bijvank M, Koster RD (2019) Integrated scheduling and assignment of trucks at unit-load cross-dock terminals with mixed service mode dock doors. Eur J Oper Res 278(3):752–771
33. Shahmardan A, Sajadieh MS (2020) Truck scheduling in a multi-door cross-docking center with partial unloading—reinforcement learning-based simulated annealing approaches. Comput Ind Eng 139:106134
34. Wang H, Alidaee B (2019) The multi-floor cross-dock door assignment problem: Rising challenges for the new trend in the logistics industry. Transp Res Part E Logist Transp Rev 132:30–47
35. Wisittipanich W, Hengmeechai P (2017) Truck scheduling in a multi-door cross docking terminal by modified particle swarm optimization. Comput Ind Eng 113:793–802

36. World Resources Institute (n.d.) Climate analysis indicators tool—report. https://www.climatewatchdata.org/ghg-emissions?end_year=2018&start_year=1990. Accessed 19 May 2020
37. Zhang J, Onal S, Das S (2019) The dynamic stocking location problem—dispersing inventory in fulfillment warehouses with explosive storage. Int J Prod Econ 224:107550
38. Zuluaga JPS, Thiell M, Perales RC (2017) Reverse cross-docking. Omega 66:48–57

Active Edible Packaging: A Sustainable Way to Deliver Functional Bioactive Compounds and Nutraceuticals

Anka Trajkovska Petkoska, Davor Daniloski, Nishant Kumar, Pratibha, and Anita T. Broach

Abstract Edible materials intended for packaging purposes could be used as carriers of bioactive and nutraceutical substances, and consequently to provide an active role in packaging. Active compounds, including a variety of antioxidants, antimicrobials, probiotics, or other functional ingredients incorporated in edible films and coatings could protect foods against deterioration during storage, reducing undesirable effects, extend their shelf life, improve the quality, as well as provide certain health benefits to consumers. Micro- or nano-encapsulation of bioactive compounds and nutraceuticals or other forms of inclusions of bioactives into edible materials aim to increase the stability of bioactive compounds, utilisation and customise their delivery mechanisms from the packaging to food item or human body, but also to extend their usage in a variety of food categories. Proper active inclusions in edible packaging can serve also as indicators, making a smart type of packaging; they could act on humidity, food changes, or deterioration in a timely manner. Applications of active edible packaging for a variety of food items remain still an open field for further investigation; it is a sustainable method of food preservation with environmental concerns for a less polluted world.

A. Trajkovska Petkoska (✉)
Faculty of Technology and Technical Sciences, St. Clement of Ohrid University of Bitola, Dimitar Vlahov, 1400 Veles, Republic of Macedonia
e-mail: anka.trajkovska@uklo.edu.mk

A. Trajkovska Petkoska · A. T. Broach
CSI: Create.Solve.Innovate. LLC, 2020 Kraft Dr., Suite 3007, Blacksburg, VA 24060, USA

D. Daniloski
Advanced Food Systems Research Unit, Institute for Sustainable Industries and Liveable Cities and College of Health and Biomedicine, Victoria University, Melbourne, VIC 8001, Australia

Food Chemistry and Technology Department, Teagasc Food Research Centre, Moorepark, Fermoy, Cork P61 C996, Ireland

N. Kumar
National Institute of Food Technology Entrepreneurship and Management, Kundli, Sonipat, Haryana 131028, India

Pratibha
National Institute of Technology, Kurukshetra, Haryana 136119, India

S. S. Muthu (eds.), *Sustainable Packaging*, Environmental Footprints and Eco-design of Products and Processes, https://doi.org/10.1007/978-981-16-4609-6_9

Keywords Active edible packaging · Intelligent packaging · Food preservation · Smart packaging · Sustainability

1 Introduction

Active edible packaging has gained a lot of attention in the past decade; the enhanced functional attributes obtained by incorporating natural ingredients, antioxidants, antimicrobials, colourants, flavourants, nutraceuticals in edible materials, are of special interest. They possess few to several functions; starting from those that packaging material should satisfy at least the food safety and improved quality to possible human health benefits, including delivery of vitamins, minerals, probiotics [33, 95, 131]. Moreover, intelligent packaging is also a raising trend in the packaging sector, there is a packaging solution that responds as a pH-indicator based on food changes, deterioration or so. The smart variants of active and intelligent packaging options usually have high sensitivity and safety attributes for packed food and users; they are with a low cost for real-time monitoring of the freshness of the food products, able to provide direct information without damaging the package integrity (e.g. by visual colour changes) [108, 151, 198].

This chapter presents active packaging based on edible materials. More particularly, bioactive additives to edible materials in context to create an active role of edible packaging (antimicrobial and antioxidant) and their application for different food categories all in context of sustainable packaging options are presented here.

2 Active and Intelligent Packaging Options

Active packaging is described as a shift of a passive packaging system towards an active role providing better food protection and safety, increased shelf life, without compromising on food quality. It serves as a medium or material that provides an interaction between the packaging, product, and the environment, including various physical, chemical, or biological activities that change the environment of the packed food [39, 111, 165]. According to European regulation (EC) No 450/2009, active packaging systems are defined as "*deliberately incorporate components that would release or absorb substances into or from the packaged food or the environment surrounding the food*" [41, 187].

Active packaging systems are attractive from an economic point of view. Namely, the spoilage of food products during processing, distribution and storage has a negative impact on the food industry. In addition, deterioration caused by the growth of microorganisms and lipid oxidation are the main problems in food losses, therefore to reduce these losses, extend the shelf life, improve the quality, and safety of food products, food engineers and scientists have designed different active packaging systems as an alternative to the traditional packaging options. The active packaging has the potential to control moisture content, delayed oxidation, controlled respiration rate, regulation of gas diffusion through the packages (oxygen, carbon dioxide, ethylene), absorption of oxygen, adding flavours, antioxidants, antimicrobial agents, etc. In general, the smart packaging systems are classified under two major categories depending upon their functional roles: compounds that absorb (scavengers) or discharge (emitters) gasses to transform the package's internal atmosphere actively [187].

In the past years, many advancements in active packaging technology are focussed on intelligent packaging (i.e. time–temperature indicators, humidity indicators, gas indicators, ripeness, biosensors, radiofrequency identification) that are responsible for the improvement in food quality, safety and shelf life [19, 85, 114, 175]. In general, smart packaging is a mutual term for active and intelligent materials, such types of packed food items could be found on the European market if they comply with the restrictions set out in Regulation (EC) 1935/2004 and the (EC) 450/2009. Smart packaging systems respond to environmental stimuli by repairing or alerting the consumer regarding the contamination or the presence of pathogens. Custom-made nanosensors are used for food analysis, detection of flavours or colours, drinking water, and clinical diagnosis. Application of nanosensors in food packaging aids in tracing the physical, chemical and biological modifications, customised designed nanodevices utilised in smart packaging could help in the detection of toxins, chemicals and food pathogens [111, 137].

The smart active packaging is able to protect the food products such as meat (beef, pork, poultry), seafood and fish products (shrimp, fish fillet) and dairy products (milk, cheese) [138, 169, 198]. Capello et al. [26] used anthocyanins extracted from jabuticaba fruit (*Plinia caulifora*) and purple sweet potato (*Ipomoea Batatas* L.) peels as a pH indicator to monitor the freshness of meat at different temperature storages. Also, the intelligent film based on gelatine/carrageenan matrix with functional fillers of propolis and shikonin has been investigated for the freshness of packaged milk. The functional fillers performed antimicrobial and antioxidant activity, their addition improved UV blocking property without disturbing the film transparency [151]. In another case, extracts of black chokeberry pomace mixed in chitosan have been used as dyes in intelligent films (pH indicators) in response to pH from 1 to 10 [80]. The anthocyanins extracted from propolis extract [108], black rye bran [181], red cabbage [100] and roselle calyxes [19] are used as antimicrobial agents and pH indicators in smart packaging options.

2.1 Antimicrobial Edible Packaging

Perishable foods like meat, dairy, fruits and vegetables are frequently spoiled by microbial growth and contamination; they can be protected by using antimicrobial packaging [17, 32, 36]. An antimicrobial agent has the ability to reduce, restrict or inhibit the growth of spoilage and pathogenic microorganisms in food products. Antimicrobial packaging could play a multifunctional role, specifically reduced harmful microbial activity in food, enhanced food safety, reduced food loss and improved food shelf life. In addition, biobased antimicrobial agents in packaging could provide extra safety for human health. Generally, the target microorganisms that cause food deterioration and loss are *Listeria monocytogenes, Pseudomonas aeroginosa, Escherichia coli O157, Salmonella, Staphylococcus aureus, Campylobacter, Clostridium perfringens, Aspergillus niger* [54, 150, 168].

According to Regulation No. 1333/2008, antimicrobials, or preservatives, are "*substances which prolong the shelf life of foods by protecting them against deterioration caused by micro-organisms and/or which protect against growth of pathogenic microorganisms*" [14].

The most used antimicrobials are essential oils (EOs) like basil, thyme, oregano, cinnamon, clove, rosemary, and then enzymes obtained from animal sources (lysozyme, lactoferrin), bacteriocins from microbial sources (nisin, natamycin), organic acids (sorbic, propionic, lauric, citric acid), or natural polymers (chitosan and its derivatives, alginates); all have been approved as food contact materials [8, 33]. For example, chitosan can inhibit the growth of bacteria (*Listeria monocytogenes, Bacillus cereus, Staphylococcus aureus, Escherichia coli, Salmonella typhimurium*), yeast and mould (*Candida lambica, Botrytis cinerea, Rhizoctonia solani, Phomopsis asparagi, Fusarium oxysporum*) [14]. Additionally, inorganic/metallic nanoparticles are used for the same purpose as well [111, 116]. Generally, the high surface energy of inorganic nanoparticles (Ag, Au, ZnO and TiO_2) allows them to act as antimicrobial agents—the mechanism of action is usually by disrupting the cell membrane. Nanoclays with large aspect ratio shapes, on the other hand, besides the main function to improve the gas and moisture barrier of polymers, are also acting as oxygen scavengers—it further lowers the survivability of microbes within the packaging, restricting access to oxygen or water. Table 1 reviews cases of antimicrobial edible packaging for different food categories.

2.2 Antioxidant Edible Packaging

Polyphenols, flavonoids, vitamins, poly-unsaturated fatty acids, curcumin, astaxanthin, resveratrol, and many others, perform antioxidant properties and are used in food science. According to Regulation No. 1333/2008, antioxidants are "*substances which prolong the shelf life of foods by protecting them against deterioration caused by oxidation, such as fat rancidity and colour changes*" [14]. They can act as chain-breaking antioxidants or as hydroperoxide-deactivating antioxidants. When antioxidant compounds are appropriately encapsulated they can provide a suitable effect through the edible packaging system. The main nanoencapsulation techniques applied to antioxidants are similar to those that are applied to antimicrobials such as: lipid-based nanoencapsulation techniques, encapsulation techniques based on

Table 1 Antimicrobial edible packaging applications on a variety of food products

Food product	Edible materials	Beneficial effect	References
Fruits and vegetables			
Cherry tomato	Chitosan with liposome encapsulation of Artemisia Annua oil	Inactivation of *Escherichia coli* O157:H7	Cui et al. [37]
Cherry tomato	Nanoemulsion-thymol-quinoa protein/chitosan	Inhibition of *Botrytis cinerea*	Robledo et al. [147]
Brocolli	Chitosan with different bioactive substances: tea tree, rosemary, pollen and propolis, pomegranate and resveratrol	Inhibitory effects on *Escherichia coli* and *Listeria monocytogenes*	Alvarez et al. [9]
Okra	Alginate coating with nanoemulsified basil (*Ocimum basilicum. L*) oil	Effective against spoilage fungi *Penicillium chrysogenum* and *Aspergillus flavus*	Gundewadi et al. [77]
Green beans	Chitosan with a nanoemulsion of mandarin EO	Reduction of *L. Innocua* over the examined storage time, owing to synergistic antimicrobial effects	Donsì et al. [53]
Blueberries	Gelatine with Persian gum	Inactivating and reducing murine norovirus	Sharif et al. [162]
Blueberries/raspberries	Carrageenan and green tea extract	Antiviral activity against murine norovirus and hepatitis A virus	Falcó et al. [60]
Strawberries/ raspberries	Coatings based on alginate-oleic acid with green tea	Antiviral effect against foodborne pathogens	Falcó et al. [60]
Strawberry	Banana starch-chitosan and Aloe Vera (AV) gel composition	Reduced fungal decay and increased shelf life up to 15 days during storage (20% AV). Maintainance of physicochemical properties, reduced weight loss and limited water vapour transfer	Pinzon et al. [132]
Strawberry (fresh cut)	Gellan-based coatings (Gel) with Geraniol (G) and Pomegranate Extract (PE)	Gel + G significantly reduced microbial counts. PE incorporation did not control microbial growth. Samples with coatings + G showed a better firmness loss than control	Tomadoni et al. [172]

(continued)

Table 1 (continued)

Food product	Edible materials	Beneficial effect	References
'Chandler' strawberries	Six types of coatings based on methyl cellulose solution with curcumin and limonene	Limonene liposomes showed significantly lower fungal growth compared to the control (on the 14th day of storage). Titratable acidity and total phenolic contents were found to be higher in limonene-coated samples compared to other coatings	Dhital et al. [45]
Fuji apples (fresh cut)	Nanoemulsion-based coatings with lemongrass EO	Inactivation of *Escherichia coli.*	Salvia-Trujillo et al. [156]
Cantaloupe (fresh cut)	Coatings based on chitosan, pectin, and *trans*-cinnamaldehyde	Extension of shelf life to 7–9 days compared to control (4 days, at 4 °C). The coating with chitosan (alone) was effective only against yeast and moulds at a low concentration, but coatings with encapsulated *trans*-cinnamaldehyde are effective against mesophilic microorganisms	Martiñon et al. [104]
Meat, fish and their products			
Sliced ham	Oregano EO (OEO) incorporated in Na-alginate films	Reduction of *Listeria* population. Presence of OEO resulted in colour differences (decrease of the product's quality, but improved aroma)	Pavli et al. [124]
Sliced cooked ham	PLA containing cellulose nanocrystals with nisin	Inhibited capacity of *Listeria monocytogenes,* better physicochemical and structural properties (stored for 14 days, 4 °C)	Salmieri et al. [155]
Salami	Whey protein isolate (WPI) films with cassava starch plant extracts and addition of ethanolic fractions of rambutan peel extract and cinnamon oil	WPI films performed low in vitro antibacterial activity, but the highest efficacy of delayed microbial growth	Chollakup et al. [34]

(continued)

Table 1 (continued)

Food product	Edible materials	Beneficial effect	References
Lamb	Whey protein isolate/cellulose nanofibre films with 1.0% TiO_2 and 2.0% rosemary EO	Inhibition of gram-positive greater than of gram-negative bacteria; increased shelf life of treated meat (15 days) compared to the control (6 days)	Sani et al. [159]
Turkey deli meat	Pullulan films with AgNPs, ZnONPs, oregano oil (2% OR) and rosemary oil (2% RO)	AgNPs and OR edible films were more active than ZnONPs and RO. AgNPs, ZnONPs, OR and RO films exhibited antibacterial activity against pathogens: *Listeria monocytogenes* and *Staphylococcus aureus*	Khalaf et al. [92]
Turkey deli meat	Chitosan, lauric arginate ester (LAE) and nisin	Reduced *Listeria innocua*. Combining antimicrobial coatings/films with flash pasteurisation further reduced *L. Innocua*	Guo et al. [78]
Fresh cut turkey pieces	Mixture of whey proteins and chitosan, with the addition of cranberry or quince juice	Microbiological deterioration and the development of *S. typhimurium, Escherichia coli* O157:H7, and *Campylobacter jejuni* in coated pieces is stopped for at least 6 days	Brink et al. [24]
Harbin red sausage	Chitosan (CTS) coatings (0, 1, 2, and 3%)	Storage stability was improved with an increase in the concentration of CTS coating. Inhibited microbial growth—total aerobic bacteria and lactic acid bacteria	Dong et al. [52]
Ham/bologna	Apple, carrot and hibiscus-based films with carvacrol and cinnamaldehyde as antimicrobials	Antimicrobial effect against *Listeria monocytogenes.* Carvacrol films showed better antimicrobial activity than cinnamaldehyde films; films were more effective on ham than on bologna	Ravishankar et al. [144]

(continued)

Table 1 (continued)

Food product	Edible materials	Beneficial effect	References
Fresh beef	Whey protein film with 1—2.5% EOs: cinnamon, cumin, thyme	All EOs reduced viable bacterial count (thyme performed the greatest antimicrobial activity)	Dohhi [50]
Beef steak	Whey protein concentrate with casein hydrolysate	Increased antimicrobial properties of the coating and decreased growth of coliform bacteria	Haque et al. [81]
Chicken thigh meat	Whey protein/alginate coating with different loadings of lactoperoxidase system (LPOS)	Increased antibacterial effect against *Enterobacteriaceae* spp. with increasing the LPOS load	Molayi et al. [106]
Sea bass fillets	A combination of liquid smoke and thymol encapsulated in chitosan nanofibers	Delayed growth of total mesophilic aerobic bacteria, psychrophilic bacteria, yeast and mould during the storage period	Ceylan et al. [30]
Rainbow trout	Whey protein + lactoperoxidase system (LPOS) against gram-negative bacteria	Reduction of total specific spoilage organisms *Shewanella putrefaciens* and *Pseudomonas fluorescens* of the fillets more than 1.5 logs by the end of storage	Shokri et al. [167]
Rainbow trout fillets	Whey protein isolate coating with EOs (ginger and chamomile)	Bacterial growth was inhibited in samples with high EOs concentrations	Yıldız and Yangılar [189]
Tuna (*Thunnus obesus*) chunks	Whey protein isolate (WPI) in MAP or vacuum-packed samples	WPI (with 8% glycerol) in MAP presented the lowest bacterial counts, thiobarbituric acid and total volatile basic-nitrogen contents	Xie et al. [182]
Dairy products			
Cheddar cheese	Chitosan-coated nisin-silica liposome	Sustained antibacterial activity against *L. Monocytogenes* without affecting the sensory properties of samples	Cui et al. [36]
Nabulsi cheese	Chitosan-and bitter vetch protein-based films	Effective at hindering microorganism growth in wrapped unsalted cheese. Maintained pH of the fresh product during storage	Sabbah et al. [154]

(continued)

Table 1 (continued)

Food product	Edible materials	Beneficial effect	References
Cheese	Coatings based on: sodium alginate, sodium alginate + *Lactobacillus acidophilus* and sodium alginate + *Lactobacillus helveticus*	The *L. acidophilus* and *L. helveticus* inclusion in coatings reduced the presence of the total coliform (at 10 days). Probiotic bacteria reduced coating flexibility. *Lactobacillus helveticus* diffuse to the cheese interior	Olivo et al. [115]
Acid-Curd Cheese (quark)	Furcellaran (FUR) and whey protein isolate (WPI) films containing pu-erh (PE) or green tea (GT) extracts	The FUR/WPI with GT revealed bacteriostatic effect against *Staphylococcus aureus.* The number of *Lactococcus* and total bacteria count decreased in almost all samples during storage except for FUR/WPI + PE The yeast count increased during storage in samples regardless of packaging used. Most of the examined films had a negative influence on organoleptic quality of cheese	Pluta-Kubica et al. [133]
Ricotta cheese	Chitosan/Whey protein (CWP) films	Lower O_2 and CO_2 permeability: CWP versus chitosan, but higher water vapour permeability of CWP versus chitosan film. Lower amount of lactic acid bacteria, mesophilic and psychotropic microorganisms. Extended shelf life of CWP versus chitosan film	Di Pierro et al. [47]
Bakery			
Bakery products	Antimicrobial activity of EOs: thyme, cinnamon, oregano, and lemongrass	Inhibit the growth of harmful microorganisms. Extended shelf life and enhanced safety	Gavahian et al. [69]

(continued)

Table 1 (continued)

Food product	Edible materials	Beneficial effect	References
Sliced wheat bread	Chitosan-carboxymethyl cellulose-oleic acid (CMC-CH-OL) incorporated with different concentrations (0.5, 1 and 2%) of ZnONPs	Increased microbial shelf life from 3 to 35 days for CMC-CH-OL-ZnO NPs (2%) compared to the control. Reduced the number of yeasts and moulds (over 15 days); improved antimicrobial properties for coatings contains 1% and 2% ZnO NPs with no fungal growth. Better maintenance of moisture content	Noshirvani et al. [113]

biologically derived polymeric nanocarriers, encapsulation techniques based on non-biological polymeric nanocarriers, electrospraying and electrospinning, nanocomposite encapsulation, etc., some of them can be followed by freeze-drying or spray-drying [64, 175].

Some examples are given in continuation; e.g. a case of anthocyanin-rich bayberry extract (with antioxidant property) in cassava starch was used to develop food packaging films with antioxidant and pH-sensitive properties; namely, this film has presented colour changes when exposed to hydrogen chloride and ammonia gases [196]. On the other hand, the effects of green tea and black tea extracts on the physical, structural and their antioxidant properties of chitosan films have been investigated by Peng et al. [127], while *Curcuma longa* L. rhizomes that contain curcuminoid pigments (curcumin)—which are phenolic compounds in gelatine-based films has provided the antioxidant behaviour towards the packed food. The antioxidant capacity of these films increased with increasing concentrations of curcuma extract [22]. Furthermore, a multicomponent film based on sodium caseinate, pectin, Zedo gum, and Poulk extract was used to produce a bioactive edible film, Poulk extract concentration had a significant effect on the antioxidant activity of these films as well [151]. The modified whey protein concentrate and isolate films with melanin (isolated from watermelon, *Citrullus lanatus* and seeds), also exhibited high antioxidant activity [101]. Table 2 presents more examples of antioxidant and combined edible packaging for a variety of food products.

2.3 Combined and Other Types of Active Edible Packaging

There is an extended list of reports for both antimicrobial and antioxidant properties of the same active packaging system [186, 195]. The film's mechanical properties of cassava-starch-based packaging have been examined after its enrichment

Table 2 Antioxidant and combined edible packaging applications on a variety of food products

Food product	Edible materials	Beneficial effect	References
Fruits and vegetables			
Tomato	Aloe vera based coating	Delayed ripening and extended shelf life (up to 39 days) compared to control sample (19 days)	Athmaselvi et al. [13]
Cucumber	Corn starch and mint (*Mentha viridis L.*) extract	Enhancement of shelf life and quality stored at room /low temperature (25 °C/10 °C)	Raghav and Saini [142]
Carrots	Protein, polyalcohol and polysaccharide coatings	Extended shelf life	Villafañe [177]
Strawberry	Prickly pear cactus mucilage (*Opuntia ficus indica*)	Extended shelf life	Del-Valle et al. [44]
Lime fruit	Pectin-based coating	Increase in shrivelling/wilting and loss in green colour during storage time. The changes are accelerated at increased temperatures	Maftoonazad and Ramaswamy [103]
Banana	Rice-starch coating blended with sucrose esters	Extended postharvest quality during ripening (at $20 \pm 2\ ^{0}C$); effective in delaying ethylene biosynthesis and reducing respiration rate. Shelf life prolonged for 12 days	Thakur et al. [171]
Apple cv. Golab kohanz	Nanochitosan coating	Significantly reduced weight loss, respiration rate, ethylene production and peroxidase activity. Softening process is slowed down; improved the flesh colour after the climacteric peak	Gardesh et al. [68]

(continued)

Table 2 (continued)

Food product	Edible materials	Beneficial effect	References
Fresh cut apples	Whey protein nanofibrils (WPNF)—coatings plasticised with glycerol (Gly) and trehalose (Tre)	Formation of WPNF increased surface smoothness, homogeneity, continuity, hydrophobicity and transparency, and decreased the moisture content and water solubility of the films. Coatings with WPNF (5%), glycerol (4%) and trehalose (3%) provided the best protection towards retarding the total phenolic content, browning and weight loss	Feng et al. [62]
Apple and potato slices	Calcium caseinate (CC) and whey protein (WP) solutions	WP showed better antioxidant capacity than CC. The addition of carboxymethyl cellulose to the formulations improved their antioxidative power (delayed browning; best scavenging of oxygen free radicals and ROS were found for films based on WP and carboxymethyl cellulose)	Yousuf et al. [192]
Papaya (fresh cut)	Psyllium gum with /without sunflower oil	Quality maintenance and shelf-life improvement	Yousuf and Srivastava [193]
Apricot	Chitosan coatings	Increased content of total phenolics and antioxidant activity	Ghasemnezhad et al. [71]
Kiwifruit slices	*Opuntia ficus-indica* mucilage coating	Higher firmness and lower weight loss for treated versus untreated slices	Allegra et al. [6]

(continued)

Table 2 (continued)

Food product	Edible materials	Beneficial effect	References
Grape berry	Coatings based on carnauba wax-lemongrass oil nanoemulsions	Effective at reducing weight loss, firmness, phenolic compounds and antioxidant activity. Inhibition of *Salmonella typhimurium* and *Escherichia coli.*	Kim et al. [94]
Tomato Bell pepper Litchi	Chitosan: pullulan composite enriched with pomegranate peel extract	Improved the shelf life of fruit and vegetables by maintaining higher phenolic, antioxidant and sensorial attributes	Kumar et al. [96, 97]
Meat, fish and their products			
Sliced ham	Na-alginate films with probiotic bacteria	Probiotic bacteria successfully delivered into the product by the edible film	Pandhi et al. [122]
Muscle foods	Collagen films	Reduced moisture loss, minimised lipid oxidation, prevented discolouration and reduced dripping	Pandhi et al. [122]
Fresh pork	Chitosan-gelatine coatings with grape seed extract and/or nisin	Effectively inhibited meat oxidation and microbial spoilage; grape seed extract further enhanced antioxidant activity against meat oxidation. Nisin in the coating did not further improve the antimicrobial and antioxidant effects	Xiong et al. [183]

(continued)

Table 2 (continued)

Food product	Edible materials	Beneficial effect	References
Fresh pork loin	Oregano EO and resveratrol nanoemulsion in pectin coating	Significantly prolonged the shelf life of pork by minimising the pH and colour change, retarding lipid and protein oxidation, maintaining meat tenderness and inhibiting microbial growth	Xiong et al. [184]
Lamb meat	Nano-encapsulated *Satureja khuzestanica* EO (SKEO) in chitosan coatings	Delayed microbial growth and chemical spoilage. Encapsulation decelerated release of SKEO—it led to prolonged antimicrobial and antioxidant activity as well as improved sensory attributes	Pabast et al. [120]
Lamb	Cellulose nanofiber/Whey protein matrix containing TiO_2 particles (1%) and rosemary EO (2%)	Reduced microbial growth, lipid oxidation and lipolysis of the lamb meat. Increased the shelf life of treated meat (15 days) compared to the control (6 days)	Alizadeh-Sani et al. [5]
Beef meat	Antioxidants present in olives, hydroxytyrosol and 3,4-dihydroxyphenylglycol were added to a pectin-fish gelatin film. A composite film with beeswax is also compared	Film with antioxidants reduced the formation of oxidation products compared to control films (without antioxidants). Suppressed lipid oxidation. The combined effect of acting as an oxygen barrier and the specific antioxidant activity of beeswax-based film and lipid oxidation was suppressed (stored for 7 days at 4 °C)	Bermúdez-Oria et al. [18]

(continued)

Table 2 (continued)

Food product	Edible materials	Beneficial effect	References
Chilled beef	WPNF-based edible coatings with glycerol (Gly) and TiO_2 nanotubes (TNTs) as antimicrobial agents	TNTs showed greater antibacterial activity than TiO_2NPs. WPNF/TNT coatings improved lipid peroxidation and antioxidant activity, limited microbial growth, reduced weight loss, and extended the shelf life of the beef	Feng et al. [61]
Beef steak/catfish fillet	Whey protein isolate (WPI) and casein hydrolysate; WPI and oolong tea extract; WPI with both casein hydrolysate and oolong tea	Substantial reduction of protein oxidation of tested samples in different packaging compositions	Mukherjee and Haque [107]
Beef muscle slices	Milk protein films with 1.0% oregano, 1.0% pimento, or 1.0% oregano-pimento (1:1) EO mix	Stabilisation of lipid oxidation (pimento was with the highest antioxidant activity). Bacteria count decreased by using oregano-based films (0.95 log reduction of *Pseudomonas spp.* and 1.12 log reduction of *Escherichia coli* O157:H7)	Oussalah et al. [119]
Salami	Whey protein concentrate (WPC) films with a blend of *Cinnamomum cassia, Cinnamomum zeylanicum*, and *Rosmarinus officinalis* EOs at different loadings	WPC films incorporated with EOs retard lipid oxidation induced by UV light in food	Ribeiro-Santos et al. [145]

(continued)

Table 2 (continued)

Food product	Edible materials	Beneficial effect	References
Portuguese sausages	Whey protein concentrate coating containing *Origanum virens* EO	Coated paínhos: higher acidity and protection against colour fading. Coated alheiras: significant reduction of the lipid peroxidation. Inhibition of the total microbial load for both coated sausages. Shelf-life extension: for paínhos (20 days) and alheiras (15 days)	Catarino et al. [29]
Rainbow trout (*Oncorhynchus mykiss*)	Whey protein concentrate (WPC) coatings with or without glycerol	Prolonged shelf life: 9–15 days depending on the WPC compositions	Yıldız and Yangılar [188]
Shrimps	Bilayer films based on agar and Na-alginate with cinnamon oil	Antioxidant activity; effective against photobacterium phosphoreum	Arancibia et al. [12]
Dairy products			
Cheddar cheese	Guar/Tragacanth-gum-based coating	Moisture decrease and protein contents increase in the samples during ripening (90 days). pH, acidity, fat in dry matter, tyrosine and tryptophan contents of samples significantly improved with these coatings	Pourmolaie et al. [135]
Bakery, nuts and powdered products			
Muffins	Triticale-based film/coating	Retard the staling process	Pandhi et al. [122]

(continued)

Table 2 (continued)

Food product	Edible materials	Beneficial effect	References
Fruit bars	Films based on sodium alginate, carboxymethyl cellulose and whey protein isolate	No significant effect on chemical properties. Maintained textural properties, limited moisture loss; loss of total phenolic content and radical scavenging activity during the storage was prevented	Eyiz et al. [58]
Fruit bars	Layer-by-layer deposition of the polycation chitosan and the polyanion alginate with ascorbic acid	Increased ascorbic acid content, antioxidant capacity, firmness and fungal growth prevention during storage period	Bilbao-Sainz et al. [21]
Bread (mini-burger buns)	Sodium alginate, pectin, whey protein concentrate used as coatings	Drying behaviour of coatings on bread surfaces	Chakravartula et al. [31]
Fresh walnut kernels	Chitosan with green tea extract	Antioxidant and antifungal effects: significant inhibition of lipid oxidation and fungal growth	Sabaghi et al. [153]
Roasted peanuts (RP)	Carboxymethyl cellulose (RP-CMC), methyl cellulose (RP-MC) and whey protein (RP-WPI) coatings	The stability of RP-CMC is twice as long compared to RP. Chemical indicator values and intensity ratings of oxidised and cardboard flavors had lower increase during storage	Riveros et al. [146]

(continued)

Table 2 (continued)

Food product	Edible materials	Beneficial effect	References
Roasted peanuts (RP)	Coatings: Whey protein isolate 11%, corn protein (zein) 15%, and whey protein isolate with carboxymethyl cellulose (CMC) 0.5%	The coating in combination with ultrasonication treatment was an effective method in delaying the formation of oxidative volatile compounds and hence inhibiting rancidity of RP compared to uncoated ones	Wambura and Yang [179]
Peanuts	Aqueous whey protein isolate (WPI) solution	WPI delayed oxygen uptake of dry roasted peanuts at intermediate (53%) and low (21%) storage relative humidity. Extended shelf life	Shendurse et al. [166]
Sunflower seed kernels	Whey protein concentrate/glycerol/carboxymethylcellulose/rosemary extract	The coating was adjusted to provide the kernels desirable colour and oxidative stability properties	Hosseini et al. [83]
Dried pistachio kernels	Whey protein concentrate (WPC)/glycerol solutions	The control samples showed a delay in oxidation. Regardless of composition or thickness, WPC coatings provided a glossy appearance	Javanmard [87]

with propolis extract. As a result of the propolis's phenolic compounds, the film presented antioxidant (Artepelin C—the antioxidant component in the propolis) and antimicrobial activity against *Stapylococus aureus* (more effective) and *Escherichia coli* (less effective) [38]. Multifunctional carboxymethyl cellulose/agar-based smart films were examined by combining cellulose nanocrystals separated from onion peel and shikonin (isolated from the roots of *Lithospermum erythrorhizon*), the composite film has performed both antimicrobial and antioxidant activity, but it also enhanced physical and functional properties. The addition of shikonin has provided functional properties like pH-responsive colour change, without significant change in the water vapor permeability and thermal stability of the film [149]. Another case has reported quercetin in three different polymers, carboxymethyl cellulose (CMC), gelatine, and PLA, which have exhibited antioxidant and antimicrobial behaviour. More specifically, the CMC/quercetin and gelatine/quercetin films showed significant inhibition function on *Listeria monocytogenes* and *Escherichia coli* [59]. Examples of combined (antioxidant and antimicrobial) edible packaging are given in Table 2.

3 Health Benefits of Active Edible Packaging

Increased consumer interest in health, nutrition, food safety, and eco-friendly materials has led to improve research and development of biobased films and coatings intended for edible packaging. The concept of active edible packaging is a relatively novel concept—the edible materials as a vehicle for active compounds can contribute to consumers' health benefits besides the basic benefits that the packaging should provide for packed foodstuffs. The health-promoting substances involved in edible matrices could be vitamins, minerals, probiotics, phytochemicals, among others [27, 105, 114, 125], more particularly, anthocyanins, betalains, lycopene, carotenoids, chlorophylls, etc. are bioactives that are responsible for the human health. In some cases, there is a synergistic relationship amongst bioactive compounds reported by many authors that are good for the prevention and treatment of some diseases, such as oxidative stress related diseases, immunity enhancers, antimicrobial or anticancer [28, 64, 164].

3.1 Probiotic, Prebiotics and Synbiotics

Probiotics are bioactive compounds that could be added to edible materials and perform specific health benefits. According to the definition given by the FAO/WHO (2002), they are "*live microorganisms which, when administered in adequate amounts, confer a health benefit on the host*". Particularly, the main genera of probiotic micro-organisms preferred by the food industry are *Lactobacillus* and *Bifidobacterium.* In this context, LAB of the genus *Lactobacillus* has been studied for their probiotic properties, preventing the deterioration of the microbiota as well as the inhibition of pathogenic microorganisms at the oral cavity and colon [48, 82, 143]. Some of the reported beneficial effects of probiotics in human health include anti-cancer, anti-allergic, anti-diabetic, anti-obesity, anti-pathogenic, immunomodulatory and anti-inflammatory activities [55, 161], but also antimicrobial effects of probiotics incorporated in edible materials have been reported [125, 134, 163].

However, the application of probiotics in food products is not an easy procedure, since these microorganisms might lose their viability before consumption (e.g. during processing, storage or during digestion) [123]. Namely, for health improving purposes, it is required that a minimum of intestinal content of probiotic bacteria are present in the final food product,e.g. a dose of 10^{8-9} viable cells of probiotics per day is recommended [134].

The use of edible materials as carriers of living microorganisms is reported by many authors. Namely, the entrapment of probiotic cells in edible materials is a favourable way to overcome the limitations related to the use of probiotics in food products. Also, the encapsulation methods of probiotics are of high importance due to their sensibility to environmental conditions, low stability in the food processing steps/storage (loss of viability), or in the GIT. Survival of probiotics depends on the strain, food characteristics, processing technologies, and storage conditions (e.g. temperature, time), to name a few [57, 75, 82, 123]. The encapsulation of probiotics in edible matrices that favour their optimal survival could be obtained by spray drying, spray freeze-drying /electrospray and cross-linking gelation [134].

In the same context, the intake of prebiotics is being considered as a healthy habit, too; this group of ingredients (inulin, lactulose, fructooligosaccharides, galactooligosaccharides or the human milk oligosaccharides) has shown properties that contribute to the well-being of the consumers. A *prebiotic* is defined as "*a non-digestible food ingredient that beneficially affects the host by selectively stimulating the growth and/or activity of one or a limited number of bacteria in the colon, and thus improves host health*" [63, 125, 143]. Their beneficial effects include control of cholesterol, inhibition of colitis, protection of colon and other organs against cancer, improvement of mineral bioavailability, constipation relief or treatment of diarrhoea and allergies, potentiation of immune defence, reduction of cardiovascular disease risk factors and antioxidant activity [123, 55].

Recently, the positive effects of the combination of probiotics and prebiotics on human health more particularly favouring beneficial microbes in the human gut, and their possible synergistic interactions, the development of suitable encapsulation systems for controlled and targeted release has been reported as well [123]. Namely, the term "*synbiotics*" is used for formulation composed of at least one probiotic microorganism and one prebiotic substance, "synbiotics" alludes to synergism—it should be reserved for products in which the prebiotic compound selectively favours the probiotic compound. Furthermore, it may help maintaining the cell viability of probiotics and incorporated within the structure of edible films/coatings, it was found to provide effective probiotic protection [134].

It is known that the synergistic combination of prebiotics with probiotic strains promotes colonisation in the intestinal tract as well as preventing several forms of cancer. For instance, the chemical structure of some oligosaccharides makes them resistant to digestive enzymes and thus, they are able to reach the large intestine where they become available for fermentation by saccharolytic bacteria [125]. Therefore, the synbiotic agents have beneficial effects on host welfare through the enhancement of activity and survival of beneficial microorganisms in the GIT, so that they can selectively provoke the growth and stimulate the metabolism of one or more health-promoting bacteria [55].

There are still some major challenges that should be overcome in order to achieve a wider industrial application of probiotics and prebiotics and their best combination with edible materials and technologies in order to tailor to specific foods or consumers' needs. Another challenge is maintaining a high cell density of probiotics during the formation process of edible films/coatings and after ingestion (still possess positive impact on human health). All these issues should be addressed in future research activities [125, 134, 173].

4 Application of Active Edible Packaging

The active edible packaging could play an important role as oxygen scavenger, moisture scavenger, antioxidant release, ethylene absorber, CO_2 emitter and/or antioxidant packaging system for different types of food products, including fruits, vegetables, meat and meat products, bakery products, etc. [187]. The application of active edible packaging reduced the undesirable effects and lipid peroxidation of food products by minimising the respiration rate, colour browning, maintain firmness, phenolic, antioxidant and sensory characteristics, while ensuring their safety, quality and integrity [121]. Figure 1 summarises the application of edible coatings and films as an active packaging on a variety of food items.

4.1 Fruits and Vegetables

Extensive research has been conducted on prolongation of the shelf life and improving the quality and safety of fruits and vegetables. Different types of edible films and coatings with addition of bioactive compounds have been reported. In most of the cases, chitosan has been chosen as a potentially viable alternative for fruit and vegetable preservation [2, 43, 96, 109, 141]. Multicomponent edible films and coatings also can be produced with suitable ingredients in order to provide desired barrier protection and to serve as a vehicle of bioactive substances that enhance the functionality, thus avoiding pathogen or foodborne microorganism growth on the surface of fruits and vegetables. At the same time, the combination of edible materials and minerals, vitamins or other nutraceutical compounds can reinforce the nutritional value of the commodities, without reducing the taste acceptability [4, 16]. There are many studies on protection and extended shelf life of particular types of fruits and vegetable, particularly grapes [158], guava [73], banana [180], mango [25], blackberries [148],

Fig. 1 Many variants of application of active edible packaging to different food categories

blueberries [1], raspberry [86], strawberry [45], fresh-cut pineapple [20], fresh figs [176], pear [65], plum [121], pomegranate [91], gooseberry [72], tomato [129], green bell pepper [97], okra [7], cherry tomatoes [158], cucumber [79], mushrooms [157] have been reported.

The compositions of alginate/chitosan-based films with functional additives phenolic, essential oils and nano-forms have been reported for prolonging the shelf life of fruits and vegetables. The functional molecules could play a synergistic role along with the chitosan and alginate-based edible coatings and retain the moisture, antioxidant potential, enhance the activity of antioxidant enzymes, reduces the activity of browning enzymes, and imparts antimicrobial properties in fresh fruits and vegetables [109]. A significant improvement in antimicrobial property of gelatine films against two food pathogens, *Escherichia coli* and *Staphylococcus aureus* was obtained in the presence of carboxymethyl cellulose (CMC) when was applied as edible packaging on cherry tomatoes (*Solanum lycopersicum var. cerasiforme*) and grapes (*Vitis vinifera*). The effectiveness of gelatine–CMC blend films to extend the shelf life of agricultural products was evaluated over a 14-day preservation study, as well as to control the weight loss and browning index of the tested fruits [158]. Similarly, CMC and pectin-based active coatings have also shown great potential, for both, protective effect and carrying functional compounds (antimicrobials, antioxidants, anti-browning agents, texture enhancers and nutraceuticals) and prevention of unwanted reactions in horticultural products [121]. Furthermore, curcumin and limonene have been used as natural antimicrobials, as they were prepared from their liposomes and were over-coated with methylcellulose. Based on the number of berries with visible mould, limonene liposomes showed significantly lower fungal growth compared to the control on the 14th day of storage. Titratable acidity and total phenolic contents were also found to be higher in limonene-coated strawberries compared to other performed coatings [45]. In another study, pear as a highly perishable climacteric fruit with a short shelf life affected by several microbial diseases, but with suitable edible coatings, they could be protected very successfully. As an example, the application of chitosan-*Ruta graveolens* EO coatings during 21 days of storage (at 18 °C) have performed microbiological protective capabilities. Aerobic mesophilic bacteria and moulds were significantly reduced without affecting consumer perception [128]. Additionally, "*Babughosha*" pear (*Pyrus communis* L.) fruit has been packed with edible coatings containing soy protein isolate in combination with additives hydroxypropyl methylcellulose and olive oil (stored at 28 ± 5 °C, 60 ± 10% RH) [42]. The examined coating could withhold the levels of ascorbic acid, chlorophyll and sugar content in the treated fruits. Activities of enzymes associated with fruit softening (b-galactosidase, polygalacturonase, pectin methyl esterase) have showed delayed peaks, consequently, there was an extension of the shelf life of pear fruits up to 15 days, compared to 8 days for untreated pear fruits [42]. Guava is another highly perishable fruit, generally attacked by pathogenic species like the fungi *Colletotrichum gloeosporioides*, which causes anthracnosis. To diminish the losses caused by pathogenic fungi, coatings of chitosan with Ruta graveolens EO have been applied in situ and their effects on the physical properties and microbiological quality of the guavas were studied. Better physicochemical behaviour and lower microbiological decay as compared to the uncoated ones were observed, the coated samples have shown a high percentage of inhibition in the development of anthracnose lesions and the proposed edible packaging could extend the stability of the guavas fruit up to 12 days [73]. Contrarily, the tomato (*Solanum lycopersicum* L.) has a shorter shelf life and postharvest losses affect its marketing. Chitosan-*Ruta graveolens* EO coatings on the postharvest quality of Tomato var. "chonto" stored at low temperature (4 °C) was tested over the period of 12 days. Aerobic mesophilic bacteria were significantly reduced, the coatings with a particular concentration of the EO completely inhibited the mould and yeast growth on tomato surfaces without negatively affecting the consumer acceptance [129]. Tomato [96], bell pepper [97] and litchi [98] fruits' and vegetables' shelf life were extended using composite edible coating functionalised by chitosan: pullulan with pomegranate peel extract by maintaining sensory attributes, firmness, antioxidant property, lowest weight loss [117]. In addition, no mould growth was observed on the sliced cherry tomatoes that were in direct contact with the films during 7 days of storage, proving the promising application of the films as active food packaging materials [117]. The use of quinoa protein, chitosan and sunflower oil as a coating of fresh blueberries has been shown as a promising packaging to control the growth of moulds and yeasts (32 days of storage), extending the shelf life colour and others during the storage period at room and cold storage conditions. Furthermore, corn starch/chitosan nanoparticles/thymol nanocomposite films were utilized for extending the shelf life of cherry tomatoes; the cherry tomatoes exhibited no significant changes in firmness and the lowest weight loss [117]. In addition, no mould growth was observed on the sliced cherry tomatoes that were in direct contact with the films during 7 days of storage, proving the promising application of the films as active food packaging materials [117]. The use of quinoa protein, chitosan, and sunflower oil as a coating of fresh blueberries has been shown as a promising packaging to control the growth of moulds and yeasts (32 days of storage), extending the shelf life of fresh blueberries stored at 4 0C and 75% RH [1]. The strawberry samples were coated with chitosan and licorice extract—the coating has maintained good quality parameters for treated samples during storage and showed good microbiological preservation in comparison with controls [141]. More examples of edible packaging of food and vegetables are reviewed in Tables 1 and 2.

Bioprotection is another very promising approach for microbiological quality and safety for postharvest storage of raw or minimally processed fruits and vegetables; namely, the use of *Lactic acid bacteria (LAB)* in these food categories helps to maintain their quality and extending the shelf life by causing a significant reduction or inhibition of foodborne pathogens. The antibacterial effect of *LAB* is owing to its ability to produce antimicrobial compounds (bacteriocins). The use of bacteriocins is considered a very suitable and promising approach for microbiological quality, and safety for postharvest storage of fruits and vegetables [4].

4.2 Dairy Products

Edible packaging could serve as a control of respiratory rate and the prevention of spoilage, reduction of weight loss in the preservation of dairy products' quality and safety. Recently, its use for cheese preservation is becoming more exploited and resulted in the commercialisation of some edible coatings and films [35]. The fungal microbiota that usually grows on the cheese surface during ripening processes promotes rind formation and the development of organoleptic characteristics, imparting positive sensory attributes to cheeses [10, 133]. In addition, undesirable moulds may occur as cheese contamination. The investigation has revealed that the main components of rind fungal communities of Tuscan pecorino cheese were *P. solitum, P. discolour* and *P. verrucosum*, to name a few. For their prevention in dairy industries usually antibiotic natamycin [3] is used, which may represent a risk factor for human health and environmental sustainability. However, agro-industrial by-products with natural antimicrobial properties, i.e. tannins and chitosan, were tested in a cheese-making trial producing Tuscan pecorino cheese. They did not significantly affect the number and composition of fungal communities developed during Pecorino Toscano cheese ripening, as well as its physical, chemical and nutritional profiles, showing that they may represent effective alternatives to the antibiotic natamycin [3]. Another case of antimicrobial edible coatings based on chitosan, sodium alginate and carboxymethyl cellulose with probiotic strains (*Bifdobacterium lactis, Lactobacillus acidophilus* and *Lactobacillus casei*) was conducted for preservations of soft cheese for a period of 45 days, instead of using traditional plastic packaging. Also, the probiotic counts in tested films were more than 8.0 log CFU/g after 45 days that is satisfactory [57]. Anti-listeria effects of a chitosan-coated nisin-silica liposome on Cheddar cheese have been examined as well [36]. Two biodegradable zein-based blend coatings were evaluated on the impact on the quality of the "Minas Padrao" cheese in a storage period of 56 days [126]. Namely, cheese samples with edible coatings exhibited ca. 30% lower weight loss and avoided microbiological contamination for more than 50 days, when compared to unpackaged cheese samples that exhibited contamination after 21 days [126]. Chitosan/PVA/TiO2-NPs were evaluated as packaging for soft white cheese, the bionanocomposites have exhibited superior antibacterial activity against gram-positive (*Staphylococcus aureus*), gram-negative (*Pseudomonas aeruginosa, Escherichia coli*) bacteria, and fungi (*Candidia albicans*). These results have indicated that the total bacterial counts, mould, and yeast and coliform decreased with the increasing storage period and disappeared at the end of the storage period compared to control samples (stored at 7 °C for 30 days) [191]. In addition, gelatin/carrageenan/propolis/shikonin composite film showed potent antimicrobial and antioxidant activity, namely [151] have proposed the intelligent film that can effectively monitor the freshness of packaged milk. This smart film can serve as a freshness indicator to extend the food shelf life of dairy products [151]. Tables 1 and 2 review applications of different edible packaging on dairy products.

4.3 Meat, Fish and Derived Products

Meat, fish and their products are perishable food items; they can spoil under improper storage conditions. Thus, the application of active and intelligent packaging systems in muscle-based food products, which are prone to contamination, has attracted big attention in the past years. The aim of packaging meat and muscle products is to suppress spoilage, bypass contamination, enhance the tenderness (by allowing enzymatic activity), decrease weight loss, and to preserve the cherry red colour in red meat products [40, 111].

Edible films and coatings with active components are a good option for the preservation of this food category. They can be used for preventing moisture loss, delaying microbial spoilage, restricting the growth of pathogenic microorganisms, slowing lipid, protein and pigment oxidation, and at the same time to extend the shelf life. Active compounds (with antioxidant and antimicrobial properties, probiotic, prebiotic, flavourants) added in edible matrices can improve the sensory and quality characteristics of packaged meat products too [46, 51, 174, 194]. In general, many cases of preservation with edible packaging are reported in the literature, particularly for chicken [67], cooked ham [11], beef steaks [102], beef patties [178], fish fillets [152], white shrimp [93], salmon products [89], etc. and are also presented in Tables 1 and 2.

Alginate-based edible films containing natural antioxidants from pineapple peel have been applied in the microbial spoilage control, colour preservation, and barrier to lipid oxidation of beef steaks (stored at 4 °C for 5 days) [102]. Bioactive films exhibited higher antioxidant activity than the alginate film without the bioactive substances. Namely, the results have shown that control films without active compounds had no significant effect on decreasing the microbial load of aerobic mesophilic and *Pseudomonas spp.*, while the films containing encapsulated hydro-alcoholic extract showed a significant inhibitory effect on microbial growth of meat over 2 days of storage, they were effective also for maintaining the colour hue and intensity of red beef meat samples. Consequently, pineapple peel antioxidants have the potential to retard lipid oxidation in meat samples [102]. Another study has been conducted on chitosan-based edible films/basil EO and characterised for their physicochemical and biological properties. These films have been tested as packaging for wrapping cooked ham samples during 10 days of storage. It was demonstrated that the active film can both control the bacterial growth of the cooked ham and markedly inhibit the pH increase of the packaged food, and therefore can extend the shelf life of packed meat products [11]. Feng et al. [61] have reported that whey protein nanofibrils/TiO_2 nanotubes coatings can retard lipid peroxidation, improve antioxidant activity, limit microbial growth, reduce weight loss, and extend the shelf life of chilled beef. These compositions are cheap and perform a high antioxidant and antimicrobial activity, they can be used for various food products, including raw and chilled meat [61]. Moreover, Villasante et al. [178] have investigated the use of gelatin films with incorporated bioactive compounds from the walnut/walnut shell—they have shown good protection against the oxidation of beef patties. The presence of *Thymus vulgaris L.* EO in edible films has reduced yeast populations, whereas aerobic mesophilic bacteria, lactic acid bacteria and enterobacteria were not affected by its presence in the films. In addition, the presence of the chitosan-thyme EO layer reduced water condensation inside the package, whereas packages containing only chitosan had evident water droplets, and thyme odour was perceived as desirable in cooked meat [140].

In addition, there are also studies conducted on edible packaging on a variety of fish and fillets [99], tilapia (*Oreoschromis niloticus*) fillets [152]. For instance, vanillin and chitosan coating has been used in combination for regulating the microbiota composition and shelf life of turbot (*Scophthalmus maximus*) fillets. After treatment with this composite coating, the relative abundance of *Pseudomonadaceae* and *Lactobacillaceae* have decreased significantly due to the growth inhibition of potential bacteria, spoilage bacteria, along with the rich bacterial diversity at the end of the storage period [99]. Another study was conducted on the guar gum coating with incorporated thyme oil on the quality of tilapia fish fillets (15 days of storage, at 4 °C), as a means to extend shelf life. The microbiological analyses have demonstrated that there was greater microbial growth in the uncoated fillets than in the coated ones, therefore, the bioactive coating with thyme oil retards microbial colonisation of fish [152]. There are also studies conducted on chitosan coatings enriched with propolis extract, it has performed an extension of shelf life of refrigerated *Nemipterus japonicus* fillets by effective retardation of the growth of total mesophilic bacteria and psychrotrophic bacteria [56]. Furthermore, Tapilatu et al. [170] have treated yellowfin tuna with chitosan in form of nanoparticle—nanochitosan, they have significantly suppressed the bacteria activity (at 28 °C). The preservation of fresh salmon fillet at cold storage (4 °C for 15 days) has been also performed by the combination of gelatine, chitosan, gallic acid (antioxidant effect) and clove oil (antimicrobial effect). The authors have shown a prolonged shelf life of the salmon fillet for at least 5 days [185].

The novel trend of application of nanosensors for this food category provides food spoilage or contamination to alarm the consumers by detecting toxins, pesticides, and microbial contamination based on flavour production or colour formation. Most of the nanoparticles used for packaging for this purpose have also had the potential of antimicrobial activity—they may act as carriers for antimicrobial polypeptides and may provide protection against microbial spoilage [111, 136].

4.4 Bakery, Nuts and Powdered Materials

Numerous studies have reported the practical applications of edible materials on bakery products [15, 23, 76, 134]. There is also a report of the use of Mediterranean herbal-based films/coatings with volatile EO (e.g. from oregano) as an active component that could preserve bread and sunflower grains. It has been shown to have a good control of the growth of fungi in bread loaves without any sensory changes, or flavour corrector (bread crust flavour) was developed to cover the off-odour from EO and reduce its sensory impacts [130]. Examples of edible coatings with antimicrobial and antioxidant properties for bakery products and nuts are given in Tables 1 and 2.

Furthermore, the food that is sold as a powder and needs to be solubilised before consumption in particular coffee, cappuccino, powdered milk, spices, tea, dehydrated fruits and vegetables is of great interest for a wide population from the aspect of proper packaging and consumption [139], edible films can be used for this purpose as well. Due to many advantages of inulin (e.g. as fibre, carbohydrate and fat replacement), it could be also used for packaging of powdered products. Incorporation of inulin with edible packaging materials improved the appearance, mechanical properties and solubility of the materials—it could replace foil for freshness due to the self-adhesive nature, gloss, transparency, homogeneity, well-defined margins in edible packaging. It is a valuable ingredient for the production of biobased edible materials intended for food packaging [139].

5 Safety, Environmental and Economic Concerns of Active Edible Packaging

5.1 Safety Concerns of Active Edible Packaging

Food safety is one of the major global health problems. Although science and technology have developed to considerable levels, food contamination is a very serious issue, since foodborne diseases are still widespread [90]. Utilisation of natural biomaterials and bioactive ingredients in edible packaging may be less toxic and have greater consumer acceptance, they have become more attractive in different food sectors, not just packaging [33]. They are eco-friendly alternatives to petroleum-based materials, nevertheless, the safety of using biobased packaging applications is starting to be questionable. Hence, insufficient scientific research has been conducted in order to measure the risks related to the occurrence of small particles, such as nanomaterials, that are biologically effective upon ingestion. Nano-based food packaging displays different activities from the same substances at the macro or micro-level, they exhibit specific biokinetic attributes due to their small size and large surface area and can be easily transported throughout the human body. Therefore, they exhibit certain activities and interactions through the cell membranes, bioabsorption and migration. A number of authors studied the migration of nanoparticles from the edible packaging of the food, however, there are still concerns about their health issues (the level of passage from the film to the packaged food, absorption in the body and/or human organs, cell infiltration in the body, digestion) [32, 111, 199]. These constant anxieties of the utilisation of nanomaterials in edible packaging are critical to determining the optimum levels of functional nanoparticles that can be safely applied in biomaterials without affecting human health. Moreover, understanding the way the nanomaterials drive performance during their pass from food to humans and their possible presence in the human body is crucial [95, 131, 190].

The regulatory bodies such as Food and Drug Administration (FDA) and European Union (EU) require more information to develop and implement safety regulations/policies for the application of nanomaterials in food contact packaging. Biobased plastic materials are also subjected to these regulations and amendments may include specifications regarding biobased materials, such as Regulation (EC) no 2019/1338 [112]. According to the "Plastic Food Contact Materials" Regulation (EU) 10/2011, only nanoparticles authorised and specifically mentioned in the specification of Annex I of the regulation can be used in plastic packaging for food. This also applies to nanoparticles that are intended to be used behind a functional barrier [160]. However, there is an immediate requirement for a well-established international organisation to oversee and regulate the use of nanomaterials in the food sector [111].

On the other hand, the general requirements stated in Regulation 1935/2004/EC for the safe use of active and intelligent packaging have been integrated by Regulation 450/2009/EC; it has been stated that active substances should be either contained in separate containers (e.g. in sachet forms) or directly be incorporated in the packaging material (e.g. oxygen-absorbing films) [32, 48].

In addition, there is another important concern that should be clarified within the regulatory policies; many edible materials are made with components that can cause allergic reactions in some consumers to like, milk, soybeans, fish, peanuts, wheat are well-known allergens. Therefore, edible packaging materials that contain allergen must also be clearly declared to the consumer [66].

Finally, the consumers' knowledge about the practical implementation, packaging requirements and additional environmental impacts are limited; consumers' behaviour is less sustainable than intended. Notably, awareness training based on scientific facts, clear product and packaging information based on labelling schemes (eco-labelling) and nudging for sustainable behaviour can potentially support consumers in their sustainable buying behaviour for the global well-being [118].

5.2 Economic Sustainability of Edible Packaging

In the past decades, a number of scientists have been working on eco-friendly solutions of bioplastics and their applications in the packaging sector; a part of the economic sustainability of biobased active packaging carries a bright future for their environmental friendliness. However, there are no precise statistics that will compare fossil-based with biocomposite materials intended for food packaging, some studies indicated that biobased packaging made from residues is 3–5 times more expensive compared to fossil-based packaging [32]. Another study explains that the cost is an important driving factor for the customer acceptance of edible films; currently, their cost is high, approx. 10–50 times higher than the petroleum-derived plastic films. As the production of edible films is in the development phase, the high cost of edible films cannot be taken as a negative point at this moment; thus, their cost should be lower than or equal to the petroleum-derived plastics [88]. Moreover, regulations offered by FDA and EU have limitations in the way of commercial production of biobased food packaging materials in terms of the implementation of standards, standards on biobased materials and food contacts, explicitly in labels and unavailability of proper waste management facilities. To overcome most of the limitations, policy support should be carefully examined and made as a function of sustainable development [32, 88].

5.3 Environmental Concerns of Edible Packaging

In relation to sustainability and the circular economy, waste management and recyclability of polymer (nano)composite materials is also of high importance. Mechanical recycling of the plastic composite materials has been reported; however, the data are limited for recycling of biobased packaging. According to Zhao et al. [197], Guillard et al. [74] only 10–14% is recycled, far below the global recycling rates for paper (58%) or iron/steel (70–90%), while 10% of plastics are incinerated, and ≈ 80% are landfilled, most of them still finishes in landfill zones and leaks into natural systems e.g. oceans [74, 160]. Therefore, by 2020 over 6 billion metric tonnes of plastic waste in total has accumulated worldwide, causing great environmental concerns according to study [197].

The concepts of recycling, reduction and reuse of food packaging could be reconsidered seriously while keeping the major functions of the packaging. Namely, packaging design can significantly contribute to this concept, if the end-of-life step is already considered during the development phase of the packaging option. Additionally, sustainable food packaging in the circular economy may only be achieved by combining efforts, and involving all stakeholders, food and packaging manufacturers, recycling companies, decision-makers, civil society, as well as consumers [49, 70, 74, 84, 110]. More specifically, by 2050, the global society has to be focussed on production of at least 50% of the food packaging materials from renewable, non-food resources by using up-cycling of organic wastes, while the rest 50% fossil-based materials being closed-loop recycled. The biobased packaging should be fully biodegradable and compostable solving the current issues of plastic waste accumulation in line with the EU circular economy strategy [74].

References

1. Abugoch James L, Tapia C, Plasencia D, Pastor A, Castro Mandujano O, López L, Escalona Contreras V (2016) Shelf-life of fresh blueberries coated with quinoa protein/chitosan/sunflower oil edible film. J Sci Food Agric 96: 619–626
2. Adiletta G, di Matteo M, Petriccione M (2021) Multifunctional role of chitosan edible coatings on antioxidant systems in fruit crops: a review. Int J Mol Sci 22:2633
3. Agnolucci M, Daghio M, Mannelli F, Secci G, Cristani C, Palla M, Giannerini F, Giovannetti M, Buccioni A (2020) Use of chitosan and tannins as alternatives to antibiotics to control mold growth on PDO Pecorino Toscano cheese rind. Food Microbiol 92:103598
4. Agriopoulou S, Stamatelopoulou E, Sachadyn-Król M, Varzakas T (2020) Lactic acid bacteria as antibacterial agents to extend the shelf life of fresh and minimally processed fruits and vegetables: quality and safety aspects. Microorganisms 8:952
5. Alizadeh-Sani M, Mohammadian E, Mcclements DJ (2020) Eco-friendly active packaging consisting of nanostructured biopolymer matrix reinforced with TiO_2 and essential oil: application for preservation of refrigerated meat. Food Chem 1–9
6. Allegra A, Sortino G, Inglese P, Settanni L, Todaro A, Gallotta A (2017) The effectiveness of *Opuntia ficus-indica* mucilage edible coating on post-harvest maintenance of 'Dottato' fig (*Ficus carica* L.) fruit. Food Packag Shelf Life 12:135–141
7. Aloui H, Jguirim N, Khwaldia K (2017) Effects of biopolymer-based active coatings on postharvest quality of okra pods in Tunisia. Fruits 72:363–369
8. Aloui H, Khwaldia K (2016) Natural antimicrobial edible coatings for microbial safety and food quality enhancement. Compr Rev Food Sci Food Saf 15:1080–1103
9. Alvarez MV, Ponce AG, Moreira MDR (2013) Antimicrobial efficiency of chitosan coating enriched with bioactive compounds to improve the safety of fresh cut broccoli. LWT Food Sci Technol 50:78–87

10. Amariei S, Norocel L, Gutt G (2020) New edible packaging material with function in shelf life extension: applications for the meat and cheese industries. Foods 9:562
11. Amor G, Sabbah M, Caputo L, Idbella M, de Feo V, Porta R, Fechtali T, Mauriello G (2021) Basil essential oil: composition, antimicrobial properties, and microencapsulation to produce active chitosan films for food packaging. Foods 10:121
12. Arancibia M, Giménez B, López-Caballero M, Gómez-Guillén M, Montero P (2014) Release of cinnamon essential oil from polysaccharide bilayer films and its use for microbial growth inhibition in chilled shrimps. LWT-Food Sci Technol 59:989–995
13. Athmaselvi K, Sumitha P, Revathy B (2013) Development of aloe vera based edible coating for tomato. Int Agrophysics 27:369–375
14. BARBOSA, C. H., ANDRADE, M. A., VILARINHO, F., FERNANDO, A. L. & SILVA, A. S. 2021. Active edible packaging. Encyclopedia 1:360-370s
15. Bartolozzo J, Borneo R, Aguirre A (2016) Effect of triticale-based edible coating on muffin quality maintenance during storage. J Food Meas Charact 10:88–95
16. Basumatary IB, Mukherjee A, Katiyar V, Kumar S (2020) Biopolymer-based nanocomposite films and coatings: recent advances in shelf-life improvement of fruits and vegetables. Critical Rev Food Sci Nutr 1–24
17. Batiha GE-S, Hussein DE, Algammal AM, George TT, Jeandet P, Al-Snafi AE, Tiwari A, Pagnossa JP, Lima CM, Thorat ND (2021) Application of natural antimicrobials in food preservation: recent views. Food Control 108066
18. Bermúdez-Oria A, Rodríguez-Gutiérrez G, Rubio-Senent F, Fernández-Prior Á, Fernández-Bolaños J (2019) Effect of edible pectin-fish gelatin films containing the olive antioxidants hydroxytyrosol and 3,4-dihydroxyphenylglycol on beef meat during refrigerated storage. Meat Sci 148:213–218
19. Bhargava N, Sharanagat VS, Mor RS, Kumar K (2020) Active and intelligent biodegradable packaging films using food and food waste-derived bioactive compounds: a review. Trends Food Sci Technol 105:385–401
20. Bierhals VS, Chiumarelli M, Hubinger MD (2011) Effect of cassava starch coating on quality and shelf life of fresh-cut pineapple (*Ananas Comosus* L. *Merril* cv "*Pérola*"). J Food Sci 76:62–72
21. Bilbao-Sainz C, Chiou BS, Punotai K, Olson D, Williams T, Wood D, Rodov V, Poverenov E, McHugh T (2018) Layer-by-layer alginate and fungal chitosan based edible coatings applied to fruit bars. J Food Sci 83:1880–1887
22. Bitencourt C, Fávaro-Trindade C, Sobral PDA, Carvalho R (2014) Gelatin-based films additivated with curcuma ethanol extract: antioxidant activity and physical properties of films. Food Hydrocoll 40:145–152
23. Bravin B, Peressini D, Sensidoni A (2006) Development and application of polysaccharide–lipid edible coating to extend shelf-life of dry bakery products. J Food Eng 76:280–290
24. Brink I, Šipailienė A, Leskauskaitė D (2019) Antimicrobial properties of chitosan and whey protein films applied on fresh cut turkey pieces. Int J Biol Macromol 130:810–817
25. Cai C, Ma R, Duan M, Deng Y, Liu T, Lu D (2020) Effect of starch film containing thyme essential oil microcapsules on physicochemical activity of mango. *LWT* 131:109700
26. Capello C, Trevisol TC, Pelicioli J, Terrazas MB, Monteiro AR, Valencia GA (2020) Preparation and characterization of colorimetric indicator films based on chitosan/polyvinyl alcohol and anthocyanins from agri-food wastes. J Polym Environ 1–14
27. Cardona F, Andrés-Lacueva C, Tulipani S, Tinahones FJ, Queipo-Ortuño MI (2013) Benefits of polyphenols on gut microbiota and implications in human health. J Nutr Biochem 24:1415–1422
28. Carvalho APAD, Conte-Junior CA (2021) Health benefits of phytochemicals from Brazilian native foods and plants: antioxidant, antimicrobial, anti-cancer, and risk factors of metabolic/endocrine disorders control. Trends Food Sci Technol 111:534–548
29. Catarino MD, Alves-Silva JM, Fernandes RP, Gonçalves MJ, Salgueiro LR, Henriques MF, Cardoso SM (2017) Development and performance of whey protein active coatings with *Origanum virens* essential oils in the quality and shelf life improvement of processed meat products. Food Control 80:273–280

30. Ceylan Z, Unal Sengor GF, Yilmaz MT (2018) Nanoencapsulation of liquid smoke/thymol combination in chitosan nanofibers to delay microbiological spoilage of sea bass (*Dicentrarchus labrax*) fillets. J Food Eng 229:43-49
31. Chakravartula SSN, Soccio M, Lotti N, Balestra F, Dalla Rosa M, Siracusa V (2019) Characterization of composite edible films based on pectin/alginate/whey protein concentrate. Materials 12:2454
32. Chawla R, Sivakumar S, Kaur H (2021) Antimicrobial edible films in food packaging: Current scenario and recent nanotechnological advancements-a review. Carbohydr Polym Technol Appl 2:100024
33. Chen W, Ma S, Wang Q, Mcclements DJ, Liu X, Ngai T, Liu F (2021) Fortification of edible films with bioactive agents: a review of their formation, properties, and application in food preservation. Critical Rev Food Sci Nutr 1–27
34. Chollakup R, Pongburoos S, Boonsong W, Khanoonkon N, Kongsin K, Sothornvit R, Sukyai P, Sukatta U, Harnkarnsujarit N (2020) Antioxidant and antibacterial activities of cassava starch and whey protein blend films containing rambutan peel extract and cinnamon oil for active packaging. LWT-Food Sci Technol 1–10
35. Costa MJ, Maciel LC, Teixeira JA, Vicente AA, Cerqueira MA (2018) Use of edible films and coatings in cheese preservation: opportunities and challenges. Food Res Int 107:84–92
36. Cui H, Wu J, Li C, Lin L (2016) Anti-listeria effects of chitosan-coated nisin-silica liposome on Cheddar cheese. J Dairy Sci 99:8598–8606
37. Cui H, Yuan L, Li W, Lin L (2017) Edible film incorporated with chitosan and Artemisia annua oil nanoliposomes for inactivation of *Escherichia coli O157: H7* on cherry tomato. Int J Food Sci Technol 52:687–698
38. Cunha GF, Soares JC, de Sousa TL, Buranelo M (2020) Cassava-starch-based films supplemented with propolis extract: physical, chemical, and microstructure characterization. Biointerface Res Appl Chem 11:12149–12158
39. Daniloski D, Gjorgjijoski D, Petkoska AT (2020) Advances in active packaging: perspectives in packaging of meat and dairy products. Adv Mater Lett 11:1–10
40. Daniloski D, Petkoska AT, Galić K, Ščetar M, Kurek M, Vaskoska R, Kalevska T, Nedelkoska DN (2019) The effect of barrier properties of polymeric films on the shelf-life of vacuum packaged fresh pork meat. Meat Sci 158:107880
41. Daniloski D, Petkoska AT, Lee NA, Bekhit AE-D, Carne A, Vaskoska R, Vasiljevic T (2021) Active edible packaging based on milk proteins: a route to carry and deliver nutraceuticals. Trends Food Sci Technol 111:688–705
42. Dave RK, Rao TR, Nandane A (2017) Improvement of post-harvest quality of pear fruit with optimized composite edible coating formulations. J Food Sci Technol 54:3917–3927
43. Deb Majumder S, Sarathi Ganguly S (2020) Effect of a chitosan edible-coating enriched with citrus limon peel extracts and *Ocimum tenuiflorum* leaf extracts on the shelf-life of bananas. Biosurface Biotribology 6:124–128
44. Del-Valle V, Hernández-Muñoz P, Guarda A, Galotto M (2005) Development of a cactus-mucilage edible coating (*Opuntia ficus indica*) and its application to extend strawberry (*Fragaria ananassa*) shelf-life. Food Chem 91:751–756
45. Dhital R, Joshi P, Becerra-Mora N, Umagiliyage A, Chai T, Kohli P, Choudhary R (2017) Integrity of edible nano-coatings and its effects on quality of strawberries subjected to simulated in-transit vibrations. LWT 80:257–264
46. Dhumal CV, Sarkar P (2018) Composite edible films and coatings from food-grade biopolymers. J Food Sci Technol 55:4369–4383
47. di Pierro P, Sorrentino A, Mariniello L, Giosafatto CVL, Porta R (2011) Chitosan/whey protein film as active coating to extend *Ricotta* cheese shelf-life. LWT Food Sci Technol 44:2324–2327
48. Díaz-Montes E, Castro-Muñoz R (2021) Edible films and coatings as food-quality preservers: an overview. Foods 10:249
49. Dieterle M, Schäfer P, Viere T (2018) Life cycle gaps: interpreting LCA results with a circular economy mindset. Procedia CIRP 69:764–768

50. Dohhi E-BS (2014) Evaluation of the antimicrobial action of whey protein edible films incorporated with cinnamon, cumin and thyme against spoilage flora of fresh beef. Int J Agric Res 9:242–250
51. Domínguez R, Barba FJ, Gómez B, Putnik P, Kovačević DB, Pateiro M, Santos EM, Lorenzo JM (2018) Active packaging films with natural antioxidants to be used in meat industry: a review. Food Res Int 113:93–101
52. Dong C, Wang B, Li F, Zhong Q, Xia X, Kong B (2020) Effects of edible chitosan coating on Harbin red sausage storage stability at room temperature. Meat Sci 159:107919
53. Donsì F, Marchese E, Maresca P, Pataro G, Vu KD, Salmieri S, Lacroix M, Ferrari G (2015) Green beans preservation by combination of a modified chitosan based-coating containing nanoemulsion of mandarin essential oil with high pressure or pulsed light processing. Postharvest Biol Technol 106:21–32
54. Du WX, Olsen C, Avena-Bustillos R, McHugh T, Levin C, Mandrell R, Friedman M (2009) Antibacterial effects of allspice, garlic, and oregano essential oils in tomato films determined by overlay and vapor-phase methods. J Food Sci 74:390–397
55. Durazzo A, Nazhand A, Lucarini M, Atanasov AG, Souto EB, Novellino E, Capasso R, Santini A (2020) An updated overview on nanonutraceuticals: focus on nanoprebiotics and nanoprobiotics. Int J Mol Sci 21:2285
56. Ebadi Z, Khodanazary A, Hosseini SM, Zanguee N (2019) The shelf life extension of refrigerated Nemipterus japonicus fillets by chitosan coating incorporated with propolis extract. Int J Biol Macromol 139:94–102
57. El-Sayed HS, El-Sayed SM, Mabrouk AM, Nawwar GA, Youssef AM (2021) Development of eco-friendly probiotic edible coatings based on chitosan, alginate and carboxymethyl cellulose for improving the shelf life of uf soft cheese. J Polym Environ 1–13
58. Eyiz V, Tontul İ, Türker S (2020) The effect of edible coatings on physical and chemical characteristics of fruit bars. J Food Meas Charact 1–9
59. Ezati P, Rhim J-W (2021) Fabrication of quercetin-loaded biopolymer films as functional packaging materials. ACS Appl Polym Mater 3:2131–2137
60. Falcó I, Randazzo W, Sánchez G, López-Rubio A, Fabra MJ (2019) On the use of carrageenan matrices for the development of antiviral edible coatings of interest in berries. Food Hydrocoll 92:74–85
61. Feng Z, Li L, Wang Q, Wu G, Liu C, Jiang B, Xu J (2019) Effect of antioxidant and antimicrobial coating based on whey protein nanofibrils with TiO_2 nanotubes on the quality and shelf life of chilled meat. Int J Mol Sci 20:1184
62. Feng Z, Wu G, Liu C, Li D, Jiang B, Zhang X (2018) Edible coating based on whey protein isolate nanofibrils for antioxidation and inhibition of product browning. Food Hydrocoll 79:179–188
63. Fernandes LM, Guimarães JT, Pimentel TC, Esmerino EA, Freitas MQ, Carvalho CWP, Cruz AG, Silva MC (2020) Edible whey protein films and coatings added with prebiotic ingredients. Elsevier, Agri-food industry strategies for healthy diets and sustainability
64. Fleming E, Luo Y (2021) Co-delivery of synergistic antioxidants from food sources for the prevention of oxidative stress. J Agric Food Res 3:100107
65. Gago C, Antão R, Dores C, Guerreiro A, Miguel MG, Faleiro ML, Figueiredo AC, Antunes MD (2020) The effect of nanocoatings enriched with essential oils on 'rocha' pear long storage. Foods 9:240
66. Galus S, Arik Kibar EA, Gniewosz M, Kraśniewska K (2020) Novel materials in the preparation of edible films and coatings—a review. Coatings 10:674
67. Garavito J, Moncayo-Martínez D, Castellanos DA (2020) Evaluation of antimicrobial coatings on preservation and shelf life of fresh chicken breast fillets under cold storage. Foods 9:1203
68. Gardesh ASK, Badii F, Hashemi M, Ardakani AY, Maftoonazad N, Gorji AM (2016) Effect of nanochitosan based coating on climacteric behavior and postharvest shelf-life extension of apple cv. Golab Kohanz. LWT 70:33–40
69. Gavahian M, Chu Y-H, Lorenzo JM, Mousavi Khaneghah A, Barba FJ (2020) Essential oils as natural preservatives for bakery products: Understanding the mechanisms of action, recent findings, and applications. Critical Rev Food Sci Nutr 60:310–321

70. Geueke B, Groh K, Muncke J (2018) Food packaging in the circular economy: overview of chemical safety aspects for commonly used materials. J Clean Prod 193:491–505
71. Ghasemnezhad M, Shiri M, Sanavi M (2010) Effect of chitosan coatings on some quality indices of apricot (*Prunus armeniaca* L.) during cold storage. Casp J Environ Sci 8:25–33
72. González-Locarno M, Maza Pautt Y, Albis A, Florez Lopez E, Grande Tovar CD (2020) Assessment of chitosan-rue (*Ruta graveolens* L.) essential oil-based coatings on refrigerated cape gooseberry (*Physalis peruviana* L.) quality. Appl Sci 10:2684
73. Grande Tovar CD, Delgado-Ospina J, Navia Porras DP, Peralta-Ruiz Y, Cordero AP, Castro JI, Chaur Valencia MN, Mina JH, Chaves López C (2019) Colletotrichum gloesporioides inhibition in situ by chitosan-*Ruta graveolens* essential oil coatings: effect on microbiological, physicochemical, and organoleptic properties of guava (*Psidium guajava* L.) during room temperature storage. Biomolecules 9:399
74. Guillard V, Gaucel S, Fornaciari C, Angellier-Coussy H, Buche P, Gontard N (2018) The next generation of sustainable food packaging to preserve our environment in a circular economy context. Front Nutr 5:121
75. Guimaraes A, Abrunhosa L, Pastrana LM, Cerqueira MA (2018) Edible films and coatings as carriers of living microorganisms: a new strategy towards biopreservation and healthier foods. Compr Rev Food Sci Food Saf 17:594–614
76. Guldas M, Bayizit AA, Yilsay TO, Yilmaz L (2010) Effects of edible film coatings on shelf-life of mustafakemalpasa sweet, a cheese based dessert. J Food Sci Technol 47:476–481
77. Gundewadi G, Rudra SG, Sarkar DJ, Singh D (2018) Nanoemulsion based alginate organic coating for shelf life extension of okra. Food Packag Shelf Life 18:1–12
78. Guo M, Jin TZ, Wang L, Scullen OJ, Sommers CH (2014) Antimicrobial films and coatings for inactivation of *Listeria innocua* on ready-to-eat deli turkey meat. Food Control 40:64–70
79. Gutiérrez-Pacheco MM, Ortega-Ramírez LA, Silva-Espinoza BA, Cruz-Valenzuela MR, González-Aguilar GA, Lizardi-Mendoza J, Miranda R, Ayala-Zavala JF (2020) Individual and combined coatings of chitosan and carnauba wax with oregano essential oil to avoid water loss and microbial decay of fresh cucumber. Coatings 10:614
80. Halász K, Csóka L (2018) Black chokeberry (*Aronia melanocarpa*) pomace extract immobilized in chitosan for colorimetric pH indicator film application. Food Packag Shelf Life 16:185–193
81. Haque ZZ, Zhang Y, Mukherjee D (2016) Casein hydrolyzate augments antimicrobial and antioxidative persistence of Cheddar whey protein concentrate based edible coatings. Food Sci Technol 17:468–477
82. Hellebois 'T, Tsevdou M, Soukoulis C (2020) Functionalizing and bio-preserving processed food products via probiotic and synbiotic edible films and coatings. Probiotic Prebiotics Foods: Challs Innov Adv
83. Hosseini H, Hamgini EY, Jafari SM, Bolourian S (2020) Improving the oxidative stability of sunflower seed kernels by edible biopolymeric coatings loaded with rosemary extract. J Stored Prod Res 89:1–11
84. Hughes R (2017) The EU circular economy package–life cycle thinking to life cycle law? Procedia CIRP 61:10–16
85. Ingle A, Philippini R, Martiniano S, Antunes F, Rocha T, Da Silva S (2021) Application of microbial-synthesized nanoparticles in food industries. Microb Nanobiotechnology: Princ Appl
86. Ishkeh SR, Shirzad H, Asghari MR, Alirezalu A, Pateiro M, Lorenzo JM (2021) Effect of chitosan nanoemulsion on enhancing the phytochemical contents, health-promoting components, and shelf life of raspberry (*Rubus sanctus Schreber*). Appl Sci 11:2224
87. Javanmard M (2008) Shelf life of whey protein-coated pistachio kernel (*Pistacia vera* L.). J Food Process Eng 31:247–259
88. Jeya Jeevahan J, Chandrasekaran M, Venkatesan SP, Sriram V, Britto Joseph G, Mageshwaran G, Durairaj RB (2020) Scaling up difficulties and commercial aspects of edible films for food packaging: a review. Trends Food Sci Technol 100:210–222

89. Jonušaite K, Venskutonis PR, Martínez-Hernández GB, Taboada-Rodríguez A, Nieto G, López-Gómez A, Marín-Iniesta F (2021) Antioxidant and antimicrobial effect of plant essential oils and sambucus nigra extract in salmon burgers. Foods 10:776
90. Ju J, Xie Y, Guo Y, Cheng Y, Qian H, Yao W (2019) The inhibitory effect of plant essential oils on foodborne pathogenic bacteria in food. Crit Rev Food Sci Nutr 59:3281–3292
91. Kawhena TG, Opara UL, Fawole OA (2021) Optimization of gum arabic and starch-based edible coatings with lemongrass oil using response surface methodology for improving postharvest quality of whole "wonderful" pomegranate fruit. Coatings 11:442
92. Khalaf H, Sharoba A, El-Tanahi H, Morsy M (2013) Stability of antimicrobial activity of pullulan edible films incorporated with nanoparticles and essential oils and their impact on turkey deli meat quality. J Food Dairy Sci 4:557–573
93. Khaledian S, Basiri S, Shekarforoush SS (2021) Shelf-life extension of pacific white shrimp using tragacanth gum-based coatings containing Persian lime peel (*Citrus latifolia*) extract. LWT 141:110937
94. Kim I-H, Oh YA, Lee H, Song KB, Min SC (2014) Grape berry coatings of lemongrass oil-incorporating nanoemulsion. LWT Food Sci Technol 58:1–10
95. Kraśniewska K, Galus S, Gniewosz M (2020) Biopolymers-based materials containing silver nanoparticles as active packaging for food applications–a review. Int J Mol Sci 21:1–18
96. Kumar N, Neeraj P, Trajkovska Petkoska A (2021) Improved shelf life and quality of tomato (*Solanum Lycopersicum* L.) by using chitosan-pullulan composite edible coating enriched with pomegranate peel extract. ACS Food Sci Technol 1–11
97. Kumar N, Ojha A, Upadhyay A, Singh R, Kumar S (2021) Effect of active chitosan-pullulan composite edible coating enrich with pomegranate peel extract on the storage quality of green bell pepper. LWT 138:110435
98. Kumar NP, Pareek S (2020) Bioactive compounds of moringa (*Moringa* species). In: Murthy HN, Paek KY (eds) Bioactive compounds in underutilized vegetables and legumes. Springer International Publishing, Cham
99. Li T, Sun X, Chen H, He B, Mei Y, Wang D, Li J (2020) Effect of the combination of vanillin and chitosan coating on the microbial diversity and shelf-life of refrigerated turbot (*Scophthalmus maximus*) filets. Front Microbiol 11:462
100. Liang T, Sun G, Cao L, Li J, Wang L (2019) A pH and NH_3 sensing intelligent film based on *Artemisia sphaerocephala Krasch.* gum and red cabbage anthocyanins anchored by carboxymethyl cellulose sodium added as a host complex. Food Hydrocoll 87:858–868
101. Łopusiewicz Ł, Drozłowska E, Trocer P, Kostek M, Śliwiński M, Henriques MH, Bartkowiak A, Sobolewski P (2020) Whey protein concentrate/isolate biofunctional films modified with melanin from watermelon (*Citrullus lanatus*) seeds. Materials 13:3876
102. Lourenço SC, Fraqueza MJ, Fernandes MH, Moldão-Martins M, Alves VD (2020) Application of edible alginate films with pineapple peel active compounds on beef meat preservation. Antioxidants 9:667
103. Maftoonazad N, Ramaswamy HS (2019) Application and evaluation of a pectin-based edible coating process for quality change kinetics and shelf-life extension of lime fruit (*Citrus aurantifolium*). Coatings 9:285
104. Martiñon ME, Moreira RG, Castell-Perez ME, Gomes C (2014) Development of a multilayered antimicrobial edible coating for shelf-life extension of fresh-cut cantaloupe (*Cucumis melo* L.) stored at 4 °C. LWT Food Sci Technol 56:341–350
105. Milea ȘA, Aprodu I, Enachi E, Barbu V, Râpeanu G, Bahrim GE, Stănciuc N (2021) β-lactoglobulin and its thermolysin derived hydrolysates on regulating selected biological functions of onion skin flavonoids through microencapsulation. CyTA-J Food 19:127–136
106. Molayi R, Ehsani A, Yousefi M (2018) The antibacterial effect of whey protein–alginate coating incorporated with the lactoperoxidase system on chicken thigh meat. Food Sci Nutr 6:878–883
107. Mukherjee D, Haque ZZ (2016) Reduced protein carbonylation of cube steak and catfish fillet using antioxidative coatings containing cheddar whey, casein hydrolyzate and oolong tea extract. Ann Food Sci Technol 17:529–536

108. Mustafa P, Niazi MB, Jahan Z, Samin G, Hussain A, Ahmed T, Naqvi SR (2020) PVA/starch/propolis/anthocyanins rosemary extract composite films as active and intelligent food packaging materials. J Food Saf 40:e12725
109. Nair MS, Tomar M, Punia S, Kukula-Koch W, Kumar M (2020) Enhancing the functionality of chitosan-and alginate-based active edible coatings/films for the preservation of fruits and vegetables: a review. Int J Biol Macromol 164:304–320
110. Niero M, Hauschild MZ (2017) Closing the loop for packaging: Finding a framework to operationalize circular economy strategies. Procedia Cirp 61:685–690
111. Nile SH, Baskar V, Selvaraj D, Nile A, Xiao J, Kai G (2020) Nanotechnologies in food science: Applications, recent trends, and future perspectives. Nano-Micro Letters 12:1–34
112. Nilsen-Nygaard J, Fernández EN, Radusin T, Rotabakk BT, Sarfraz J, Sharmin N, Sivertsvik M, Sone I, Pettersen MK (2021) Current status of biobased and biodegradable food packaging materials: Impact on food quality and effect of innovative processing technologies. Comprehensive Reviews in Food Science and Food Safety 20:1333–1380
113. Noshirvani N, Ghanbarzadeh B, Mokarram RR, Hashemi M (2017) Novel active packaging based on carboxymethyl cellulose-chitosan-ZnO NPs nanocomposite for increasing the shelf life of bread. Food Packag Shelf Life 11:106–114
114. Oliveira Filho JG, Braga ARC, De Oliveira BR, Gomes FP, Moreira VL, Pereira VAC, Egea MB (2021) The potential of anthocyanins in smart, active, and bioactive eco-friendly polymer-based films: A review. Food Res Int 110202.
115. Olivo PM, Da Silva Scapim MR, Maia LF, Miazaki J, Rodrigues BM, Madrona GS, Bankuti FI, Dos Santos Pozza MS (2020) Probiotic coating for ripened cheeses with *Lactobacillus Acidophilus* and *Lactobacillus Helveticus* inclusion. J Agric Stud 8:152-–70
116. Omerović N, Djisalov M, Živojević K, Mladenović M, Vunduk J, Milenković I, Knežević NŽ, Gadjanski I, Vidić J (2021) Antimicrobial nanoparticles and biodegradable polymer composites for active food packaging applications. Compr Rev Food Sci Food Saf 1–27
117. Othman SH, Othman NFL, Shapi'i RA, Ariffin SH, Yunos KFM (2021) Corn starch/chitosan nanoparticles/thymol bio-nanocomposite films for potential food packaging applications. Polymers 13:390
118. Otto S, Strenger M, Maier-Nöth A, Schmid M (2021) Food packaging and sustainability–consumer perception vs. Correlated scientific facts: a review. J Clean Prod 126733
119. Oussalah M, Caillet S, Salmiéri S, Saucier L, Lacroix M (2004) Antimicrobial and antioxidant effects of milk protein-based film containing essential oils for the preservation of whole beef muscle. J Agric Food Chem 52:5598–5605
120. Pabast M, Shariatifar N, Beikzadeh S, Jahed G (2018) Effects of chitosan coatings incorporating with free or nano-encapsulated *Satureja* plant essential oil on quality characteristics of lamb meat. Food Control 91:185–192
121. Panahirad S, Dadpour M, Peighambardoust SH, Soltanzadeh M, Gullón B, Alirezalu K, Lorenzo JM (2021) Applications of carboxymethyl cellulose- and pectin-based active edible coatings in preservation of fruits and vegetables: a review. Trends Food Sci Technol 110:663–673
122. Pandhi S, Kumar A, Alam T (2019) Probiotic edible films and coatings: concerns, applications and future prospects. J Packag Technol Res 3:261–268
123. Pateiro M, Gómez B, Munekata PE, Barba FJ, Putnik P, Kovačević DB, Lorenzo JM (2021) Nanoencapsulation of promising bioactive compounds to improve their absorption, stability, functionality and the appearance of the final food products. Molecules 26:1547
124. Pavli F, Argyri AA, Skandamis P, Nychas G-J, Tassou C, Chorianopoulos N (2019) Antimicrobial activity of oregano essential oil incorporated in sodium alginate edible films: Control of *Listeria monocytogenes* and spoilage in ham slices treated with high pressure processing. Materials 12:3726
125. Pavli F, Tassou C, Nychas G-JE, Chorianopoulos N (2018) Probiotic incorporation in edible films and coatings: bioactive solution for functional foods. Int J Mol Sci 19:150
126. Pena-Serna C, Penna ALB, Lopes Filho JF (2016) Zein-based blend coatings: impact on the quality of a model cheese of short ripening period. J Food Eng 171:208–213

127. Peng Y, Wu Y, Li Y (2013) Development of tea extracts and chitosan composite films for active packaging materials. Int J Biol Macromol 59:282–289
128. Peralta-Ruiz Y, Grande-Tovar CD, Navia Porras DP, Sinning-Mangonez A, Delgado-Ospina J, González-Locarno M, Maza Pautt Y, Chaves-López C (2021) Packham's triumph pears (*Pyrus communis* L.) post-harvest treatment during cold storage based on chitosan and rue essential oil. Molecules 26:725
129. Peralta-Ruiz Y, Tovar CDG, Sinning-Mangonez A, Coronell EA, Marino MF, Chaves-Lopez C (2020) Reduction of postharvest quality loss and microbiological decay of tomato "Chonto"(*Solanum lycopersicum* L.) using chitosan-e essential oil-based edible coatings under low-temperature storage. Polymers 12:1822
130. Pérez-Santaescolástica C, Munekata PE, Feng X, Liu Y, Bastianello Campagnol PC, Lorenzo JM (2020) Active edible coatings and films with Mediterranean herbs to improve food shelf-life. Critical Rev Food Sci Nutr 1–13
131. Petkoska AT, Daniloski D, D'cunha NM, Naumovski N, Broach AT (2021) Edible packaging: sustainable solutions and novel trends in food packaging. Food Res Int 140:109981
132. Pinzon MI, Sanchez LT, Garcia OR, Gutierrez R, Luna JC, Villa CC (2020) Increasing shelf life of strawberries (*Fragaria* ssp) by using a banana starch-chitosan-aloe vera gel composite edible coating. Int J Food Sci Technol 55:92–98
133. Pluta-Kubica A, Jamróz E, Juszczak L, Krzyściak P, Zimowska M (2021) Characterization of furcellaran-whey protein isolate films with green tea or pu-erh extracts and their application as packaging of an acid-curd cheese. Food Bioprocess Technol 14:78–92
134. Pop OL, Pop CR, Dufrechou M, Vodnar DC, Socaci SA, Dulf FV, Minervini F, Suharoschi R (2020) Edible films and coatings functionalization by probiotic incorporation: a review. Polymers 12:12
135. Pourmolaie H, Khosrowshahi Asl A, Ahmadi M, Zomorodi S, Naghizadeh Raeisi S (2018) The effect of guar and tragacanth gums as edible coatings in Cheddar cheese during ripening. J Food Saf 38:e12529
136. Prajapati S, Padhan B, Amulyasai B, Sarkar A (2020) Nanotechnology-based sensors. Elsevier, Biopolymer-based formulations
137. Primožič M, Knez Ž, Leitgeb M (2021) (Bio) Nanotechnology in food science—food packaging. Nanomaterials 11:292
138. Priyadarshi R, Ezati P, Rhim J-W (2021) Recent advances in intelligent food packaging applications using natural food colorants. ACS Food Sci Technol 1:124–138
139. Puscaselu R, Gutt G, Amariei S (2019) Rethinking the future of food packaging: biobased edible films for powdered food and drinks. Molecules 24:3136
140. Quesada J, Sendra E, Navarro C, Sayas-Barberá E (2016) Antimicrobial active packaging including chitosan films with *Thymus vulgaris* L. essential oil for ready-to-eat meat. Foods 5:57
141. Quintana SE, Llalla O, García-Zapateiro LA, García-Risco MR, Fornari T (2020) Preparation and characterization of licorice-chitosan coatings for postharvest treatment of fresh strawberries. Appl Sci 10:8431
142. Raghav P, Saini M (2018) Development of mint (*Mentha viridis* L.) herbal edible coating for shelf life enhancement of cucumber (*Cucumis sativus*). Int J Green Herb Chem 7:379–391
143. Rashidinejad A, Bahrami A, Rehman A, Rezaei A, Babazadeh A, Singh H, Jafari SM (2020) Co-encapsulation of probiotics with prebiotics and their application in functional/synbiotic dairy products. Critical Rev Food Sci Nutr 1–25
144. Ravishankar S, Jaroni D, Zhu L, Olsen C, McHugh T, Friedman M (2012) Inactivation of *Listeria monocytogenes* on ham and bologna using pectin-based apple, carrot, and hibiscus edible films containing carvacrol and cinnamaldehyde. J Food Sci 77:377–382
145. Ribeiro-Santos R, de Melo NR, Andrade M, Azevedo G, Machado AV, Carvalho-Costa D, Sanches-Silva A (2018) Whey protein active films incorporated with a blend of essential oils: characterization and effectiveness. Packag Technol Sci 31:27–40
146. Riveros CG, Mestrallet MG, Quiroga PR, Nepote V, Grosso NR (2013) Preserving sensory attributes of roasted peanuts using edible coatings. Int J Food Sci Technol 48:850–859

147. Robledo N, Vera P, López L, Yazdani-Pedram M, Tapia C, Abugoch L (2018) Thymol nanoemulsions incorporated in quinoa protein/chitosan edible films; antifungal effect in cherry tomatoes. Food Chem 246:211–219
148. Rodríguez MC, Yépez CV, González JHG, Ortega-Toro R (2020) Effect of a multifunctional edible coating based on cassava starch on the shelf life of *Andean* blackberry. Heliyon 6:1–8
149. Roy S, Kim H-J, Rhim J-W (2021) Synthesis of carboxymethyl cellulose and agar-based multifunctional films reinforced with cellulose nanocrystals and shikonin. ACS Appl Polym Mater 3:1060–1069
150. Roy S, Rhim J-W (2020) Effect of CuS reinforcement on the mechanical, water vapor barrier, uv-light barrier, and antibacterial properties of alginate-based composite films. Int J Biol Macromol 164:37–44
151. Roy S, Rhim J-W (2020) Preparation of gelatin/carrageenan-based color-indicator film integrated with shikonin and propolis for smart food packaging applications. ACS Appl Bio Mater 4:770–779
152. Ruelas-Chacon X, Aguilar-González A, De La Luz Reyes-Vega M, Peralta-Rodríguez RD, Corona-Flores J, Rebolloso-Padilla ON, Aguilera-Carbo AF (2020) Bioactive protecting coating of guar gum with thyme oil to extend shelf life of tilapia (*Oreoschromis niloticus*) fillets. Polymers 12:3019
153. Sabaghi M, Maghsoudlou Y, Khomeiri M, Ziaiifar AM (2015) Active edible coating from chitosan incorporating green tea extract as an antioxidant and antifungal on fresh walnut kernel. Postharvest Biol Technol 110:224–228
154. Sabbah M, di Pierro P, Dell'Olmo E, Arciello A, Porta R (2019) Improved shelf-life of Nabulsi cheese wrapped with hydrocolloid films. Food Hydrocoll 96:29–35
155. Salmieri S, Islam F, Khan RA, Hossain FM, Ibrahim HM, Miao C, Hamad WY, Lacroix M (2014) Antimicrobial nanocomposite films made of poly (lactic acid)–cellulose nanocrystals (PLA–CNC) in food applications—part B: effect of oregano essential oil release on the inactivation of *Listeria monocytogenes* in mixed vegetables. Cellulose 21:4271–4285
156. Salvia-Trujillo L, Rojas-Graü MA, Soliva-Fortuny R, Martín-Belloso O (2015) Use of antimicrobial nanoemulsions as edible coatings: Impact on safety and quality attributes of fresh-cut Fuji apples. Postharvest Biol Technol 105:8–16
157. Sami R, Elhakem A, Alharbi M, Benajiba N, Almatrafi M, Abdelazez A, Helal M (2021) Evaluation of antioxidant activities, oxidation enzymes, and quality of nano-coated button mushrooms (*Agaricus Bisporus*) during storage. Coatings 11:149
158. Samsi MS, Kamari A, Din SM, Lazar G (2019) Synthesis, characterization and application of gelatin–carboxymethyl cellulose blend films for preservation of cherry tomatoes and grapes. J Food Sci Technol 56:3099–3108
159. Sani MA, Ehsani A, Hashemi M (2017) Whey protein isolate/cellulose nanofibre/TiO2 nanoparticle/rosemary essential oil nanocomposite film: Its effect on microbial and sensory quality of lamb meat and growth of common foodborne pathogenic bacteria during refrigeration. Int J Food Microbiol 251:8–14
160. Sarfraz J, Gulin-Sarfraz T, Nilsen-Nygaard J, Pettersen MK (2021) Nanocomposites for food packaging applications: an overview. Nanomaterials 11:10
161. Schütz F, Figueiredo-Braga M, Barata P, Cruz-Martins N (2021) Obesity and gut microbiome: review of potential role of probiotics. Porto Biomed J 6:1–6
162. Sharif N, Falcó I, Martínez-Abad A, Sánchez G, López-Rubio A, Fabra MJ (2021) On the use of Persian gum for the development of antiviral edible coatings against *Murine* norovirus of interest in blueberries. Polymers 13:224
163. Sharifi-Rad J, Rodrigues CF, Stojanović-Radić Z, Dimitrijević M, Aleksić A, Neffe-Skocińska K, Zielińska D, Kołożyn-Krajewska D, Salehi B, Milton Prabu, S (2020) Probiotics: versatile bioactive components in promoting human health. Medicina 56:433
164. Sharma M, Usmani Z, Gupta VK, Bhat R (2021) Valorization of fruits and vegetable wastes and by-products to produce natural pigments. Critical Rev Biotechnol 1–42
165. Sharma S, Singh RK (2020) Cold plasma treatment of dairy proteins in relation to functionality enhancement. Trends Food Sci Technol 102:30–36

166. Shendurse A, Gopikrishna G, Patel A, Pandya A (2018) Milk protein based edible films and coatings–preparation, properties and food applications. J Nutr Health Food Eng 8:219–226
167. Shokri S, Ehsani A, Jasour MS (2015) Efficacy of lactoperoxidase system-whey protein coating on shelf-life extension of rainbow trout fillets during cold storage (4 C). Food Bioprocess Technol 8:54–62
168. Siripatrawan U, Vitchayakitti W (2016) Improving functional properties of chitosan films as active food packaging by incorporating with propolis. Food Hydrocoll 61:695–702
169. Sun G, Chi W, Zhang C, Xu S, Li J, Wang L (2019) Developing a green film with pH-sensitivity and antioxidant activity based on κ-carrageenan and hydroxypropyl methylcellulose incorporating *Prunus maackii* juice. Food Hydrocoll 94:345–353
170. Tapilatu Y, Nugraheni PS, Ginzel T, Latumahina M, Limmon GV, Budhijanto W (2016) Nano-chitosan utilization for fresh yellowfin tuna preservation. Aquat Procedia 7:285–295
171. Thakur R, Pristijono P, Bowyer M, Singh SP, Scarlett CJ, Stathopoulos CE, Vuong QV (2019) A starch edible surface coating delays banana fruit ripening. LWT 100:341–347
172. Tomadoni B, Moreira MDR, Pereda M, Ponce AG (2018) Gellan-based coatings incorporated with natural antimicrobials in fresh-cut strawberries: microbiological and sensory evaluation through refrigerated storage. LWT 97:384–389
173. Tomasik P, Tomasik P (2020) Probiotics, non-dairy prebiotics and postbiotics in nutrition. Appl Sci 10:1470
174. Umaraw P, Munekata PE, Verma AK, Barba FJ, Singh V, Kumar P, Lorenzo JM (2020) Edible films/coating with tailored properties for active packaging of meat, fish and derived products. Trends Food Sci Technol 98:10–24
175. Vasile C, Baican M (2021) Progresses in food packaging, food quality, and safety—controlled-release antioxidant and/or antimicrobial packaging. Molecules 26:1263
176. Vieira TM, Moldão-Martins M, Alves VD (2021) Composite coatings of chitosan and alginate emulsions with olive oil to enhance postharvest quality and shelf life of fresh figs (*Ficus carica* L. cv.'*Pingo De Mel*'). Foods 10:718
177. Villafañe F (2017) Edible coatings for carrots. Food Rev Int 33:84–103
178. Villasante J, Martin-Lujano A, Almajano MP (2020) Characterization and application of gelatin films with pecan walnut and shell extract (*Carya illinoiensis*). Polymers 12:1424
179. Wambura P, Yang WW (2010) Ultrasonication and edible coating effects on lipid oxidation of roasted peanuts. Food Bioprocess Technol 3:620–628
180. Wang H, Qian J, Ding F (2018) Emerging chitosan-based films for food packaging applications. J Agric Food Chem 66:395–413
181. Wu C, Sun J, Zheng P, Kang X, Chen M, Li Y, Ge Y, Hu Y, Pang J (2019) Preparation of an intelligent film based on chitosan/oxidized chitin nanocrystals incorporating black rice bran anthocyanins for seafood spoilage monitoring. Carbohydr Polym 222:115006
182. Xie J, Tang Y, Yang S-P, Qian Y-F (2017) Effects of whey protein films on the quality of thawed bigeye tuna (*Thunnus obesus*) chunks under modified atmosphere packaging and vacuum packaging conditions. Food Sci Biotechnol 26:937–945
183. Xiong Y, Chen M, Warner RD, Fang Z (2020) Incorporating nisin and grape seed extract in chitosan-gelatine edible coating and its effect on cold storage of fresh pork. Food Control 110:107018
184. Xiong Y, Li S, Warner RD, Fang Z (2020) Effect of oregano essential oil and resveratrol nanoemulsion loaded pectin edible coating on the preservation of pork loin in modified atmosphere packaging. Food Control 114:107226
185. Xiong Y, Kamboj M, Ajlouni S, Fang Z (2021) Incorporation of salmon bone gelatine with chitosan, gallic acid and clove oil as edible coating for the cold storage of fresh salmon fillet. Food Control 125:107994
186. Yang H, Zhang Y, Zhou F, Guo J, Tang J, Han Y, Li Z, Fu C (2021) Preparation, bioactivities and applications in food industry of chitosan-based maillard products: a review. Molecules 26:166
187. Yildirim S, Röcker B, Pettersen MK, Nilsen-Nygaard J, Ayhan Z, Rutkaite R, Radusin T, Suminska P, Marcos B, Coma V (2018) Active packaging applications for food. Compr Rev Food Sci Food Saf 17:165–199

188. Yıldız PO, Yangilar F (2016) Effects of different whey protein concentrate coating on selected properties of rainbow trout (*Oncorhynchus mykiss*) during cold storage (4 C). Int J Food Prop 19:2007–2015
189. Yildiz PO, Yangilar F (2017) Effects of whey protein isolate based coating enriched with *Zingiber officinale* and *Matricaria recutita* essential oils on the quality of refrigerated rainbow trout. J Food Saf 1–8
190. Youssef AM, El-Sayed SM (2018) Bionanocomposites materials for food packaging applications: concepts and future outlook. Carbohyd Polym 193:19–27
191. Youssef AM, El-Sayed SM, Salama HH, El-Sayed HS, Dufresne A (2015) Evaluation of bionanocomposites as packaging material on properties of soft white cheese during storage period. Carbohyd Polym 132:274–285
192. Yousuf B, Qadri OS, Srivastava AK (2018) Recent developments in shelf-life extension of fresh-cut fruits and vegetables by application of different edible coatings: a review. LWT-Food Sci Technol 89:198–209
193. Yousuf B, Srivastava AK (2015) Psyllium (*Plantago*) gum as an effective edible coating to improve quality and shelf life of fresh-cut papaya (*Carica papaya*). Int J Biol Biomol Agric Food Biotechnol Eng 9:702–707
194. Yu D, Regenstein JM, Xia W (2019) Bio-based edible coatings for the preservation of fishery products: a review. Crit Rev Food Sci Nutr 59:2481–2493
195. Yuan G, Chen X, Li D (2016) Chitosan films and coatings containing essential oils: the antioxidant and antimicrobial activity, and application in food systems. Food Res Int 89:117–128
196. Yun D, Cai H, Liu Y, Xiao L, Song J, Liu J (2019) Development of active and intelligent films based on cassava starch and Chinese bayberry (*Myrica rubra Sieb. et Zucc.*) anthocyanins. RSC Adv 9:30905–30916
197. Zhao X, Cornish K, Vodovotz Y (2020) Narrowing the gap for bioplastic use in food packaging: an update. Environ Sci Technol 54:4712–4732
198. Zia J, Mancini G, Bustreo M, Zych A, Donno R, Athanassiou A, Fragouli D (2021) Porous pH natural indicators for acidic and basic vapor sensing. Chem Eng J 403:126373
199. Zielińska A, Costa B, Ferreira MV, Miguéis D, Louros J, Durazzo A, Lucarini M, Eder P, Chaud MV, Morsink M (2020) Nanotoxicology and nanosafety: Safety-by-design and testing at a glance. Int J Environ Res Public Health 17:4657

Possibilities for the Recovery and Valorization of Single-Use EPS Packaging Waste Following Its Increasing Generation During the COVID-19 Pandemic: A Case Study in Brazil

Felipe Luis Palombini and Mariana Kuhl Cidade

Abstract In Brazil, from capitals to small municipalities, urban solid waste treatment deeply relies on the sorting of collected residues by local associations. Those Sorting Units are composed of low-income workers that rely on the selling of the waste as their main revenue source. However, the trading of the sorted residues hinges on the commercial interest from buyers, therefore only a few materials of interest are potentially recyclable, while the rest ends up being discarded in landfills or open-air dumps all over the country. Due to the COVID-19 pandemic, the country has observed a significant increase in the consumption and generation of polymeric packaging waste throughout 2020 and 2021. Particularly, expanded polystyrene (EPS) packaging has its use risen mostly due to lockdown measures and the consequent increase in delivery services and takeout meals. Despite the low cost and practicability of single-use packaging in terms of food safety, some materials are considered problematic regarding their recyclability, such as the case of EPS. This chapter addresses the case study of EPS packaging waste in two cities of southern Brazil, over recyclability issues considering social, environmental, economic, and political spheres. The possibilities for increasing the recyclability of EPS in Brazil are discussed as well as the main challenges of making the waste more interesting for potential buyers. In general, being EPS a notable low-cost and lightweight commodity, its transport as a residue for reuse purposes often is hampered by logistics costs, as well as the low intrinsic value of the recycled material for application in different products. Finally, we propose a new application of chemically recycled EPS as a gemstone for application in contemporary jewelry, using silver recovered from scrap sources. The material recycling and the gemstone manufacturing processes are described using simplified techniques as a way to permit its replication by both jewelry professionals and untrained workers, aiming at valorizing the residue. Parting from an extremely inexpensive material with a low-quality perception, we discuss the opportunities of the trash-to-treasure transformation as a way to potentially increase the material

F. L. Palombini
Design and Computer Simulation Group, School of Engineering, Federal University of Rio Grande do Sul — UFRGS, Av. Osvaldo Aranha 99/408, Porto Alegre, RS 90035-190, Brazil

M. K. Cidade (✉)
Department of Industrial Design, Federal University of Santa Maria — UFSM, Av. Roraima, n° 1000, Prédio 40, Sala 1136, Santa Maria, RS 97105-900, Brazil

S. S. Muthu (eds.), *Sustainable Packaging*, Environmental Footprints and Eco-design of Products and Processes, https://doi.org/10.1007/978-981-16-4609-6_10

value and consequently the revenue of the Sorting Units workers, while reducing its environmental impact.

Keywords Sustainability · Problematic materials · COVID-19 · Packaging · Contemporary jewelry · Recycling · Gemstones · Plastic waste

1 Introduction

Brazil's current urban solid waste treatment system deeply relies on the action of collecting and selecting residues by local associations, called Sorting Units [14, 52]. In the country's current waste treatment system, the Municipal Solid Waste (MSW) generated is distributed by each municipality among those associations. The Sorting Units encompass a series of registered facilities that are designated by the respective sanitation departments of each city, where personnel may be hired to work. The main income of the workers from those units relies on the probability of the sorted MSW to be accepted by potential buyers [28, 52]. If the collected and sorted residue has some economic interest, it is then compacted and packed into bales to be sold as a source of material for further reprocessing and recycling procedures. However, if a certain type of waste has difficulties in finding potential buyers, or if it is already identified by the workers to be considered almost virtually impossible to sell, the residue is discarded. Besides, following the discarded treatment policies of MSW given by many cities in the country, a significant amount of waste is landfilled or even tossed into open dumps [52]. Despite a relative substantial gathering coverage of municipal solid waste in the country, where over 92% of the waste is estimated to be collected, according to the Brazilian Association of Special Waste and Public Cleaning Companies [1], Brazil still lacks important measures to fight the waste problem, and some materials are considered a key part in this matter. For instance, only 1.3% of the nearly 11 million tons of plastic generated annually in the country is estimated to be recycled and recovered [71]. Giving the lower price of most virgin, commodity-based polyolefins, recycling and using them as a secondary-sourced material becomes less lucrative, thus hampering the possibilities of finding potential buyers for the material. Sorting Units' associates are generally low-income workers from peripheral regions of major urban centers, and every increase in the selling of the sorted materials makes a big difference in their revenue. In this sense, finding newer ways of increasing the interest in certain types of the so-called "problematic residues" may contribute not only environmentally, but also in the economic and social spheres.

From a social point of view, by considering "problematic" in the Sorting Units [52] one material must be qualified as having little or no economic interest. According to the associations, residues that are made of those materials are less likely to be sold, thus being discarded. In that category, they include several types of film-like polymeric packaging, glasses and different ceramics, thermosets, and others. In addition to not being commercialized, these problematic materials also slow down the overall sorting procedure. Once the selection of residues is performed manually, workers need to open the litter bags, spread the dry waste into the conveyor belt, and sort them according to the type o residue (aluminum, polypropylene, polyethylene, glasses, etc.) by either its appearance or, even, sound (*e.g.,* plastic resins make different sounds when crushed) [48]; then when a residue is not used, not only it is discarded but the workers also waste their time sorting it. Particularly one of the materials with the lower expectancy of being recovered due to commercialization is expanded polystyrene (EPS). In addition to being considered a commodity with a lower intrinsic value by itself as a virgin material, EPS foam also has a remarkable low density [25], which means that even more material is needed for a determined amount of weight. This leads to several disadvantages in their acceptance as a secondary material, from the viewpoint of the Sorting Units, giving that most sales are done by setting a price per weight instead of volume. First, the associations would have to reserve a significant space in their facilities just to storage this type of material, in order to accumulate a significant amount of EPS worthy of commercialization. Second, since most of the Sorting Units do not have any sophisticated equipment for recycling or pre-processing polymeric residues, EPS would have to be deposited in its original form. Most residues are sold without any pre-processing, which also includes cleaning procedures, thus hindering the value of the secondary material. At the very most, some associations may use some hydraulic presses to form bales of residues for stocking. Third, even if the Units could store compacted EPS in large facility areas, the material would still be too light to be transported by buyers in a substantial quantity to be considered profitable to some extent. Furthermore, in major Brazilian cities, EPS has to be transported to far located facilities in order to be reprocessed, which makes the costs impracticable. For instance, there is the case of Porto Alegre, the capital city of the southernmost state of Rio Grande do Sul. According to the city's Municipal Department of Urban Cleaning (DMLU, in the Portuguese acronym), the closest EPS recycling facility is in the state of Espírito Santo, located at approximately 2100 km (or 1300 miles) away [18]; not mentioning the costs of filling up a whole truck with EPS and traveling that distance for all the material to be sold at just BRL 20 to BRA 50 (approximately USD 4 to USD 9, as in March 2021), according to the department [18]. Fourth, considering that someone has the possibility to acquire and transport secondary EPS, the buyer would still be in charge of cleaning and recycling the residue, prior to using it as source material. Once again, giving the lower price of opting for a virgin styrene resin to manufacture EPS in the first place, some tax subsidy policies would be required to favor the use of recycled EPS. Fifth, regarding the application of recycled polymeric material in the country, according to the current regulation from the Brazilian Health Regulatory Agency [3], no secondary plastic material can be used in direct contact with food, regardless of its origin, recycling treatment, and chemical constitution or condition. Therefore, even if a clean and reprocessed EPS—which could have been originally used as food packaging—is finally recycled, there are fewer applications to which the waste can be destined. And six, even that EPS can be technically considered a recyclable material [61, 65, 72], with all the practical difficulties faced by most municipal urban cleaning departments, the material's recovery ends up being virtually unfeasible. Brazil's National Policy on Solid Waste (PNRS, Portuguese acronym), defined by Federal Law nº 12.305/10 [9] establishes an integrated management for USW as a "set of mechanisms and procedures that guarantee society's access to information and participation in the formulation, implementation and assessment processes of public policies related to solid waste". In the law, the control, management, and treatment measures of the generated solid waste should start from several spheres considering all stakeholders [52]. That means that in the case of EPS, both the manufacturer of the material and cleaning responsible agents, as the municipal departments, are in charge of dealing with it responsibly. On the other hand, the same Policy states that a "solid waste that, after having exhausted all the possibilities for treatment and recovery for available and economically viable technological processes, show no other possibility than the final disposal environmentally sound" [9]. Meaning that in the case of the city of Porto Alegre, giving the fact that EPS cannot have an economically viable recycling, the waste can be legally landfilled, as commented by DMLU [18]. All those points intensify the difficulty of finding buyers that could be potentially interested in acquiring and recycling EPS waste, thus increasing the concern this type of residue has in the environmental and in the social spheres.

Recently, another issue regarding the use of expanded polystyrene has gained the headlines in the country due to the increasing utilization of single-use polymeric products. The consumption of single-use plastics, such as cups, straws, plates, bags, or bottles, has been a current and urgent global environmental matter for many years [19, 68, 73], primarily because of the great discrepancy between the time at which the product is used and how long it actually takes to decompose. Meanwhile, studies have also highlighted the impact single-use plastic debris have including during their decomposition on multiple environments; particularly in marine ecosystems [59], where the polymeric chains are disintegrated very slowly, leading to the formation of macro- and microplastics in the sediment and sea, that are then mistaken by the fauna as being food [42, 53, 67]. Another case where single-use plastics have raised recent concern is in medic- and healthcare-related products [39, 43]. Once polyolefins are also a common choice material for the manufacturing of several disposable medical supplies, such as syringes, gloves, face masks, among others, their extensive use automatically implies the generation of a large amount of waste, that needs to be treated even more carefully in terms of post-use treatment, recovery, recycling, or landfilling, due to the risk of contamination. More recently, since the beginning of the COVID-19 pandemic, the consumption of single-use plastics has grown in several segments, due to a number of reasons, mostly related to the health condition of the country as well as the "stay at home" circumstance defined in the majority of cities. This scenario has led to a significant rise in the generation of USW, primarily related to single-use polymeric products. All over Brazil, the manufacturing and waste generation of single-use plastics, locally referred to as "disposable materials", has increased from 25 to 30% in the second quarter of 2020 [27], compared to the same period of 2019, even though the generation of overall USW has decreased in about 9% in the same period [37]. A similar increase—from 15 to 25%—was expected to be registered throughout 2020 in the country, according to another report from the Brazilian Association of Special Waste and Public Cleaning Companies [2]. This consumption pattern has been noticed and reported in many major Brazilian cities, both capital and inland. For instance, in São Paulo, the city administration pointed out a 39% increase in the selective collection of USW in 2020 when compared to 2019 [37]. In Porto Alegre, despite lockdown measures, the selective collection of USW increased over 9%, ever since the beginning of the COVID-19 pandemic, according to a survey from the Municipal Department of Urban Services [38]. In Santa Maria, an inland city of the center region of the Rio Grande do Sul state, in southern Brazil, the selective collection has increased up to 18% in 2020, according to the Municipal Department of Environment [74]. All this growth can be derived from two main reasons. The first one is related to the obvious increase in the consumption and the consequent waste generation of medical supplies due to the COVID-19 pandemic, in view of the fact that Brazil is one of the most affected countries in the world. According to the report from ABRELPE [2], the generation of this type of waste in 2020 is expected to be f10 to 20 times higher than the year before. Another recent report from the Heinrich Böll Foundation—Brazil [33] points out that the country would potentially be generating over 10.5 thousand metric tonnes of plastic waste per month with disposable face masks alone. In the same report, the medical waste generation of a COVID-19-infected, hospitalized patient (about 7.5 kg/day) is estimated to be almost seven times higher than the average USW generation per regular inhabitant in the country. But the increase in single-use plastics in Brazil is not only noticed in the medical and healthcare segment, directly. Several recent studies have approached the concerning increase of single-use plastics during COVID-19, however most are focused on waste derived from personal protective equipment [36, 54, 57, 63]. Therefore, in the second reason, following lockdown and more general restriction measures for the movement of individuals, when people are largely working at home and are—depending on the city regulations or health status—either avoiding, being unable or prohibited to going to restaurants or dinners, the usage of delivery and takeout meals has grown significantly [62]. The money spent on delivery apps raised almost 95% between January and May 2020, compared to the same period of 2019 [11]. As an example, in the city of Porto Alegre, the delivery sector grew up to 75% in 2020 [17], largely used by the transport of meals. Added to the fact that almost all types of meals are delivered using single-use plastics, this represents a large amount of polymeric packaging being employed. Giving the known characteristic properties of EPS of having an exceptionally low density, a great thermal insulation capability, a relatively high glass transition temperature (suitable for use with food), and a very affordable price [5], it becomes a preferred way of packaging hot meals for delivery. Adding up with the employment of polyethylene bags, as a consequence, the consumption of plastics with less than one year of usage has turned them into the type of product that has grown the most during the pandemic [33]. Despite the local prohibition of single-use plastics in many cities and states in Brazil in recent years, which impacted multiple branches of the food and service industries, from local restaurants to major supermarket chains, the Justice has been forced to revoke laws due to the pandemic. For instance, in the city of São Paulo, an injunction from the São Paulo Court of Justice suspended the municipal law that prohibited the supply of disposable plastic cups, plates, and cutlery as of 1 January 2021 [45], following a lawsuit of Direct Action of Unconstitutionality filed by the Union of the Plastic

Material Industry, Transformation and Recycling of Plastic Material of the State of São Paulo [64]. And not only Brazil is passing through the same controversy. In the USA, COVID-19 is also changing how the country consumes those products, given that single-use plastics are returning to the daily life of consumers. For decades, these types of products had been restricted or even eliminated from many cities and states, varying from taxes and fees to complete bans [32]. For instance, the states of New York and California have introduced measures for reducing single-plastic usages, particularly plastic bags. However, due to the COVID-19 shutdowns, single-use plastics have returned to grocery stores, restaurants, and cafes of most cities and states, reaching about 50 items that have reduction policies revered across the country in 2020 [32]. The same resurgence has been observed all over Europe during 2020, despite recent efforts from the Single-Use Plastics Directive [55]. Just like Brazil, not only the usage of single-use medical and personal protective equipment has increased, but also disposable plastic cutlery, plates, cups, transport packaging, and many others in the continent. A recent report by the European Environment Agency [50] affirms that as most European restaurants were closed for on-site dining, many have shifted to takeout and delivery options, thus forcing them to once again adopt single-use plastic products. As the report also comments, the generation of this type of waste is particularly difficult to fight due to the great difference in the price of the virgin resin when compared to recycled alternatives. Additionally, a recent drop in oil prices has contributed to even widen the manufacturing and reprocessing cost differences between new and recycled plastics during the pandemic [10].

Following all the above-mentioned circumstances, it can be seen that the current scenario of the significant increase in the usage of EPS packaging and other single-use plastics globally is due to multifactorial reasons. However, considering Brazil a country with a traditionally poor reprocessing culture of EPS, even before the COVID-19 pandemic, finding alternatives for the recovery of this type of waste is crucial for environmental reasons in the country, as well as in many other regions facing the same difficulties. Since the waste reprocessing mechanisms in the country largely depend upon the selling success of a particular type of USW, as well as considering that it also serves as the main revenue source of a number of low-income families, this issue is also a social and economic matter. Additionally, the poor recyclability of EPS in Brazil is predominantly due to the lower intrinsic value the residue has, facing the high costs it demands on a large number of cities to transport and recycling it, especially when compared with the lack of options for applying the recycled material, as well as with the price of the virgin resin as a regular alternative. As a consequence, one of the ways of addressing the problem of EPS packaging waste in the country can be summarized as the need to increase the added value of the residue, to such an extent that it can become interesting to potential buyers. In this manner, this chapter presents one practical possibility to recover this type of waste by means of increasing its value on the market via its application in contemporary jewelry. Section 2 presents an overview of contemporary jewelry, briefly describing its origins and recent history, in addition, to present how it can be employed as a way to improve sustainability issues, by either be presented as an ecological method of producing newer goods, as well as a social mechanism to benefit from it. The next section describes an experimental procedure proposed to chemically recycle valueless post-consumer EPS packaging resulting in application in contemporary jewelry. Finally, the last section discusses how simplified practices and different design-based approaches can contribute to a more sustainable society.

2 Contemporary Jewelry—A More Sustainable Approach

Over the past centuries, jewelry has developed a variety of different purposes, being able to carry symbolic, social, economic, and religious meanings, besides, obviously, having the function of adorning, beautifying. In addition, jewelry has also been considered a voiceless way of communicating between individuals, whether to express the bearer's thoughts, origins, or culture, for instance, or to identify peers with specific characteristics. Following this broad meaning, jewelry has been the object of numerous interpretations, being currently considered a multidisciplinary area, which encompasses art, materials, techniques, processes, anthropology, crafts, and design. Since the early period of human history (ca. 40,000 BC), materials and handling techniques have been discovered, developed, and improved by civilizations, starting with the wonder and curiosity about natural materials, such as bones, canines, tusks, and fangs, hides, skins, and leathers from hunting, as well as seeds, rocks, woods, shells, vegetable fibers, etc., and later evolved into the insertion of metal alloys and gems found in their habitats [8]. Traditional noble materials used in jewelry, such as gold, platinum, silver, diamond, ruby, emerald, among others, gained this name due to their general low reactivity (resistance to corrosion, oxidation, and deterioration), in addition to aesthetic and symbolic aspects. Gold, for example, was discovered around 5000 BC, where its aesthetics, malleability, and oxidation resistance, made it the symbolic material of the divine [4, 8, 29, 40]. Its beauty and resemblance to the maximum divinity, the Sun, both in terms of color and brightness, caught the attention of those who handled it. In addition to gold, other noble metallic materials traditionally used for the manufacture of jewelry are platinum and silver, in which the introduction of alloy elements with lower melting points, such as copper, brass, palladium, among others, was required for their manipulation [40]. These elements were introduced in pieces of jewelry mainly to provide greater mechanical resistance and hardness when compared to traditional materials. However, some of these materials have also been used as the basis for costume jewelry pieces, generally inexpensive or imitation materials, which were principally made of brass and copper.

According to Cappellieri et al. [15], on the one hand, there is art, with the pride of its authorship, on the other, fashion, with the transience of its present, and in between, jewelry with the defense of its noble materials as bastions of eternity and symbolism. If the value of a jewel has for a long time being synonymous with preciousness and nobility, today that idea is definitely being modified [6]. Materials and the tangible preciousness of the pieces are no longer the only characterizing elements to define whether or not an object belongs to the sphere of luxury [15].

The applied material, regardless of whether it is gold, platinum, diamonds, emeralds, wood, polymers, rocks, or even residues from waste, is a design choice, as for the techniques, technologies, finishes, and concepts employed in jewelry pieces are developed focusing on the changes that society has been presenting, in terms of choices, aptitudes, sustainable visions, social, cultural, and economic issues, among others. Including in this new jewelry design mode, referring to the creative processes, unusual material choices and suitable manufacturing processes are employed along with efforts to fabricate pieces with a high-quality finish, were shifted due to changes in several spheres in our societies over the centuries. Some authors mention that the initial boom occurred in the Industrial Revolution [12, 15, 29, 34], with the transition from artisanal manufacturing methods to machinery production, in the eighteenth century, passing through the so-called Second and Third Revolutions until Contemporary times. It is also important to highlight the parallelism between the contemporary development of the jewelry industry and that of product design, from modifications in the production processes to materials selection, and even the choices of aesthetics and style in the pieces. With industrialization, the consumption of gold and diamond-studded jewelry became relatively more accessible to a larger portion of society due to the advent of series production and the economic stage of society, which made it possible to buy them [29]. This stimulated the development of jewelry industries in Europe and the United States. However, with the industrialization of the production process, the driven force of all factories has turned into the need to obtain increasing profits through a growing volume of sales to the middle class, the new economic strength of the cities [58]. With this objective, the quality of life of a large part of the population has decreased, driven by the scope of the industrial districts and their productive styles. According to Heskett [34], with the need for sales growth, manufacturers started to further increase the division of labor, with the hiring of so-called "style consultants", who sought new concepts that could be adopted by the jewelry market. Under his orders, untrained designers, and with almost no knowledge, produced cheap copies of other products, being manufactured with materials of ever-lower quality, in addition to a less significant aesthetic appeal, all manufactured at the expense of long and unhealthy working hours, with few safety conditions and even lower pay [34]. Dissatisfied with the paths industrial production was taking, some thinkers with reformist ideologies proposed changes in the production chains, aiming at the recovery of style and good taste [12]. Essentially, they sought a return to production based on the Middle Ages, believing that artisanal manufacturing would bring better living conditions for workers. This type of production with a high appreciation for aesthetics and function, culminated in the Arts & Crafts movement, in the late nineteenth century, in England [34]. The ideas and production models of the movement soon spread to other European countries and the United States, exerting a significant influence on the emergence of the first modernist actions aimed at jewelry and design [16, 49]. Even though the mass production of jewelry was set in a smaller volume, in comparison to that of traditional consumer goods, the jewelry industry has developed even more associated with handcrafted and distinguished processes [29], such as, for example, techniques that used filigree details, comprised of the use of very thin metallic wires, which are twisted and shaped, generating drawings on the pieces [49]; and stamped with chiseling, in which reliefs, textures, and fine details are made with a chisel. Still, the work of notable artists such as René Lalique is important for this period, with the stylization of the forms of nature, represented by asymmetric and organic models, in which underappreciated or even neglected gems were used in jewelry with a great value, as long as they satisfied the intended effect [26, 29]. This represents one of the first applications of a low-valued decorative element into a much more valorized piece of jewelry, where the selling price of it outperforms by a large margin that of the sum of the materials employed costed at first.

The Contemporary Age has been marked by both the discovery and the first uses of important metals and alloys for jewelry, such as platinum, nickel, tin, titanium, palladium, and rhodium [29, 35]. The period also led to the development of multicolored gold, through alloys with the application of different elements [35]. Still, manufacturers started to use electroplating processes of noble metals, such as gold and silver, which would later contribute to the development of bathing techniques in costume jewelry, aiming at the reduction of costs of making jewelry. The same period brought to the jewelry market the first pearls produced in captivity, which also started to be massified [35]. With the renewed appreciation of aesthetics and good quality in industrialized products, a movement known in Europe as *Art Nouveau* starts to grow. The period marked a sense of artistic life that should be reflected in everyday products [12]. The style also fought mass industrialization, associating art and handcraftsmanship in main objects, with a great emphasis on ornamentation, through more organic and natural shapes [58]. Also, as a reflection of the massification of products at this time, European jewelers began to design adornments following the style known as *Belle Èpoque*, as a reaction to the "ordinariness" of diamond-covered jewelry inherited from the period of the Industrial Revolution. *Art Nouveau* and *Art Decó* also emerge strongly and have an effect on jewelry, since they sought to break with the usual lifestyle of the period [29, 44, 49]. The emergence of these artistic styles also contributed to the reintroduction of different materials in jewelry, such as iron, bronze, glass, ivory, and nacre, in addition to consecrating less valued gems [49]. During the period, regarding the use of innovative materials for jewelry the development of zirconia stands out, being used later as a synthetic replacement for diamonds, in jewelry pieces of lesser value [35]. In Brazil, the development of the jewelry industry took place in the middle of World War II, bringing characteristics of *Art Decó* and the Industrial Era [29]. For this reason, jewelry pieces of the 1940s became known as "cocktail" (or "*coquetel*" in Portuguese), due to the mixture of worldwide themes and inspirations from the twentieth century. Created in a context of crises and social changes, they found in light geometries—made with forged gold—a way to continue to reproduce ostentatious effects during the crisis, since they simulated thick and heavy pieces [29]. In Europe, at the end of World War II, which brought an end to deprivation, metals and gems were in short supply. Jewelers who managed to keep their businesses started to experiment and build pieces with gold-plated materials [13]. At the same time, during this post-war financial scenario, which inhibited the exhibition of fortune and luxury as jewelry, raises the design of pieces that imitated jewelry, the so-called costume jewelry. These are known as pieces of little intrinsic value and are currently characterized by not using noble materials in their production, but metallic alloys such as brass (copper and zinc) and Zamak (family of zinc-based alloys with aluminum, magnesium, and copper); they are also characterized by being gilded or plated with thin layers of gold, silver, or nickel, as well as by not having an exceptional finish or even significant design variations. The 1950s and 1960s were marked by a rupture in the standards prevailing until then, by a strong reaction to the traditional and commercial form of jewelry production, as well as by social and moral transformations [15, 69]. But it was only in the 1970s that the imitation—or costume—jewelry reached its peak, bringing to the jewelry industries the need to innovate in their creations [29]. During the period, a new generation of jewelry designers were brought to the production ideas and concepts, in addition to materials, such as resins and polymers, both in view of the new aesthetic standards and the increase in the price of gold, which has since started to transform jewelry until the present day [23].

Until the end of the 1990s, the Brazilian jewelry industries followed international trends, and the pieces, in their majority, were considered copies of these versions [29]. Jewelry industries followed the big international fairs and reproduced their pieces, through catalogs and magazines, as well as invested in the import of machinery to manufacture them or even trained their goldsmiths to plagiarize them. With the market opening in Brazil, during the same decade, the trade-ins also brought competition, which led to a reformulation in the Brazilian jewelry market. Thus, the need necessity of acquiring technology and local materials arises, as well as the call for greater knowledge of market trends for the development of its aesthetic designs. With this in mind, many Brazilian industries sought new forms of obtaining profits, thus introducing design-trained workers to their teams for the development of exclusive jewelry. In this way, not only new trends in materials and manufacturing processes for jewelry appeared but also the emergence of styles and concepts of their own.

Unlike traditional jewelry, in which noble and rare materials are ever-present, such as gold, platinum, diamonds, and other precious gems, in the so-called contemporary jewelry, new and unusual materials and technologies have been introduced, such as wood, fabrics, seeds, ornamental rocks, fibers natural, multiple metallic alloys, rubbers, polymers, among others, applied with water jet, laser engraving additive manufacturing and others [6, 7, 15, 20–22, 24, 47]. The application of unusual alternatives has also often been used as the only material in the jewelry piece, as long as it has a high finish quality, through different technologies. In addition, contemporary jewelry has also been designed and manufactured by contrasting the presence of unconventional materials along with the addition of traditional and noble metals, such as silver. Nowadays, the metal is being widely used due to its interesting cost–benefit ratio combined with the countless ways of manipulation during its production process, for example, with the inclusion of textures and intentional differentiated oxidations (e.g., with the use of sulfur), making it also a contemporary material. By challenging the prejudices of traditional jewelry and its association with wealth, social status, in addition to potential cultural and religious positioning, contemporary jewelry employs new shapes and techniques to emphasize the value of the piece, rather than simply depending on the gross costs of the materials utilized [69]. New sources of inspiration, creative processes, shapes and compositions, materials, and manufacturing processes are allowed to be extrapolated by the designer to unconventional methods, which can bring some distinction to the jewelry pieces developed, besides giving greater freedom to those who produce and use them. According to Gong and Yuan [30], the designer sees jewelry as an expression of emotions or personality, giving them a new concept of life through materials, colors, techniques, etc. With the insertion of this new way of imagining, designing, and manufacturing jewelry, a new concept is being increasingly used in this field. Sustainability, in the context of selecting the most appropriate materials and processes—aiming at both environmentally sound, economically feasible, and socially fair—has been rethought to include aspects of preservation, extraction, processing, commercialization, recovery, and recycling. Also starting from popular demand, in which society as a whole has demanded fair alternatives through all the ways of the productive process, the jewelry industry has been constantly adapting for this objective. Still, even in an initial way, designers, jewelers, and industries have been sought efforts to design and manufacture their pieces following these biases. Many project trends involving design and multidisciplinary experimentation, considered a result of a mixture of sectors, policies, processes, ideas, materials, among others, are integrating and creating something new. Today, experiments with alternative materials, which were once considered "problematic" or even derived from waste, for example, are becoming more frequent. Along with practices of artisanal jewelry manufacturing, which can also reflect on sustainable criteria, the use of newer sources of materials can increase the value of the pieces, although they also require a persistent search for higher quality standards on a product's finishing. In jewelry, actions toward the merge of unusual materials with artisanal techniques can be increasingly observed in the latest years, and the employment of residues with a considerably low recovery rate may present even better benefits from a sustainable point of view.

3 Experimental Procedure

As commented before, the COVID-19 pandemic has led to a significant increase in the consumption of single-use plastics due to its disposable characteristic, particularly related to takeaway meal packaging. However, in developing countries with an incipient waste treatment system, such as Brazil, dealing with Urban Solid Waste is even more difficult. One type of post-consumption material that has one of the worst recovery rates in most parts of the country is EPS, due to the low economic interest to deal with this kind of residue. Therefore, newer practices of adding value to some problematic source material may be a way to diminish its recycling drawbacks, by reintroducing it into a circular economy. To demonstrate the potential jewelry has on transforming the intrinsic value of some disposed of material, with no direct commercial interest, this chapter presents the process of utilizing post-consumer EPS food packaging waste as source material.

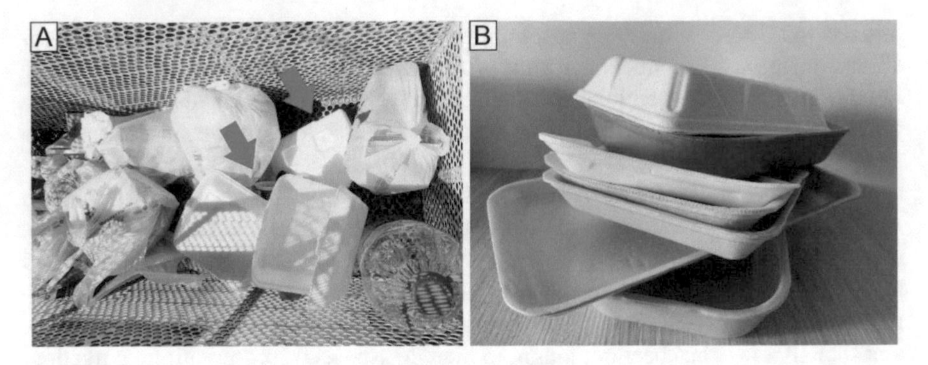

Fig. 1 Expanded polystyrene (EPS) packaging collected from waste: **A** dry waste with post-consumption single-use EPS for takeout meals; **B** examples of EPS packaging collected with different colors

3.1 EPS Waste Material

EPS post-consumer, food-packaging samples were collected from waste in the Porto Alegre and Santa Maria, the capital and an inland city, respectively, of the southernmost Brazilian state of Rio Grande do Sul. Samples with different EPS colors were sought, to investigate the possibility to use them as a gemstone replacement intended for contemporary jewelry. Almost all samples were originated from delivery or take-away meals from local restaurants, which represents a fast consumption of a product that leads to a slow decomposition residue, characteristic of a polymeric single-use packaging. The collected samples were first cleaned and divided by colors, prior to being cut into a smaller size (around 3 × 3 cm), as seen in Fig. 1.

3.2 Mold Fabrication

Samples selected for recycling were molded into the shape of real gemstones, in order to be used as a replacement alternative for jewelry application. The mold was created using gemstones with different gem cuts. For this research, gemstones with cabochon marquise, cabochon cushion antique, brilliant marquise, and a mixed cut, which combines cabbing and faceting techniques, were used as original models. The cabochon cut, derived from the French "*caboche*" ("head") consists of stones that have a flat back and a domed top [49]. The curved dome is smoothly polished and unfaced, mainly used on translucent and opaque stones, to highlight their color and pattern [66]. The cabochon marquise, also known as "*navette*" (derived from French to "incense box") comprehends within the group with unusual outlines and faces called fancy cuts [49] and consists of symmetrically curve-sided, elliptical boat shape with equally shaped points at both ends [66]. The cabochon cushion antique is a somewhat rectangular, rounded corner shape, maintaining the original curved,

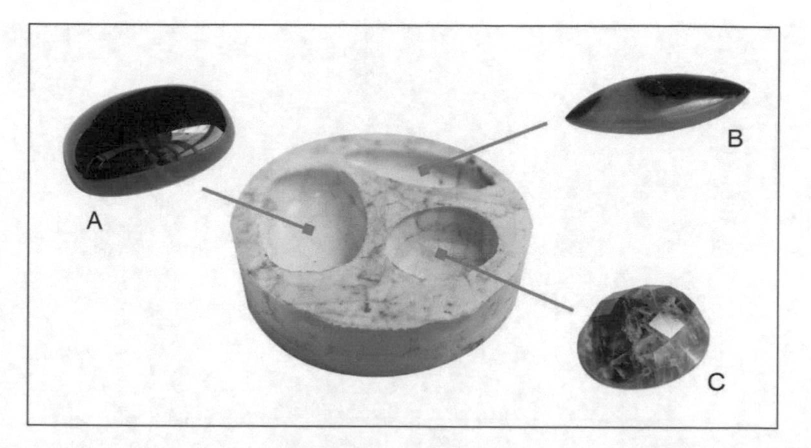

Fig. 2 Silicone rubber mold created with gemstones with different cuts: **A** cabochon cushion antique cut; **B** cabochon marquise cut; and **C** mixed cut

cabochon domed top [66]. Brilliant is a group of faced cuts that maximize the brilliance of gems and are often used on colored stones to deepen their color [49]. In the contemporary jewelry scenario, mixed cut styles have been increasingly used, and consist of the merge of more than one traditional cutting techniques [60], and are often applied with the fancy cuts group, due to multiple outlines, shapes, and facets employed [49]. The manufacture of the molds followed the placing of the selected gems inside a round stainless-steel container, which was covered with a compound silicone rubber MA4000E (Kinner® Silicone Rubber Indústria e Comércio Ltda., Ribeirão Pires, SP, Brazil). Next, the silicone was thermoset in a vulcanizer (Zezimaq® Indústria e Comércio de Máquinas e Fornos Ltda., Belo Horizonte, MG, Brazil) for 1 h at 180 °C. After cooling, the gemstones were removed from the mold. Figure 2 shows the finished mold and the gemstones used to create it, where the cabochon cushion antique cut is shown in Fig. 2A, the cabochon marquise cut is in Fig. 2B, and the mixed cut is presented in Fig. 2C.

3.3 Recycling and Pieces Obtained

Then, the cut samples with the same color were put in individual ceramic containers and processed in a fume hood for chemical dissolution [61] for the recycling procedure. Acetone $(CH_3)_2CO$, 99% (Anidrol® Produtos para Laboratórios Ltda., Diadema, São Paulo, Brazil) was used as the solvent. After dissolving, EPS was placed into prepared molds, and were left to evaporate for 48 h inside the mold to retain their shape. Finally, the pieces were removed from the mold and were left aside for around 1 week, for the acetone to completely evaporate and until the recycled gems become dry and rigid enough to be manipulated. Figure 3 illustrates the process of obtaining the polymeric gemstone pieces, via the recycling procedures followed.

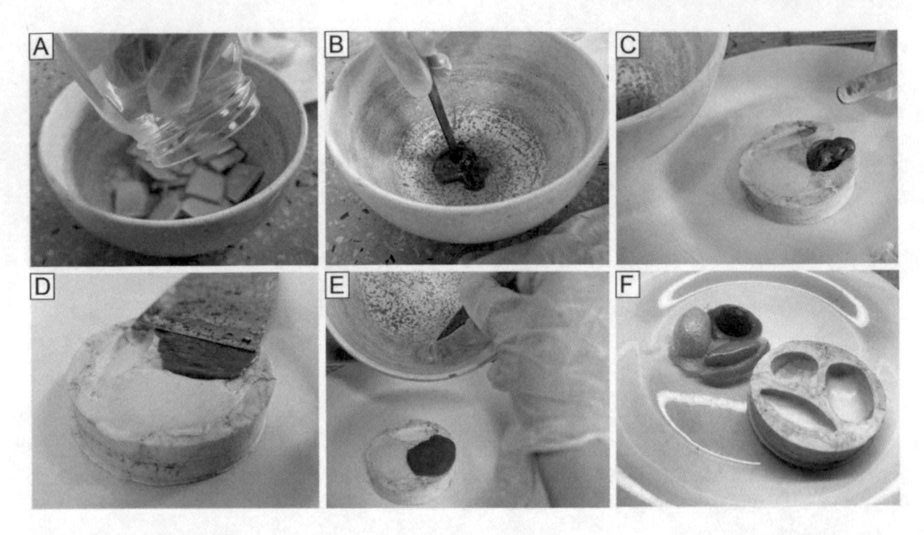

Fig. 3 Proposed experimental procedure: **A** pouring of acetone into the cut pieces of EPS; **B** mixing of the EPS until dissolved into a paste; **C** placing the dissolved EPS paste into the silicone mold; **D** flattening the top portion of the molded EPS; **E** repeating the process with a different EPS color; **F** pieces removed from the mold

Several pieces were created with the mold, using different EPS packaging colors, and the results are shown in Fig. 4. Additionally, experiments with different color combinations were followed, either by mixing cut EPS pieces that have distinct dyes, completely or partially with color gradients or by modifying the color shade with the addition of white or dark packaging pieces. Examples of the obtained pieces resulted from the process are seen in Fig. 4A. Generally, the obtained pieces retained the original mold shape fairly well. Even with non-cabochon gems, where faced cuts are employed, the recycled pieces maintained the original characteristics. It is noteworthy that the chemical recycling process of EPS resulted in the formation of air bubbles in the pieces during the solidification process, similar to reported by Gutiérrez and Colorado [31]. Despite reducing the volume of the material approximately 20 times [51], the formation of air bubbles was consistently found, regardless of how much the material was mixed during the process, or how long it was kept inside the mold. On the other hand, chemically recycled EPS of clearer colors, particularly white, produced significantly larger air bubbles after the acetone evaporates, in the solidification process. However, for the application as a gemstone replacement for contemporary jewelry, the formation of bubbles presented no issue and even gave character and individuality to each sample.

To illustrate the possibilities of the application of EPS chemically recycled into gemstones, two pieces of jewelry were developed. The pieces were manufactured using artisanal jewelry methods that are included during the addition of sterling silver (Ag 925). This type of silver alloy is produced using 92.5-wt% Ag and 7.5-wt% Cu, to improve the ductility of the silver and to make the metal more malleable. The sterling silver used in this process was obtained from scrap, being recycled from

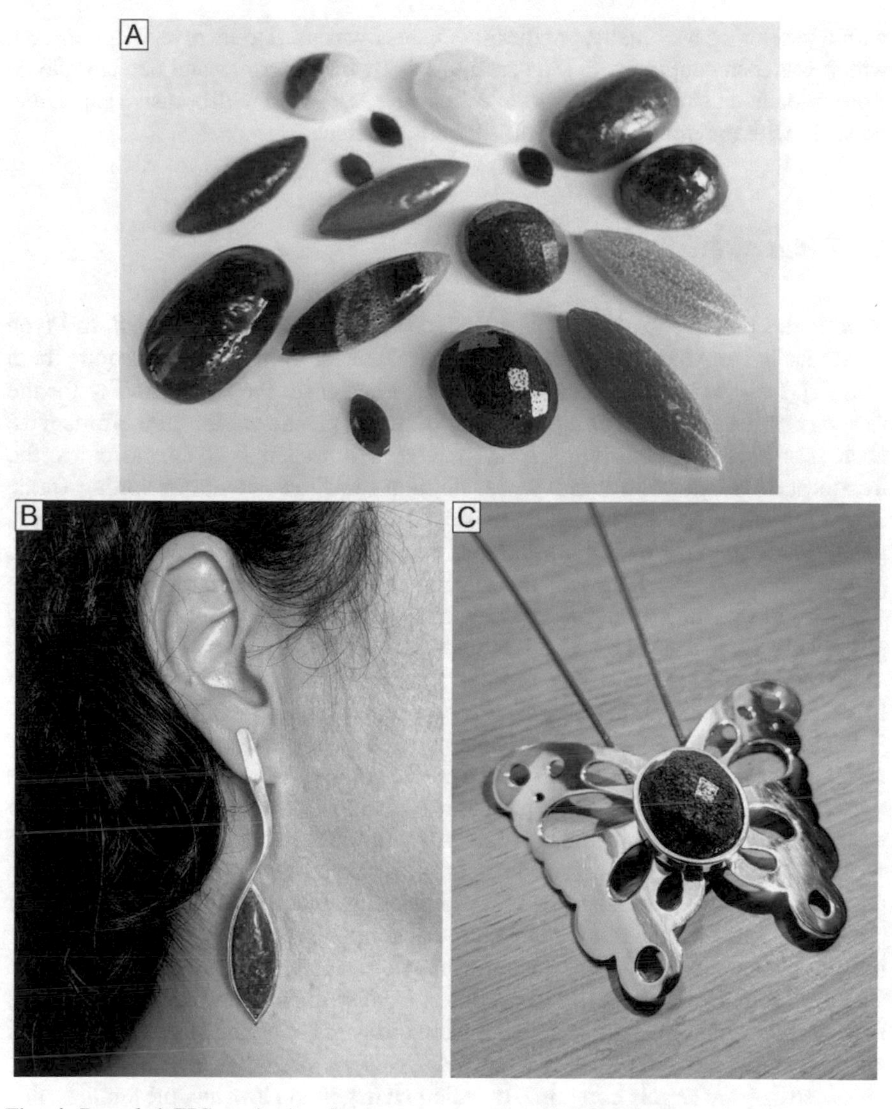

Fig. 4 Recycled EPS packaging for contemporary jewelry: **A** examples of obtained plastic gemstone pieces with different material colors; **B** application of a cabochon marquise cut piece with recycled silver into earrings; **C** application of a mixed cut piece into a pendant with recycled silver

other pieces. The first jewelry set produced consisted of a pair of earrings (Fig. 4B), where the silver was laminated with an electric rolling mill into a wire and then into a sheet until the desired thickness and length are obtained. The sheet was then shaped with jewelry pliers by shaping the earring and the gem support. The second produced piece was a pendant (Fig. 4C), following the fabrication of a silver sheet with the electric rolling mill and then with the shaping of a butterfly pattern by cutting with a jeweler's saw. Finally, another silver sheet was used to involve the gemstone, which was then coupled with the pendant. Lastly, both earrings and pendant pieces were smoothed using flies with different coarseness, sanded with abrasive paper, and polished with natural bristle and muslin buffing wheels.

4 Discussion

In countries with a poor recyclability rate of plastic waste, such as Brazil, this type of residue heavily depends on the commercial interest from buyers to acquire them [28, 52]. Despite years of national policies to designate the responsibility for the recovery or the definition of a proper end of life of plastic waste, little efforts have been seen that may contribute to this matter. From capitals to inland cities, the treatment of Urban Solid Waste passes through selection steps from Sorting Units, where low-income workers separate residues according to their type and sell them to third-party companies, who eventually recycle the waste and reinsert it into the production chain [52]. However, this situation is worrying on a sustainable, holistic scale due to several issues. The first one, obviously, is that potential buyers are only going to be interested in a secondary material that could be economically feasible when reused. If the residue cannot offer a minimum level of profit, it is not going to be commercialized. This creates an economical issue given that less revenue is being generated to the Sorting Units, as well as a great amount of secondary material that could be reprocessed and reused ends up being wasted, despite having a great usage potential. The second one regards the destination of the residue that is not sold. Brazilian legislation, in fact, accepts waste to be destined to landfilling when no other recovery choices are feasible. Furthermore, even open dumps are used to dispose of it irregularly in many cities. This directly contributes to a greater environmental impact, particularly when considering the time polymer-based residue takes to decompose and all the contaminants it releases during the process [41, 46]. The third one directly affects the welfare of the workers from the Sorting Units, because their main revenue is sourced from the gains obtained from the selling of waste. By reducing the volume of waste being sold, inevitably their salaries decrease, and since they are largely composed of underprivileged and low-income families, this represents a fair reduction in wages. Therefore, the issue of the recovery and reuse of secondary plastic material affects all spheres in the development of the country, and measures for improving their acceptance in the market are essential for keeping a sustainable consumption of polymer-based products.

There are currently several types of plastic waste that are not commercialized by most Sorting Units, varying from aluminum-metalized, bi-oriented polypropylene (BOPP) films, primarily used in the packaging of snacks, to composite materials, where different sorts of fibers are added to plastic resins for mechanical or aesthetical purposes, also largely used on food packaging, principally in organic products [52]. Despite being considered examples of multi-material products [70], known for being considered difficult to separate into their original materials, these types of residues can be successfully recovered in some recycling centers. For instance, newer programs of reverse logistics from the private sector are being settled in Brazil for the recovery of BOPP packaging [56], aiming to make this type of waste profitable by means of increasing the volume of reprocessed residues. However, there is one type of polymer waste that has become more difficult to recover due to lack of interest, even though it is consumed on a daily basis. Expanded polystyrene (EPS) is a globally known problematic material regarding its post-use treatment due to several difficulties on recycling. When considering factors like the low density of the secondary material and the low costs of the virgin styrene commodity as well as its manufacturing processes, gathering a significant volume of EPS in order to make its recycling viable leads to major logistic obstacles. In many parts of Brazil, for instance, no near EPS recycling plant can be found, making the process of collecting, storing, and transporting simply too expensive to be feasible. As a result, several major municipalities have no other choice but to landfill a large quantity of EPS residues. In addition, despite national policies forbidding inadequate treatments of several types of waste, in that case, the disposal of EPS waste is even legally supported precisely due to the costs of following more sustainable alternatives. Being one type of waste largely used in many industries, from cushioning of electronics to food packaging, EPS is a common choice. It is worth mentioning that even products that have a "sustainable" approach, such as organics, also widely employ the use of EPS, despite being considered one of the worst options in terms of recyclability [52]. According to the systematics of USW treatment in Brazil, the higher the economic interest of a given type of residue, the higher it is likely to be sold for reprocessing. As in the case of most polymeric residues, the value of the secondary material tends to be low, particularly by comparing it to its virgin equivalent. With the exception of secondary plastics that have a more established recycling process in the country, such as polyethylene terephthalate (PET)—where PET bottles from soft drinks are recycled and extruded as fibers for the manufacturing of brooms—polymeric residues like EPS are almost never used as source material. Therefore, finding alternatives for the recycling of EPS waste, either by means of increasing its intrinsic value or by discovering newer application options is a necessity in the reality of Brazil.

In addition to the already difficult scenario of plastic waste recovery in Brazil, a newer challenge has recently arisen. The consumption of single-use plastics has seen its usage growth in the past year during to the COVID-19 pandemic and its consequent social restrictions, which decrease the presence of people in local bars and restaurants. All over Brazil, this led to a considerable growth in delivery services, mainly related to the meals. As a preferred way to package food in takeout meals, thanks to its thermal insulation, low cost, and low density, the use of EPS is set as one of the main types of problematic materials for post-consumer waste treatment. This new escalation of single-use plastics during the COVID-19 pandemic has been a global matter of investigation and concern, given that scientists and environmentalists have been struggling to reduce its consumption for years. It is clear that single-use plastics may present a number of advantages during a pandemic. From the product design point of view, they are cheap, lightweight, easy to manufacture, and mold to any shape, and have sufficient mechanical properties to endure a typical usage, whether as a package, cup, plate, straw, or cutlery. Adding that to health and safety benefits, people trust a never-before-used utensil as a way to ensure a contaminant-free object, assuming that it is going to be discarded shortly after. As a result, single-use plastics have gained powerful arguments with public opinion to ensure a continued existence, even by finding breaches within the law in several spheres. Silva et al. [63] commented on the priorities that are urged to be assessed due to the COVID-19, such as finding bio-based alternatives to oil-based plastics and improving the recycling systems. Likewise, Parashar and Hait [54] highlighted that plastics would not be seen as a villain material if appropriate recycling procedures in a circular economy are defined, giving the fact that they are contributing to public health safety during COVID-19. The authors also pointed out that using not only mechanical but also chemical recycling for the creation of valuable products is a way to facilitate the obtention of circularity and a more sustainable economy.

As previously seen in the development of contemporary jewelry, the exploration of alternative materials to gold and diamonds is a practice followed since modern industrialization, as an option to circumvent the lack of aesthetic appreciation of the massification of traditional jewelry. Even though containing noble materials, jewelry pieces produced were then mostly copies, in addition to having a lower finish quality. To counterbalance, jewelers started to look for ways to retrieve one of the original characteristics of what defines one object as a jewelry piece, its uniqueness. By this means, employing different materials and manufacturing processes that can support the distinctiveness of a jewel has been a key feature of contemporary jewelry. Within this realm, many materials have been explored, still always being supplemented by what is trending in terms of style, symbolism, culture, and social importance. In a relatively new field, both in product design and in jewelry, is the social awareness related to sustainable approaches from manufacturers, particularly when environmentally found [15]. In this sense, people are more concerned with commercial practices that consider our society on a holistic scale—social, environmental, and economic—notwithstanding the company's size. From small, local jewelers to a large, international diamond producer, in contemporary jewelry, the need for a sustainability-driven and ethical commercial approach is becoming mandatory for it to be well received by the public. In particular, one way that contemporary jewelry could benefit from this demand is by employing materials considered problematic under some circumstances. In this chapter, we exemplified this idea by addressing it toward the reality in Brazil, by presenting the manufacture of parts with materials obtained from disposal and with no apparent market value, as in the case of Urban Solid Waste. By considering the country's current growing supply of valueless EPS waste, mainly derived from packaging and takeaway meals due to the COVID-19, added to the fact that restrictions that once banned this type of single-use plastic in many places are being revoked, this material has never been so problematic. We followed a chemical recycling procedure of EPS waste and its application as an alternative to gemstones in jewelry pieces. By exploring the array of colors these packages can be found in the waste, different shades and gradients can be obtained. The proposed procedure included the manufacturing of a silicone mold, where the dissolved EPS could be shaped for the latter being applied into recovered sterling silver jewelry pieces. Regarding this method of chemically reprocessing EPS for application in jewelry design, two resulting characteristics are worth underlining. The first one is related to the fact that the EPS pieces contain air bubbles that originated in the solidification process, by the evaporation of acetone. Generally, the clearer the EPS color, *i.e.*, the less color dye the material originally has, the more bubbles it is going to generate after being solidified. It has been noticed that despite the generation of these pores, the shaped pieces retained the mold geometry pretty well, therefore they mostly do not develop in the surfaces that are in contact with the mold. Another issue from this approach is that if the recycled EPS is going to be employed in a product where mechanical properties are of importance, one may consider using a more traditional mechanical recycling method of the thermoplastic resin, *i.e.*, by melting it in the reprocessing procedure. Giving the fact that these pores make the material more fragile, as this procedure would contribute to further densify the material by reducing the number and size of bubbles. On the other hand, by applying it in jewelry, there is no particular need to improve the structural integrity of the gemstone. The second characteristic worth mentioning is intrinsic to EPS, as the material is extremely lightweight. Even by chemically dissolving it with acetone, which manages to reduce its original volume approximately 20 times [51], the resulted piece is still very light. In terms of jewelry applications, this means that larger or more numerous pieces can be produced and applied without the need for excessive concerns regarding their final weight. This is a major issue in jewelry due to ergonomic factors, particularly related to earrings, where excessively heavy pieces can cause tissue damages. Lastly, employing USW-recovered materials into contemporary jewelry, particularly plastics, can be considered a safe alternative regarding ergonomic factors.

5 Conclusions

COVID-19 pandemic and the consequent lockdown measures and social distancing restrictions have led to recent growth in the consumption of delivery and takeaway meals, most of them packaged with single-use plastic. In many cities and countries, laws prohibiting this type of plastic have been revoked, once again authorizing their use, under the justification of being safer for public health. However, in countries with incipient Urban Solid Waste (USW) treatment, like Brazil, this also represents a considerable increase in residues that are mostly not going to be recovered. Among the types of single-use plastics normally found in USW that have the lowest recovery rate is expanded polystyrene (EPS), besides being one of the most employed materials for food packaging, particularly for takeaway meals. Therefore, the current situation of EPS waste in Brazil has reached a critical point, and measures for its recovery need to be discussed.

In order to be recovered in Brazil, a certain type of residue must have some economic interest from potential buyers, and the logistics that allow its reprocessing must be viable. Giving the fact that EPS has numerous disadvantages in this regard—primarily related to its low density and low intrinsic value—gathering, storing, and transporting sufficient quantities to make it profitable when compared to the virgin material is a difficult challenge and needs to be addressed. Many large Brazilian urban centers do not have easy access to EPS recycling plants, thus making its recovery not feasible. Having a residue that has so little economic interest although is so widely used, one way of addressing this matter is by increasing its perceived value via its use as some source material for new applications. One recent field of product design that has shifting paradigms of conventional materials is contemporary jewelry. In this chapter, we proposed an experimental procedure of applying post-consumer EPS waste as a base material for usage as a replacement of gemstones in traditional jewelry. Different colors of the waste material were chemically recycled and shaped into gemstones in fabricated silicone molds. The produced pieces showed an interesting pattern that could be included in the design of pieces made from recovered sterling silver. One of the main advantages of this procedure was the maintaining of EPS with low density despite the significant volume reduction, which contributes to a lightweight jewelry piece obtained.

Governments and local authorities have been facing many difficulties derived from the COVID-19 in Brazil. From public health to the economy, the pandemic forced many people to revise their habits and face new challenges. However, even though some studies have been reporting the implications of the recent growth in plastic waste pollution from personal protective equipment, this research aimed to expose the matter of one type of residue that is silently but alarmingly increasing, despite already being a well-known example of problematic materials. USW treatment in Brazil is a matter that involves several spheres and reaches social, environmental, and economic aspects of the society since low-income workers are responsible for most of the dry waste destinations in the country. EPS waste needs to be dealt with by public policies that incentive the reprocessing and recovery of this residue, either by expanding the network of recycling facilities throughout the country or by offering ways for workers to increase the value of it. In this sense, contemporary jewelry studies should be seen as one important and sustainable way of contributing to finding better solutions for problematic materials.

Acknowledgements The authors thank the "National Council for Scientific and Technological Development—CNPq" for supporting this study through the project "Chamada Universal MCTIC/CNPq 2018". This study was financed in part by the Coordenação de Aperfeiçoamento de Pessoal de Nível Superior—Brasil (CAPES)—Finance Code 001.

References

1. ABRELPE (2020) Panorama dos resíduos sólidos no Brasil 2020 [Panorama of solid waste in Brazil 2020]. ABRELPE, São Paulo
2. ABRELPE (2020) Recomendações para a gestão de resíduos sólidos durante a pandemia de coronavírus (COVID-19) [Recommendations for solid waste management during the coronavirus pandemic (COVID-19)]. São Paulo
3. ANVISA (1999) Resolução N° 105, de 19 de maio de 1999 [Resolution N° 105, from may 19, 1999]. Brasília, DF
4. Ashby MF, Johnson K (2011) Materiais e design: arte e ciência na seleção de materiais em projeto de produto, 2ª. CAMPUS, Rio de Janeiro
5. Azapagic A, Emsley A, Hamerton L (2003) Polymers, the environment and sustainable development. John Wiley & Sons, Ltd, Chichester, UK
6. Ba'ai NM, Hashim HZ (2015) Waste to wealth: the innovation of areca catechu as a biomaterial in esthetics seed-based jewelry. In: Proceedings of the international symposium on research of arts, design and humanities (ISRADH 2014). Springer Singapore, Singapore, pp 373–381
7. Balaguera YLS (2013) La influencia de los materiales en el significado de la joya. Cuad del Cent Estud en Diseño y Comun Ensayos 115–153
8. Baysal EL (2019) Personal ornaments in prehistory: an exploration of body augmentation from the Palaeolithic to the early bronze age. Oxbow Books, Oxford
9. Brasil (2010) Lei nº 12.305 de 2 de agosto de 2010. Institui a Política Nacional de Resíduos Sólidos; altera a Lei no 9.605, de 12 de fevereiro de 1998; e dá outras providências. [Law Nº. 12.305, of August 2, 2010, establishing the National Policy on Solid Waste, amending Law Nº. 9,605, of February 12, 1998 and other measures]. Brasília, DF
10. Brock J (2020) The plastic pandemic: COVID-19 trashed the recycling dream. In: REUTERS. https://www.reuters.com/investigates/special-report/health-coronavirus-plastic-recycling/. Accessed 26 Mar 2021
11. Büll P (2020) Gastos com delivery crescem mais de 94% na pandemia [Delivery costs grow more than 94% in the pandemic]. In: NOVAREJO. https://www.consumidormoderno.com.br/2020/07/08/gastos-com-delivery-crescem-mais-de-94-durante-a-pandemia/. Accessed 25 Mar 2020
12. Bürdek BE (2015) Design: history, theory and practice of product design, 2nd edn. Birkhäuser, Basel, SZ
13. de Campos AP (2011) Pensando a joalheria contemporânea com Deleuze e Guattari. Rev Trama Interdiscip 2:167–179
14. Campos HKT (2014) Recycling in Brazil: challenges and prospects. Resour Conserv Recycl 85:130–138. https://doi.org/10.1016/j.resconrec.2013.10.017
15. Cappellieri A, Tenuta L, Testa S (2020) Jewellery between product and experience: luxury in the twenty-first century. In: Gardetti MÁ, Coste-Manière I (eds) Sustainable luxury and craftsmanship. Springer, Singapore, pp 1–23
16. Cardoso R (2008) Uma introdução à história do design. Edgard Blücher, São Paulo
17. Carrizo P (2021) Quem faz o mercado da reciclagem em Porto Alegre? [Who does the recycling market in Porto Alegre?]. In: J. do Comércio. https://www.jornaldocomercio.com/_conteudo/cadernos/empresas_e_negocios/2021/02/778543-quem-faz-o-mercado-da-reciclagem-em-porto-alegre.html. Accessed 25 Mar 2020
18. Castro A (2020) Com separação incorreta, baixo preço de venda e 'coleta clandestina', Porto Alegre só recicla 6% do lixo [With incorrect separation, low sales price and 'clandestine collection', Porto Alegre only recycles 6% of the waste]. In: Sul21. https://www.sul21.com.br/caminhos-do-lixo/2020/02/com-separacao-incorreta-baixo-preco-de-venda-e-coleta-clandestina-porto-alegre-so-recicla-6-do-lixo/. Accessed 25 Mar 2020
19. Chen Y, Awasthi AK, Wei F, Tan Q, Li J (2021) Single-use plastics: production, usage, disposal, and adverse impacts. Sci Total Environ 752:141772

20. Cidade MK, Duarte L da C, Palombini FL, Barp DRA (2016) Inovações tecnológicas em gemas: aplicação de microcápsulas aromáticas em ágata gravada a laser. In: Donato M, Duarte L da C (eds) Gemas, Joias e Mineração: Pesquisas Aplicadas no Rio Grande do Sul. UFRGS/IGEO, Porto Alegre, pp 38–47
21. Cidade MK, Palombini FL, Duarte L da C, Paciornik S (2018) Investigation of the thermal microstructural effects of CO 2 laser engraving on agate via X-ray microtomography. Opt Laser Technol 104:56–64. https://doi.org/10.1016/j.optlastec.2018.02.002
22. Cidade MK, Palombini FL, Kindlein Júnior W (2015) Biônica como processo criativo : microestrutura do bambu como metáfora gráfica no design de joias contemporâneas. Rev Educ Gráfica 19:91–103
23. Corbetta G (2007) Joalheria de arte. AGE, Porto Alegre
24. de Abreu e Lima CE, Lebrón R, de Souza AJ, Ferreira NF, Neis PD (2016) Study of influence of traverse speed and abrasive mass flowrate in abrasive water jet machining of gemstones. Int J Adv Manuf Technol 83:77–87. https://doi.org/10.1007/s00170-015-7529-9
25. Ebewele RO (2000) Polymer science and technology. CRC Press, New York
26. Faggiani K (2006) O poder do design: da ostentação à emoção. Thesaurus, Brasília
27. Forster P (2020) Aumenta o consumo de descartáveis por causa da pandemia, diz associação [Consumption of disposables increases due to pandemic, says association]. In: CNN Bras. https://www.cnnbrasil.com.br/business/2020/07/29/aumenta-o-consumo-de-descartaveis-por-causa-da-pandemia-diz-abrelpe. Accessed 25 Mar 2020
28. Fuss M, Vasconcelos Barros RT, Poganietz WR (2018) Designing a framework for municipal solid waste management towards sustainability in emerging economy countries - an application to a case study in Belo Horizonte (Brazil). J Clean Prod 178:655–664. https://doi.org/10.1016/j.jclepro.2018.01.051
29. Gola E (2013) A joia: história e design, 2ª. Editora Senac São Paulo, São Paulo
30. Gong B, Yuan R (2017) Study of contemporary jewelry design emotional expression skills. J Arts Humanit 6:57. https://doi.org/10.18533/journal.v6i2.1113
31. Gutiérrez EI, Colorado HA (2020) Development and characterization of a luminescent coating for asphalt pavements. In: Minerals, metals and materials series. Springer, pp 511–519
32. Heiges J, O'Neill K (2020) COVID-19 has resurrected single-use plastics – are they back to stay? In: Conversat. https://theconversation.com/covid-19-has-resurrected-single-use-plastics-are-they-back-to-stay-140328. Accessed 25 Mar 2020
33. Heinrich Böll Stiftung (2020) Atlas do Plástico [Atlas of Plastic]. Rio de Janeiro
34. Heskett J (2005) Design: a very short introduction. Oxford University Press, New York
35. Hesse RW (2007) Jewelrymaking through history: an encyclopedia. Greenwood Press, Westport, Connecticut, USA
36. Hiemstra A-F, Rambonnet L, Gravendeel B, Schilthuizen M (2021) The effects of COVID-19 litter on animal life. Anim Biol 1–17. https://doi.org/10.1163/15707563-bja10052
37. Honorato L (2020) Quarentena faz aumentar coleta de lixo [Quarantine increases garbage collection]. In: O Estado São Paulo. https://sustentabilidade.estadao.com.br/noticias/geral,quarentena-faz-aumentar-coleta-de-lixo,70003345246. Accessed 25 Mar 2020
38. Isaías C (2020) Isolamento e trabalho em casa aumentaram produção de lixo doméstico e reciclável em Porto Alegre [Isolation and work at home increased production of household and recyclable waste in Porto Alegre]. In: Corr. do Povo. https://www.correiodopovo.com.br/notícias/geral/isolamento-e-trabalho-em-casa-aumentaram-produção-de-lixo-doméstico-e-reciclável-em-porto-alegre-1.545518. Accessed 25 Mar 2020
39. Jahnke A (2020) A discussion of single-use plastics in medical settings. Reinf Plast 64:190–192. https://doi.org/10.1016/j.repl.2019.12.002
40. Kliauga AM, Ferrante M (2009) Metalurgia básica para ourives e designers: do metal à jóia. Blücher, São Paulo
41. La Mantia FP (2002) Handbook of plastics recycling. Rapra Technology Ltd, Shrewsbury
42. Law KL, Thompson RC (2014) Microplastics in the seas. Science (80-) 345:144–145. https://doi.org/10.1126/science.1254065

43. Leissner S, Ryan-Fogarty Y (2019) Challenges and opportunities for reduction of single use plastics in healthcare: a case study of single use infant formula bottles in two Irish maternity hospitals. Resour Conserv Recycl 151:104462. https://doi.org/10.1016/j.resconrec.2019.104462
44. Magtaz M (2008) Joalheria Brasileira: do descobrimento ao século XX. Mariana Magtaz, São Paulo
45. Mello D (2020) Liminar suspende lei que proíbe copos e talheres de plástico em SP [Injunction suspends law prohibiting plastic cups and cutlery in SP]. In: Agência Bras. https://agenciabrasil.ebc.com.br/geral/noticia/2020-04/liminar-suspende-lei-que-proibe-copos-e-talheres-de-plastico-em-sp. Accessed 25 Mar 2020
46. Merrington A (2017) Recycling of plastics. In: Applied plastics engineering handbook: processing, materials, and applications, 2nd edn. Elsevier Inc., pp 167–189
47. Milewski JO (2017) Additive Manufacturing metal, the art of the possible. In: Springer series in materials science. Springer Verlag, pp 7–33
48. Millar K (2008) Making trash into treasure: struggles for autonomy on a Brazilian garbage dump. Anthropol Work Rev 29:25–34. https://doi.org/10.1111/j.1548-1417.2008.00011.x
49. Miller J (2016) Jewel: a celebration of earth's treasures. Dorling Kindersley Ltd., London
50. Mortensen LF, Tange I, Stenmarck Å, Fråne A, Nielsen T, Boberg N, Bauer F (2021) Plastics, the circular economy and Europe′s environment - A priority for action
51. Noguchi T, Miyashita M, Inagaki Y, Watanabe H (1998) A new recycling system for expanded polystyrene using a natural solvent. Part 1. A new recycling technique. Packag Technol Sci 11:19–27. https://doi.org/10.1002/(SICI)1099-1522(199802)11:1%3c19::AID-PTS414%3e3.0.CO;2-5
52. Palombini FL, Cidade MK, de Jacques JJ (2017) How sustainable is organic packaging? A design method for recyclability assessment via a social perspective: a case study of Porto Alegre city (Brazil). J Clean Prod 142:2593–2605. https://doi.org/10.1016/j.jclepro.2016.11.016
53. Palombini FL, Demori R, Cidade MK, Kindlein W, de Jacques JJ (2018) Occurrence and recovery of small-sized plastic debris from a Brazilian beach: characterization, recycling, and mechanical analysis. Environ Sci Pollut Res 25:26218–26227. https://doi.org/10.1007/s11356-018-2678-7
54. Parashar N, Hait S (2021) Plastics in the time of COVID-19 pandemic: protector or polluter? Sci Total Environ 759:144274. https://doi.org/10.1016/j.scitotenv.2020.144274
55. Pedicini P (2020) With the COVID-19 pandemic, littering from single-use plastics has seen a dramatic resurgence. In: Parliam. https://www.theparliamentmagazine.eu/news/article/drastic-actions-on-plastics. Accessed 26 Mar 2021
56. Pinheiro L (2021) Logística reversa para embalagens [Reverse logistics for packaging]. In: IstoÉ Dinheiro. https://www.istoedinheiro.com.br/logistica-reversa-para-embalagens/. Accessed 26 Mar 2021
57. Prata JC, Silva ALP, Walker TR, Duarte AC, Rocha-Santos T (2020) COVID-19 pandemic repercussions on the use and management of plastics. Environ Sci Technol 54:7760–7765. https://doi.org/10.1021/acs.est.0c02178
58. Schneider B (2010) Design - uma introdução: o design no contexto social, cultural e econômico. Edgard Blücher, São Paulo
59. Schnurr REJ, Alboiu V, Chaudhary M, Corbett RA, Quanz ME, Sankar K, Srain HS, Thavarajah V, Xanthos D, Walker TR (2018) Reducing marine pollution from single-use plastics (SUPs): a review. Mar Pollut Bull 137:157–171. https://doi.org/10.1016/j.marpolbul.2018.10.001
60. Schumann W (2009) Gemstones of the world. Sterling Publishing Company Inc., New York
61. Schyns ZOG, Shaver MP (2021) Mechanical recycling of packaging plastics: a review. Macromol Rapid Commun 42:2000415. https://doi.org/10.1002/marc.202000415
62. Sharma HB, Vanapalli KR, Cheela VS, Ranjan VP, Jaglan AK, Dubey B, Goel S, Bhattacharya J (2020) Challenges, opportunities, and innovations for effective solid waste management during and post COVID-19 pandemic. Resour Conserv Recycl 162:105052. https://doi.org/10.1016/j.resconrec.2020.105052

63. Silva ALP, Prata JC, Walker TR, Campos D, Duarte AC, Soares AMVM, Barcelò D, Rocha-Santos T (2020) Rethinking and optimising plastic waste management under COVID-19 pandemic: policy solutions based on redesign and reduction of single-use plastics and personal protective equipment. Sci Total Environ 742:140565. https://doi.org/10.1016/j.scitotenv.2020.140565
64. SINDIPLAST (2020) SINDIPLAST derruba lei municipal que proibia fornecimento de descartáveis em São Paulo [SINDIPLAST overturns municipal law that prohibited the supply of disposables in São Paulo]. http://www.sindiplast.org.br/sem-categoria/sindiplast-derruba-lei-municipal-que-proibia-fornecimento-de-descartaveis-em-sao-paulo/. Accessed 25 Mar 2020
65. Tolinski M (2011) Plastics and sustainability: towards a peaceful coexistence between bio-based and fossil fuel-based plastics. John Wiley & Sons Inc., Hoboken, NJ
66. Untracht O (2011) Jewelry concepts & technology. Doubleday, New York
67. Van Cauwenberghe L, Devriese L, Galgani F, Robbens J, Janssen CR (2015) Microplastics in sediments: a review of techniques, occurrence and effects. Mar Environ Res 111:5–17. https://doi.org/10.1016/j.marenvres.2015.06.007
68. Wagner TP (2017) Reducing single-use plastic shopping bags in the USA. Waste Manag 70:3–12. https://doi.org/10.1016/j.wasman.2017.09.003
69. Wallace J, Dearden A (2005) Digital jewellery as experience. Future interaction design. Springer-Verlag, London, pp 193–216
70. Wargnier H, Kromm FX, Danis M, Brechet Y (2014) Proposal for a multi-material design procedure. Mater Des 56:44–49. https://doi.org/10.1016/j.matdes.2013.11.004
71. Wit W de, Hamilton A, Scheer R, Stakes T, Allan S (2019) Solving plastic pollution through accountability, a report for WWF. Gland, Switzerland
72. Worrell E, Reuter MA (eds) (2014) Handbook of recycling. Elsevier
73. Xanthos D, Walker TR (2017) International policies to reduce plastic marine pollution from single-use plastics (plastic bags and microbeads): a review. Mar Pollut Bull 118:17–26
74. Xavier CS, Neves L (2020) Para onde vai tanta embalagem? [Where does so much packaging go?]. In: Diário St. Maria. https://diariosm.com.br/mix/vídeos-de-quem-é-a-responsabilidade-e-qual-é-o-destino-do-lixo-gerado-em-santa-maria-1.2236764. Accessed 25 Mar 2020

Consumers' Purchase Intention and Willingness to Pay for Eco-Friendly Packaging in Vietnam

Anh Thu Nguyen, Nguyễn Yến-Khanh, and Nguyen Hoang Thuan

Abstract Global concern for pollution from plastic packaging has been growing. In practice, packaging for consumer goods is mostly produced using plastic, which is used once and discarded. Given the huge volume of global plastic production, single-use plastic packaging has posed a massive threat to the environment. The good news is that there are initiatives around the world to reduce the production and consumption of single-use plastic. Businesses are under increasing pressures not only from consumers but also from governments and investors to phase out single-use plastic. Food manufacturers are starting to adopt the use of eco-friendly packaging for their products. However, responses from different consumer groups to eco-friendly packaging are still unclear. Questions exist on how demographics are associated with consumers' intention to buy eco-friendly packaging, especially in a developing market where plastic use is prominent. In this study, we examined the relations between consumers' demographics and their reactions towards eco-friendly packaging. We focused on packaged instant noodles in Vietnam as this emerging market has been experiencing rapid developments of convenient consumption patterns. We used choice-based experiment embedded in a self-administered online consumer survey. Demographic data were collected to explore the relationship with intention to buy and consumer responses to price levels with regard to eco-friendly packaging. There were 308 usable responses for analysis. Simple statistical analysis such as descriptive statistical analysis, frequency distribution, cross tabulation, analysis of variance and factor analysis were undertaken to assess consumers' intention to buy and willingness to pay with regard for eco-friendly packaging. T-tests and analysis of variance statistical techniques were also used to analyse the collected data. The

A. T. Nguyen (✉) · N. Yến-Khanh · N. H. Thuan
School of Business and Management, RMIT University Vietnam, 702 Nguyen Van Linh street, Ho Chi Minh City, Vietnam
e-mail: thu.nguyen@rmit.edu.vn

N. Yến-Khanh
e-mail: khanh.nguyenyen@rmit.edu.vn

N. H. Thuan
e-mail: thuan.nguyenhoang@rmit.edu.vn

S. S. Muthu (eds.), *Sustainable Packaging*, Environmental Footprints and Eco-design of Products and Processes, https://doi.org/10.1007/978-981-16-4609-6_11

findings will help packaged food businesses in managerial decisions for eco-friendly packaging strategies.

Keywords Consumers · Eco-friendly packaging · Packaged instant noodles · Purchase intention · Willingness to pay · Vietnam

1 Introduction

Since green production and consumption become a pressing issue in the global context [1], it deems more crucial to understand how consumers form intention to buy products with eco-friendly characteristics. As an emerging market with a big population of almost 98 million people [2], Vietnam has a critical role to play in the global green consumption movement. As plastic consumption in Vietnam increases at a 10 per cent annual growth rate to over 41 kg per person per year [3], Vietnam is among the top five countries, including China, Indonesia, Thailand and the Philippines, that generate 50 per cent of plastic waste in the world [4]. Thus, promoting consumer green consumption behaviour should be a priority from both government policies and manufacturers' strategies in Vietnam [5, 6]. To grow a sustainable consumption culture in this emerging market, a better understanding of Vietnamese consumer dynamics towards green consumption in general and eco-friendly packaging in particular is important [7]. Hence, this research is hoped to shed light on Vietnamese consumers' reactions to eco-friendly packaging, which can also be applied to other markets experiencing similar emerging consumer trends.

Even though policies on green production and consumption in Vietnam have not been effectively implemented [5, 8], the good news is that there are initiatives from Vietnam-based manufacturers to reduce the production and consumption of single-use plastic. Businesses are under increasing pressures not only from the government but also from consumers and investors to phase out single-use plastic and be more environmentally responsible [9–11]. Manufacturers are setting reduction goals and facilitating the use of eco-friendly packaging for their products. However, responses from different consumer groups to eco-friendly packaging are still unclear. Questions exist on how demographics are associated with consumers' purchase intention and willingness to pay for eco-friendly packaging, especially in the complex context of a developing market of Vietnam where plastic use is prominent. With this in mind, we aim to explore consumer purchase intention and willingness to pay for eco-friendly packaging in association with demographic characteristics inherent in the consumer.

The chapter is organised as follows. It will first discuss major theoretical trends and existing knowledge in eco-friendly packaging and consumer behaviour in the global context and in the emerging market of Vietnam so as to point out the research gap and propose research questions. After presenting the research method and results, the chapter discusses the findings and the implications of social contexts and situational specifics on consumers' intention to buy and willingness to pay for eco-friendly packaged food products. We then conclude with theoretical and practical contributions and outline future research directions.

2 Literature Review

Green consumption is an important field of research as it has an impact on the ecological system through consumption activities. Green consumption refers to the purchase and usage of environment-friendly products with minimal bad effects on the ecosystem [12]. As consumers buy packaged products for convenience, consumer behaviour associated with packaging is becoming an important component of green consumption [13, 14].

Sustainable packaging is defined by Sustainable Packaging Coalition [15] as beneficial, safe and healthy for individuals and communities; however, it needs to meet market criteria for performance and cost throughout its life cycle. Renewable energy and healthy materials with clean production technologies and procedures should be implemented from sourcing, manufacturing, transportation to recycling. Various terminologies have been used by researchers to refer to environmental packaging, such as green packaging, environmentally friendly packaging, eco-packaging, eco-friendly packaging, sustainable design, eco-design, design for the environment and environmentally conscious design [16]. For consistency, in this chapter, we use the term eco-friendly package (and/or eco-friendly packaged products). Eco-friendly packaged products can be defined as ones which cause fewer impacts on the environment with its packaging.

Packaging is one of the product attributes that consumers consider when they buy green products [13, 17, 18]. For example, packaging has been perceived by consumers in the United Kingdom (UK) as posing the greatest environmental and ethical concern [19]. In Sweden, green or eco-friendly packaging is ranked by 30 per cent of consumers as the most important criterion in their purchase of a drink product [20]. Nguyen et al. [21] find that Vietnamese consumers perceive eco-friendly packaging centres around three tenets, including packaging materials, market appeal and manufacturing technology. Consumer knowledge is more related to whether the materials are biodegradable or recyclable, as well as the attractiveness of graphic design and good price, rather than manufacturing technologies. In Vietnamese consumers' perceptions, paper packaging is more eco-friendly than plastic [21], yet consumers are often not aware of the life cycle assessment between paper and plastic to determine a package is more or less eco-friendly [22]. Nevertheless, generally, consumers have a more favourable attitude towards paper than plastic when it comes to packaging materials [23]. Thus, the perception of what materials are more environment-friendly can affect their decision to buy particular packaged products.

In an effort to rank the importance of eco-friendly packaging criteria in consumers' willingness to pay in China, Hao et al. [17] find four major factors, including environment-friendliness, packaging quality, commodity and package price as having the strongest weight. As of 2021, there is no prior study reported on any quantitative evaluation of customers' criteria for eco-friendly packaging in Vietnam, but a qualitative study by Nguyen et al. [21] suggests that Vietnamese buyers expect eco-friendly packaging to be non-toxic, decomposable, biodegradable and recyclable. The cost to be eco-friendly should not be too high, and at the same time, the package must be appealing in graphic designs and satisfy protective functionality. As indicated by Seo et al. [24], because consumers can evaluate the level of packaging eco-friendliness before their purchase, this product attribute can add values or vice versa in consumer perceptions. Excessive packaging can be seen as undesirable and drives consumers to other alternatives, with such features as being reusable, recyclable and biodegradable.

In the consumer decision-making process for green purchases, price is a key and complex criterion [17, 25]. In consumers' perceptions, packaging eco-friendliness has the reputation to be more expensive [26]. While higher prices are accepted by some consumers, more expensive eco-packaging do not appeal to others [17, 18]. Moreover, a study by Liobikienė et al. [27] on determinant factors influencing green purchase behaviour in European countries has pointed out the differences in consumer behaviour across different cultures. Noticeably, economic development does not substantially correlate with the frequency of green purchase behaviour among European citizens. In other words, income and price importance play variable roles in green purchase behaviour in European markets. Instead, subjective norms is reported to significantly determine green purchase behaviour, including 'perceived consumer effectiveness, attitude towards the behaviour; or variables concerned with perceived behavioural control' [27].

In addition to product attributes, key demographic variables play a vital role in how manufacturers and marketers mobilise different strategies in targeting different consumer segments. ElHaffar et al. [28] has identified different responses from different groups of consumers towards green consumption appeals and interventions. It might be efficient for marketers to target those who incline towards green living before expanding to others. Of the demographic variables, researchers have investigated education, age and gender regarding their influence on consumers' willingness to pay for green products and have depicted a diverse picture of green consumer behaviour [18, 29, 30, 31]. In the American market, 82 per cent of respondents perceive the green movement as feminine [29]. In the Netherlands, research reports that educated females more frequently value green lifestyles, and while age predicts willingness to reduce carbon emission, income does not determine the willingness for green trade-off [30]. In China, age and income have positive correlations with green purchase attitudes, while education and income are predictors of behavioural intention [31]. Still, males and females differ significantly in their green purchase attitudes and intention to buy [31]. In Vietnam, Luu [18] find that those with higher education and bigger family size are more likely to buy organic products regularly. Given the importance of consumers' demographic characteristics, the current study aims to further examine roles of demographic variables associating with consumer behaviour with regard to eco-friendly packaging.

One key issue that has been identified in the global green consumption research agenda is the gap between consumers' attitude, intention and behaviour. For example, Durif et al. [32] cite a survey in Canada that reports a 40 per cent green gap between self-perception and actual behaviours in relation to environment protection. An Ogilvy's green report points out that while 82 per cent of Americans have 'good green intentions', only 16 per cent live up to their intentions by their daily behaviours [29]. Similarly, 30 per cent of consumers in the UK express their concern for the environment but the market share of ethical food products accounts for only 5 per cent of the total food market [33]. In South Korea, marketers are encouraged to make their green messages truthful so as to persuade those who already hold a strong commitment in environment-friendly consumption so as to narrow the attitude–behaviour gap [34]. The major limitation of existing literature used to support most studies in green consumption is the reliance on Theory of Reasoned Action and Theory of Planned Behaviour, which often neglect contextual factors [1].

Some meta-analyses have examined hundreds of studies on the gap between self-reported intention to buy and actual green behaviours, as well as methodological limitations in green consumption research [1, 28, 35, 36]. The review by ElHaffar et al. [28] has summed up the following personal/intrapsychic factors as the major barriers to green behaviour: lack of awareness, lack of justification, personal values and norms, perceived self-efficacy, perceived compromise and utility. The studies argue that when consumers prioritise self-interest values over altruistic or environment values, the green gap will evidently expand [37]. Meanwhile, marketing mix and contextual/demographic factors such as product attributes, price, lack of availability of green products, lack of communication, limited income, education, social context, physical condition, temporal risk and trade-offs are major barriers to realise green intention [38, 39]. In this same external dimension, the literature collected by Joshi and Rahman [36] also elaborates that product attributes and quality, brand image, store characteristics as well as eco-labelling and certification have been found to motivate or hurdle green behaviours. To cope with the contradiction between their green attitudes, intention and behaviours, consumers turn to overlook the issue and rationalise their non-green behaviours in order to neutralise their guilty feeling and cognitive discomfort [28, 40].

With the insight on these motivators and barriers, researchers recommend various ways to bridge the green gap. ElHaffar et al. [28] call for interventions on personal factors such as reinforcement of attitudes, subjective norms and perceived behavioural efficacy and control. In order to address the attitude–behaviour inconsistency, marketers must trigger self-awareness and frame green behaviour as the consumers' self-interests and personal gains, rather than as a solely pro-social endeavour [37, 41, 42]. Part of the solutions is to make the green products' competitive advantages, product attributes, price and packaging standards superior to those of the non-green alternatives [25, 26, 39, 43]. In response to contextual/situational influences, improvement should focus on store cues and point of sale information, dynamic interaction and experience in store, healthiness attribute, influencers, social solidarity as well as digital applications and fun games [28, 38]. Joshi and Rahman

[36] highlight the importance of policymakers to build credibility and trust in eco-labelling and certification. Groening, Sarkis and Zhu [35] call for regulators to set environmental standards and performance measures in order to build stronger consumer perceptions of environmental effectiveness.

From methodological perspectives, a research approach examining both personal and contextual factors is promoted [1, 28] to overcome the social desirability and over-estimation bias when consumers tend to exaggerate their intentions to buy green products [44, 45]. Green consumer behaviour is so complex that comprehensive models are untestable [1], as consumers behave differently to different products. Thus, researchers must mobilise reductionist models to provide pieces and puzzles of the big picture. Moving in this direction, this study narrows the examination to focus on demographic characteristics relating to consumer responses to eco-friendly packaging in the packaged food market. This can help produce more useful insights to help packaged food businesses target the right consumer segments.

To close the gap due to the absence of contextual factors in understanding consumers' behaviour, this study examines the intention to buy together with willingness to pay because both of them are determinants of consumer behaviour [46, 47]. In this sense, this study is responding to the methodological shift in the green consumption research by considering personal and contextual/situational factors in understanding green consumer behaviour.

3 Research Context: Vietnam

This study focuses on Vietnam because of the nature of its emerging economy. The country is geographically located in Southeast Asia, with a large population of almost 98 million people [2]. Vietnam's GDP per capita was US$2.700 in 2019, at a growth rate of 7 per cent before the Covid-19 pandemic [48]. Even though the call for environmental-friendly consumption has been promoted by various communities and organisations, the awareness and acceptance level is still low among consumers [49, 50]. The obstacles vary, but importantly, there has not been a very effective policy to promote green production and consumption [5, 11]. Local industries still use outdated technology, regardless of their environmental damages [5, 11]. In addition, consumerist lifestyles with convenient shopping and consumption of packaged foods has escalated [51]. Even though the consumption level in Vietnam is still much lower than in developed countries [7], its large populations contribute to a widespread environmental dilemma [52]. Consequently, to build a sustainable consumption culture should be seen as an important agenda by many developing countries, including Vietnam [53]. In short, due to its big population and the emerging consumption trends, Vietnam needs to take action to minimise the footprint and contribute to global sustainability [7]. This is the main reason why this study chooses Vietnam as its research context.

3.1 *Packaged Instant Noodles in Vietnam*

The growth of packaged food products is predictable as a result of fast and convenient lifestyles. Food packaging stands as the biggest contributor in global retail packaging with 2.2 trillion units in 2019 [54]. Vietnamese consumers love instant noodles and consume them on a frequent basis [55]. The market size of instant noodles reached 27.2 trillion Vietnam Dongs, which is equivalent to US$1.2 billion in 2020 [56]. On average, each Vietnamese consumer uses 56 packets of instant noodle per year, making per capita consumption in the country top of the world and the national consumption reaches a total of 5.4 billion packets per year [57]. Because of social distancing and stockpiling due to the Covid-19 pandemic, value and volume purchases of packaged noodles increase by 11 per cent in the country in 2020 [56].

Fifty instant noodle manufacturers with hundreds of brands are competing fiercely in Vietnam's market, in which Vina Acecook, Masan and Asia Foods make up to 70 per cent of market share [55]. As a market leader, Vina Acecook alone accounts for 50 per cent of market share [55]. Most instant noodles are packaged with plastic. Only Colusa-Miliket Foodstuff and Thien Huong Foods protect their instant noodles with paper material; however, the brand Miliket by Colusa-Miliket Foodstuff only occupies 4 per cent of market share and Vi Huong by Thien Huong Foods holds a small presence in the market [55].

Packaging is essential for instant noodles because of its protection function and marketing appeal [21]. Bag, cup or bowl are the three main forms of packaging, in which the most popular material for the bag type is plastic. The single-serve bag-type instant noodles dominate Vietnam's market, accounting for 85 per cent of market share [56]. In sum, given the high consumption volume and negative environmental impacts, this study focuses on instant noodle single-use, bag-type packaging in relation to consumers' demographic characteristics and purchase intention and willingness to pay.

3.2 *What Do We Know About Green Consumption in Vietnam?*

Prior Vietnam-based research commonly used the Theory of Planned Behaviour (TPB), Behavioural Reasoning Theory (BRT) and Motivation, Opportunity and Ability (MOA) model to examine personal factors and/or contextual/situational factors relating to green consumption. Findings from the current literature have pointed out that consumers' awareness of environmental issues and eco-friendly consumption is limited [7, 58–60]. In a Vietnam-based commercial survey, 77 per cent of respondents claim that they try to take pro-environment actions in their daily life, far higher than the global average of 65 per cent [51]. However, pro-environment claims do not seem to be consistent with purchase intention and behaviour [13, 61], which can be observed in both hedonic and utilitarian purchases [62].

Prior studies have reported mixed findings on green consumption displayed in different product categories. Among young consumers, environmental self-identity is the most prominent predictor of green purchase behaviour [58, 59], in addition to environmental attitudes and subjective norms [63]. Collectivism and long-term cultural orientation are seen as a determinant of green behaviour [63]. Interestingly, young consumers feel that their 'significant others' do not influence their green purchase behaviour, possibly due to low levels of awareness and knowledge of sustainable consumption in the population [58, 59]. Nevertheless, another study in the recycled fashion industry in Vietnam finds that community influence has a positive correlation with decision to buy recycled fashion products [64]. Noticeably, Vietnamese female consumers highly value the environmental implications of green products on the community, while male consumers are more prone to green barriers such as higher price, extra time, additional efforts and unclear labelling [65]. Hence, personal characteristics of consumers may affect their intention and behaviour to some extent.

In addition to personal factors, situational factors can exert some influence on green consumption. For instance, external/situational factors also affect consumers' willingness to pay and purchase decisions for organic foods [18]. Those factors range from such demographic characteristics as family size, education and income to product attributes like processing, packaging and labelling, certification and product supply. Marketing stimuli factors, including sales promotion and advertising, coupled by green reputation of the brand also urge consumers to buy green products at retailers [14]. Furthermore, the correlation between intention and behaviour in green consumption can be improved when green products are more readily available and when consumers hold a strong belief that their green behaviour can make a difference [61]. In the market context of Vietnam, consumers have started to engage in green lifestyles as reported in some market research reports [51]. Therefore, consumers need more systemic facilitation to take green consumption actions [66]. At the same time, the consumer confidence in eco-friendly products need to be improved with more transparent and reliable green labelling criteria and procedures [7, 67]. However, different approaches are needed to communicate green characteristics of products to different consumer segments with different demographic backgrounds [49].

Although some Vietnam-based studies have attempted to span the bridge between personal and contextual factors [18, 49, 64], none of them explicitly examine consumer demographics in regard to consumers' intention and willingness to pay for eco-friendly packaged products. This study is one of a few [18, 49] to examine demographic factors in relation to Vietnamese consumers' willingness to pay for green products. This study adopts a fresh perspective by investigating consumers' demographics and product attribute factors to examine potential associations with consumer behaviour. The scope of our study is more specific as to how demographics are related to consumer purchase intention and willingness to pay for eco-friendly packaging. As consumers may respond to different products differently due to the nature of the involved product [8, 13, 64, 68], in our study, a specific product category, packaged instant noodles, was selected for the investigation. The intent was to shine light on specific consumer responses regarding eco-friendly packaging in the selected product category. Insights into consumers' green behaviour in this daily consumed product of Vietnam might suggest general consumer trends and marketing strategy implications.

4 Research Questions

The existing literature shows a diverse picture on green consumption across the globe. The knowledge gap is mainly due to a focus on consumers' internal factors such as personal norms and attitude, without a balance with contextual factors. In Vietnam, green consumption studies have investigated different aspects but have not examined eco-friendly packaging in its correlation with demographic factors. Therefore, this study seeks to help packaged food businesses in segmentation, targeting and pricing strategies in the promotion of their eco-friendly packaging. Thus, this study aims to answer the following research questions (RQ):

RQ1: What types of packaging do consumers perceive as eco-friendly?

RQ2: How do consumers react to price factors regarding eco-friendly packaging?

RQ3: How is educational background related to consumer purchase intention for eco-friendly packaging?

RQ4: How is gender related to consumer purchase intention for eco-friendly packaging?

RQ5: How is age related to consumer purchase intention for eco-friendly packaging?

5 Materials and Method

The study investigates the relationships between consumers' demographic backgrounds and consumers' purchase intention for eco-friendly packaging. Thus, we adopt a survey method in the frame of quantitative approach [69], which can help to explore and evaluate the relationships between factors and environmental behaviours [17, 47]. Furthermore, the survey method has its strength from collecting large sample sizes and thus has capability to produce generalisable results [69, 70].

The survey also employs a choice experiment, which has been popularly used in marketing and consumer research (Chen, Q, Anders and An 2013; [71].

Choice experiment is a suitable technique to investigate specific product attributes and their respective effects on price, providing estimates of willingness to pay [72]. Hence, in this research, we used a series of questions for a choice experiment which was incorporated into the online survey. The intention was to examine consumers' knowledge of packaging types and their willingness to pay a premium price for eco-friendly packaging.

To guide the survey design, we followed Rea and Parker [47]'s guidance, consisting of measurement scale, questionnaire design, pilot study, data collection and data analysis.

5.1 Measurement Scale

The purchase intention (PI) measurement scale, which was proposed by MacKenzie et al. [73], with Cronbach's alpha of 0.90, was applied in the survey instrument. Prior studies have used the scale, with high reliability. For instance, Kulviwat et al. [74] used this scale in their study and reported a Cronbach's alpha at 0.92. In this scale, a 7-point semantic differential scale was applied to measure PI through three items: (1) 'probable vs. improbable,' (2) 'likely vs. unlikely' and (3) 'possible vs. impossible'. Table 1 shows three generated items for measuring consumers' intention to purchase instant noodles in eco-friendly packages, reproduced from Nguyen [59].

In this study, we adopted a self-administered online survey. According to Sichtmann [75], it is difficult to measure willingness to pay (WTP) in a self-administered survey. In this type of survey, participants are asked to explicitly indicate their WTP for the studied product. Although the advantage is that it is simple and straightforward, the disadvantage is that external validity could be affected if participants over-report their WTP [76]. Therefore, the measurement instrument should adopt well-established scales to avoid the risk. In prior literature, Biswas and Roy [77] used the WTP scale with three items adapted from Chen [78] and Wang et al. [79].

Table 1 Measurement items for purchase intention (PI)

Adapted from	Construct	Original items	Items modified for the survey instrument
MacKenzie et al. [73]	Behavioural green purchase intention (PI)	Rate the probability that you would purchase the product (in question) 1. from 1 Unlikely to 7 likely 2. from 1 Improbable to 7 Probable 3. from 1 Impossible to 7 Possible	Please rate the probability that you would purchase instant noodles packaged environmentally friendly PI1. from 1 Unlikely to 7 likely PI2. from 1 Improbable to 7 Probable PI3. from 1 Impossible to 7 Possible

Source Reproduced from Nguyen [59]

Table 2 Measurement items for willingness to pay (WTP)

Adapted from	Construct	Original items	Items modified for the survey instrument
Laroche et al. [80]	Willingness to pay (WTP)	1. It is acceptable to pay 10 per cent more for groceries that are produced, processed, and packaged in an environmentally friendly way	WTP1. It is acceptable to pay some extra money for instant noodles packaged environmentally friendly
		2. I would accept paying 10 per cent more taxes to pay for an environmental clean-up program	WTP2. I would accept paying more taxes as a consumer to pay for eco-friendly packaging
		3. I would be willing to spend an extra $10 a week to buy less environmentally harmful products	WTP3. I would be willing to spend extra money to buy instant noodles that are packaged in a less environmentally harmful way

Source Reproduced from Nguyen [59]

However, the reliability was only medium at 0.66. We did not choose this scale for our study. We selected the WTP scale adapted from Laroche, Bergeron and Barbaro-Forleo [80] because it was reported a high reliability at 0.8. Krystallis and Chryssohoidis [81] also used this scale to measure WTP in their research on organic food. Table 2 list three adapted items for WTP scale, reproduced from Nguyen [59].

As mentioned earlier, we also incorporated questions in a choice experiment to measure consumers' WTP at different price levels for different packaging types. The two selected product attributes for examination were price and packaging. Brand and product quality were controlled in the survey.

In the experimental questions, participants were expected to make repeated choices for three products which were different in terms of price and packaging. There were three price levels: below, at and above market price. With regard to packaging, we included three options for choice: plastic, paper and biodegradable. Participants were also asked to indicate their knowledge of packaging materials based on their choice.

5.2 *Questionnaire Design*

Our questionnaire was structured into three parts. The first part consisted of background questions on personal demographic characteristics and questions on consumer purchase experiences, followed by questions relating to consumers' behavioural intention and willingness to pay. The second part was about the choice experiment. A series of questions were given so that participants could make their repeated choices based on price levels and packaging types. There were questions asking participants to identify what they thought to be eco-friendly packaging, thereby investigating consumers' knowledge of eco-friendly packaging to some extent (see Appendix 1).

As this study did not consider the effects of brand and product quality on purchase intention, brand and product quality were controlled in the questions. Regarding brand awareness, our previous research reported that the brand with the highest awareness in Vietnam was Hao Hao which had an average market price at 3,000 VND ~12 cents US [21]. Thus, we embedded Hao Hao brand in our questionnaire.

The questionnaire and relevant documents were translated into Vietnamese, as the research was conducted in the Vietnam market. To ensure the authenticity of the translation, a parallel translation technique was applied. First, one researcher and one professional translator independently translated the English questionnaire into Vietnamese versions. The two versions were compared and any differences were consensus. Then, the consensus version was reviewed and revised by another bilingual academic to ensure the accuracy of the translation. Furthermore, the translating process also took into account the aspect of semantic validity [82]. For some cases, there was no precise term that exists to convey meaning across English and Vietnamese. In these cases, we followed Brennan et al. [82]'s suggestion to provide long descriptions that can counter the lack of clarity [59].

5.3 Pilot Study

To mitigate bias response, we conducted a pre-test and a pilot survey to gather feedback about the survey questionnaire. Pilot tests are often used to check clarity and content validity of the survey questions [83]. In this research, two sub-tests were undertaken. The first pre-test was given to 25 Vietnamese university students in paper form to check if the questions were understandable and the online pilot survey was performed on 33 acquaintances from different ages and professions [59]. Both the pre-test and pilot survey used the Vietnamese version of the survey questionnaire. As a result of the pre-test and pilot survey, minor revision was made in the survey instrument.

5.4 Data Collection

The study gained institutional ethical approval. Cimigo, which is a UK-based market research company with a large online consumer panel of 43,300 members in Vietnam, was contracted to recruit participants for the self-administered online survey. A random-sampling strategy was applied to the online consumer panel across different age groups, educational backgrounds and geographic locations in Vietnam (e.g. Ho Chi Minh City, Hanoi, Danang). Research participants were active consumers of the packaged instant noodles in Vietnam.

The online survey link via Qualtrics was sent to potential participants' emails. In compliance with the confidentiality agreement, participants' email contacts were managed by Cimigo and not available to the research team. The online survey went on for 2 weeks after all invitations were sent out to the online consumer panel to recruit participants. Follow-up emails were sent to confirmed participants to ask them to activate the survey link to complete the online survey.

5.5 Data Analysis

The analysis starts with a data screening procedure. There were 585 recorded responses on Qualtrics system. We removed 192 unfinished responses and only considered 393 fully completed responses in the preliminary dataset. Accordingly, the response rate is 67.1 per cent.

5.6 Data Export

Out of 393 completed responses, we found 85 invalid cases and removed them from further analysis. First, we removed cases showing the same scored responses for the full survey. Second, we deleted cases showing that respondents completed the full questionnaire in less than 2 minutes, as it was evidenced that they did not pay sufficient attention to their responses. After the preliminary scrutiny, there were 308 usable responses in the final dataset, which was exported into SPSS 20 for statistical analysis.

5.7 Sample Demographic and Behavioural Profiles

The profiles of surveyed participants comprise 48.9 per cent of males and 51.6 per cent of females, reflecting the gender breakdown of the Vietnamese population (males 49 per cent, females 51 per cent) [84]. With regard to age, 51.3 per cent of participants were from 20 to 35 years, with the average age of 31, thus well reflecting the median age of the Vietnamese population at 31.9 [84]. In terms of education, 53.6 per cent of participants obtained university degrees; 20.1 per cent finished senior high school; 8.4 per cent gained a vocational certificate and only 7.5 per cent had postgraduate education. Hence, the sample majority reflected a highly educated population, which might not well represent the Vietnamese population and which is considered as a research limitation. The survey participants were residents of three big cities in

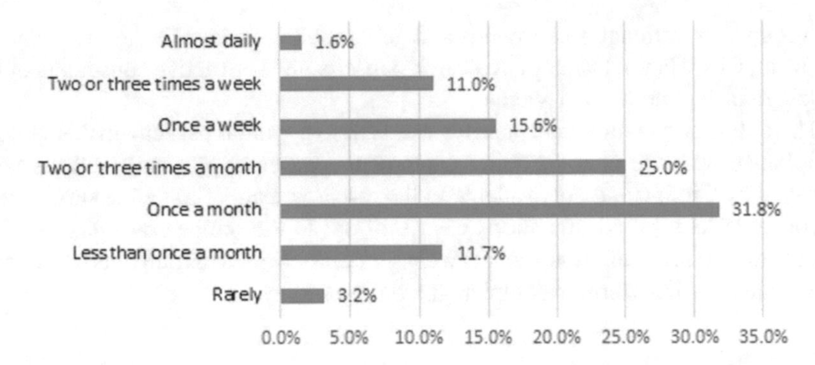

Fig. 1 Purchase frequency of packaged instant noodles in Vietnam

Vietnam—Hanoi capital in the North, Danang City in the Centre and Ho Chi Minh City in the South. Thus, the sample well represented active consumers of packaged instant noodles from the three biggest consumer markets of Vietnam.

We also examined consumers' purchase frequency in the packaged instant noodle product category. Figure 1 shows that the combined groups (who buy packaged instant noodles two to three times a week, once a week and two and three times a month) is 51.6 per cent (11% + 15.6% + 25%). As such, most Vietnamese consumers frequently make purchases in the packaged instant noodle category, on average two to ten times a month.

With respect to purchase volume, it is noted that 31.8 per cent of surveyed consumers stated that they buy 5–10 packets per time. The second largest group of consumers purchase from two to five packets (28.6 per cent). Some consumers buy a lot per time (48 packets), accounting for 12.3 per cent of the surveyed sample. This indicates that Vietnamese consumers are medium and big volume buyers of packaged instant noodles (see Fig. 2).

5.8 *Analysis of Demographic Variables*

We wanted to examine the relationships between consumers' demographics and their purchase intention concerning eco-friendly packaging. We used t-tests and ANOVA statistical techniques in this analysis. The results of t-tests are shown in Table 3 to compare the mean scores of purchase intention across two gender groups (male vs female).

It is noted that Levene's test was non-significant (sig. = 0.383), which is greater than the cut-off of 0.5, showing that equal variances can be assumed. Therefore, we found no significant differences in relation to gender when it comes to purchase intention for eco-friendly packaging.

Next, we conducted ANOVA test to see if age differences could lead to discrepancies in purchase intention concerning eco-friendly packaging. According

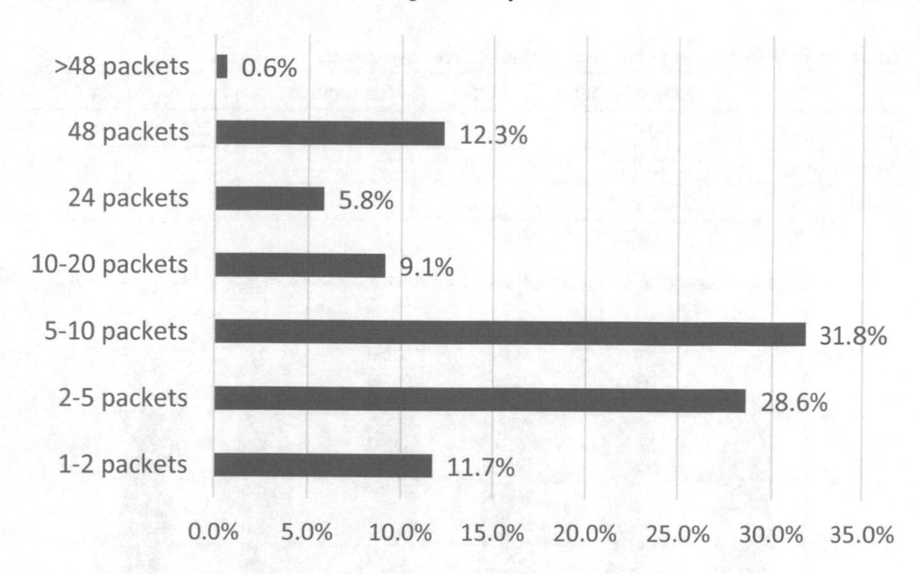

Fig. 2 Purchase volume of packaged instant noodles in Vietnam

Table 3 Independent samples *t*-test for male and female consumers' purchase intention

	Levene's test for equality of variance		t-test for equality of means				
	F	Sig	T	df	Sig. (2-tailed)	Mean difference	Std. error difference
Equal variances assumed	0.765	**0.383**	0.338	306	**0.736**	11,696	0.34643
Equal variances not assumed			0.338	306.000	**0.735**	11,696	0.34571

to Pallant (2007), the ANOVA test can be used to check for statistically significant differences between the means of two or more independent groups. In our study, age group was considered to be an independent variable, comprising of six age groups: 20–25, 26–30, 31–35, 36–40, 41–45 and 46–50. A Levene's test for homogeneity of the six age groups was undertaken with the difference of the variance of purchase intention across age groups shown in Table 4.

As can be seen in Table 4, the statistical value of sig. (0.054) is higher than the 0.05 cut-off value. This shows that there are no considerable differences in the mean scores for purchase intention across age groups. We also found that the means for purchase intention among age groups were nearly equal, ranging from 5.06 to 5.57 (see Fig. 3).

Table 4 ANOVA test for purchase intention between age groups

	Sum of squares	Df	Mean square	F	Sig
Between groups	99.497	5	19.899	2.204	0.054
Within groups	2726.383	302	8.028		
Total	2825.880	307			

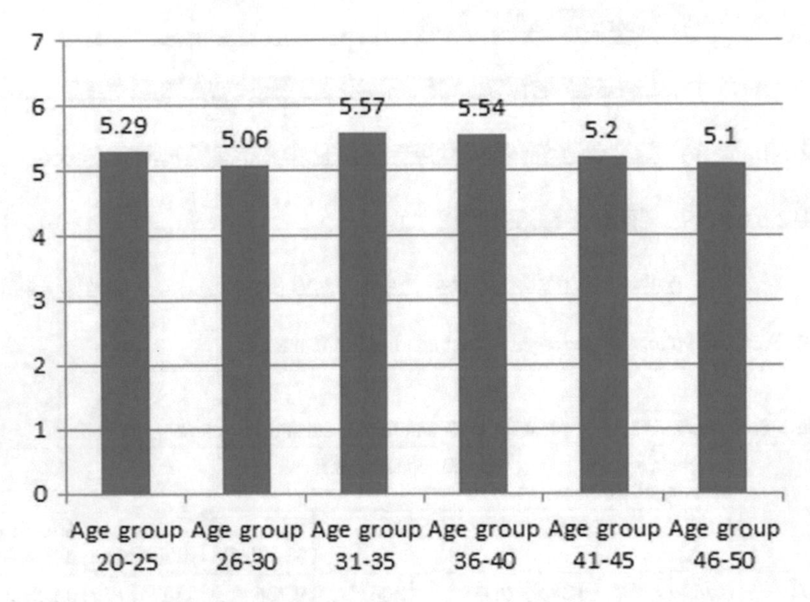

Fig. 3 Mean comparisons for purchase intention construct between age groups

Then, we did another ANOVA test to compare the means of purchase intention among six education groups. Table 5 illustrates the results of Levene's test for homogeneity across education groups.

As shown in Table 5, a statistical value of sig (0.002) is below 0.05, indicating that there is a considerable difference in the mean purchase intention scores between education groups. The effect size calculated using Eta square, which is a rough estimate of effect size in ANOVA, reflects the proportional variance in the dependent variable related to the level of an independent variable [85]. Eta squared = Sum of squares between groups/Total sum of squares = 175.398/2825.880

Table 5 ANOVA test for purchase intention between education groups

	Sum of squares	Df	Mean square	F	Sig
Between groups	175.398	5	35.080	3.997	0.002
Within groups	2650.482	302	8.776		
Total	2825.880	307			

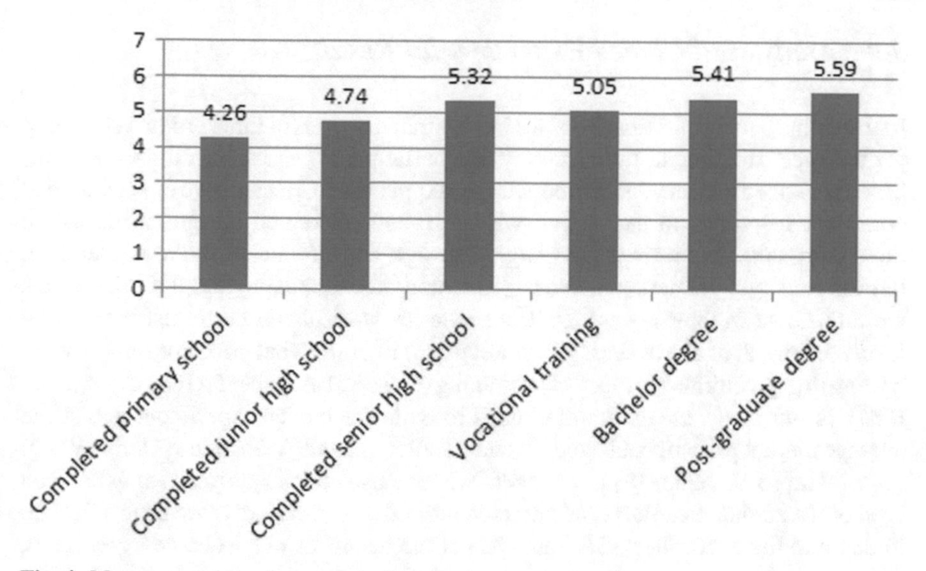

Fig. 4 Mean comparisons for purchase intention between education groups

= 0.06. This means, 6 per cent of the variance in purchase intention (dependent variable) is accounted for by education (independent variable). It can be said that consumers' education background has a medium effect on their purchase intention for eco-friendly packaging.

We also examined the mean comparisons in participants' responses concerning purchase intention (see Fig. 4). We noted that the sample size for primary school group is very small (only 5). Therefore, we did not take this group in consideration. Upon reviewing the mean scores shown in Fig. 4, we concluded that the lowest mean score (4.74) of junior high school group with sample size of 27 moderately differed from other education groups with mean scores ranging from 5.05 to 5.59.

To summarise, t-test and ANOVA test results indicate that there are no significant differences in purchase intention concerning age and gender. Moreover, education does have a medium effect on purchase intention. Hence, it can be said that there are no associations between gender and age with purchase intention for eco-friendly packaging. However, the relationship between consumers' educational background and their intention to buy eco-friendly packaging is found as medium in the studied sample: 6 per cent of the variance in purchase intention can be explained by education level.

5.9 Analysis of Choice Experimental Results

In the online survey questionnaire, we incorporated a series of three product scenarios for a choice experiment. In the survey, the participants were shown three scenario-based packaged instant noodle products (see Appendix 1). Because we only examined consumer responses to packaging, we controlled packaging design and brand in the experiment. Accordingly, packaging design and product brand was identical, but the packaging materials (paper, plastic and biodegradable) and the price levels were different in three scenarios. The scenarios were also used to test three price levels (below, at or above average market price) to see at what price consumers were willing to pay for different types of packaging (plastic PLA, paper PAP, biodegradable BIO). In our study, the assumption used to build the choice experiment is that the average market price of packaged instant noodles is 3,000 Vietnamese Dong (VND) (equivalent to 12 cents US) per packet. We also used this experiment to ask which type of packaging materials consumers identified as eco-friendly. We used SPSS 20 to describe the data collected for analysis of the results from the choice experiment.

5.9.1 Willingness to Pay for Eco-Friendly Packaging

The results show that, for different types of packaging, consumers are different in their willingness to pay above the market price. Figure 5 presents this difference through the percentage of surveyed consumers willing to pay above market price for three types of packages being studied: plastic, paper-based and biodegradable packaging. From this figure, we note that consumers are very willing to pay for biodegradable packaging, and 75.3 per cent of them are willing to pay above market price. Another packaging material that consumers are willing to pay above the market price is paper based (54.9 per cent). Consumers are less likely to pay above the market price for

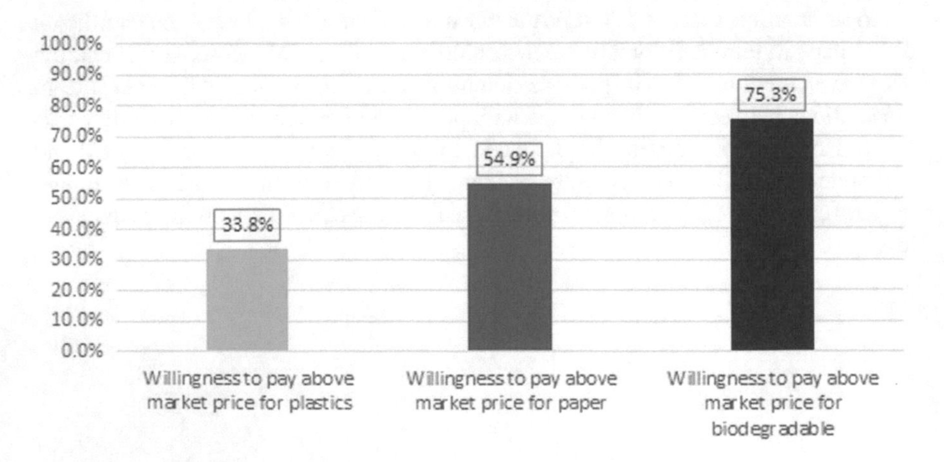

Fig. 5 Willingness to pay above market price

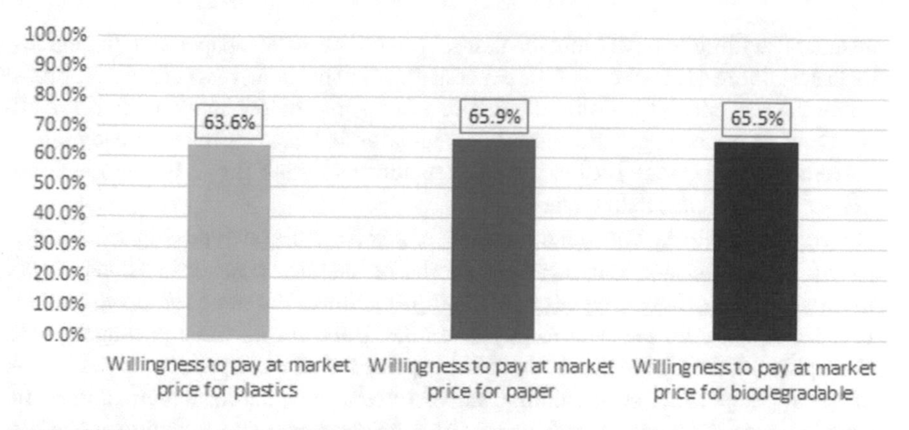

Fig. 6 Willingness to pay at market price

plastics (33.8 per cent). Our results suggest that consumers place higher value and thus willing to pay more on biodegradable than on paper-based or plastic packaging.

Moving to the next price level in our experimental choice—at market prices (3,000 VND—12 cents US), consumers show little difference in their willing to pay for different types of packaging. Figure 6 presents a more balanced percentage of willingness to pay at market prices for three types of packaging (63.6 per cent for plastics, 65.9 per cent for papers and 65.5 per cent for biodegradable packaging). This suggests that the market prices for packaged instant noodles should be reasonable of regardless the types of packaging materials.

When considering consumers' willingness to pay below market prices, the results show that consumers less likely agree with this choice. Figure 7 indicates that

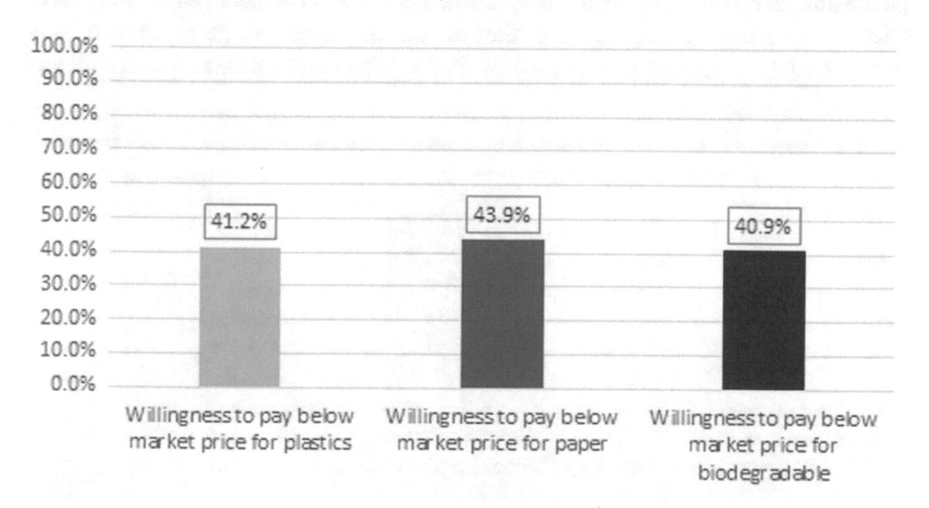

Fig. 7 Willingness to pay below market price

consumers' willing to pay below the market price for plastics, paper and biodegradable materials are 41.2 per cent, 43.9 per cent and 40.9 per cent, respectively. Approximately 60 per cent of consumers choose not to pay below market price for all three types of packaging. One possible explanation is that consumers already know the average market price for instant noodles and may think that it is impossible to purchase any products below that price.

Overall, we note that consumers' willing to pay for different types of packaging is only different at the above market price level. For the other price levels, consumers' willing to pay for different types of packaging are almost the same for three package types, i.e. roughly 60 per cent to pay at market price for all three package types, and roughly 40 per cent to pay below market price for all three package types. From a practical perspective, these findings suggest useful insights for manufacturers. In particular, manufacturers should carefully consider types of packaging materials for products that consumers consider paying higher prices for. With packaged food products, most consumers are willing to pay a premium for biodegradable packaging.

5.9.2 Consumer Perceptions of Eco-Friendly Packaging

The results also reveal different consumers perceptions regarding different types of packaging. Figure 8 presents this difference through percentages of consumers perceiving plastic, paper-based and biodegradable packaging as eco-friendly. In particular, 90.6 per cent and 77.6 per cent of consumers consider biodegradable and paper packaging as eco-friendly, respectively. Only a small portion of consumers regard plastics as eco-friendly packaging.

In sum, our results suggest that most consumers perceive paper and biodegradable packaging as eco-friendly, while only a few consumers consider plastic to be eco-friendly. This finding helps us to further understand consumer perceptions of eco-friendly packaging, which is important in the Vietnamese instant noodle context

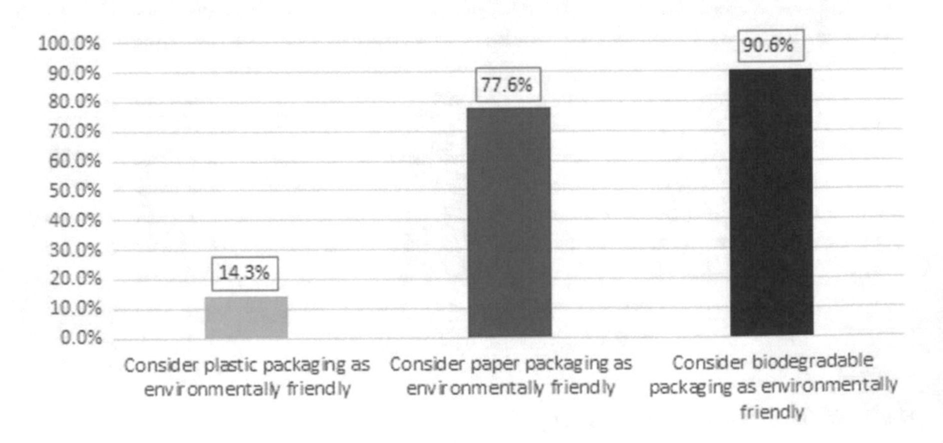

Fig. 8 Consumer perceptions of different types of packaging materials

where most of the existing packaged products using plastic packaging. Consequently, manufacturers can use this finding to form a basis for their management decisions related to eco-friendly packaging strategies, which should move towards paper and biodegradable packaging.

6 Discussion

Given the importance of promoting green purchase [50, 60], the current chapter seeks to understand consumer behaviours for eco-friendly packaging in the research context of Vietnam. This chapter presents one of the earliest studies in Vietnam to explain how demographics are associated with consumers' purchase intention and willingness to pay for eco-friendly packaging. The findings confirm that levels of education do influence purchase intention with regard to eco-friendly packaging, while gender and age show no significant effects. The findings also show that consumers' willingness to pay for eco-friendly packaging are different at different price levels. We now discuss the research findings, structured as three main themes.

6.1 Demographics' Role in Green Purchase Intention for Eco-Friendly Packaging

Existing studies have investigated the relationships between demographics and green behaviour [7, 50, 60]. In particular, they showed that age, gender, education level, occupation, income level and family size have a significant relationship with green behaviour [86, 87]. In the current chapter, we choose to analyse three demographic characteristics: age, gender and education level and how these demographics relate to consumer purchase intention for eco-friendly packaging (RQ3, 4 and 5).

In our statistical analysis, the results show no significant discrepancies in the mean scores for green purchase intention between gender groups (sig. = 0.383) and age groups (sig. = 0.054). They also show significant associations between education level and green purchase intention for eco-friendly packaging (sig. = 0.002). Our findings are firmly aligning with a recent study regarding energy consumption in Vietnam, which suggests that sustainable purchase behaviour is strongly associated with education levels, and weakly associated with gender and age [49]. The alignment can further be highlighted through the products being researched (i.e. packaged food products and energy), which belong to product categories with high frequency of consumption and thus have large impacts on promoting sustainable consumption. This alignment, to some extent, increase the generalisation of our findings regarding green purchase intention in Vietnam.

Regarding gender, existing literature shows mixed results about its associations with green purchase intention. Some studies have found that a particular gender, e.g. women, has stronger intention and behaviours towards green purchase intention [26, 88]. However, other studies have found weak associations between genders and green purchase intention [49, 89]. The current chapter positions in the latter group of studies. We further note the common nature of food products being examined in our study, which may balance product knowledge related to green knowledge in the two genders.

With respect to age, existing literature also shows mixed results regarding its associations to green purchase intention. While some studies have found old age groups with more environmental knowledge and financial resources are more green consumers [90], others have reported that young age groups with broad exposure to environmental issues are more green consumers [49]. The current study reflects this mixed nature. On the one hand, the results show no significant difference for green purchase intention among age groups. The p-value (0. 054) is really close to the cut-off of 0.05, and thus in case of using the cut-off of 0.10, the difference can be considered as significant. In other words, the results show a provisional trend towards the difference for green purchase intention among age groups. Consequently, the mixed nature is an interesting point that should be further investigated. Future research can use qualitative approaches to gain an in-depth understanding of consumer perceptions and the similarities or differences between males and females on green purchase intention.

Regarding education level, we find the effects of education on green purchase intention. In particular, we find that 6 per cent of variance in green purchase intention in respect to eco-friendly packaging can be explained by education level. We now position this finding into the related literature. While some studies have reported no relationships between education levels and eco-friend behaviour [91, 92], others have found that education levels can influence eco-friend behaviour, including green consumption [93], organic food consumption [94], environmental awareness [95] and green purchase behaviour [26]. Our findings can be positioned in the latter group, showing the important role of education levels on green consumption. Furthermore, our study extends the research context of green purchase intention into packaged food products, and thus it can provide practical implications to respective food manufacturers.

Given the educational effects on consumer behaviour across different consumer segments, the study sheds light on a long-term solution to contribute to the development of a sustainable consumption culture in Vietnam, starting from raising awareness of the effects of daily consumption habits. The shift in educational policy should be made to place more emphasis on curriculum development to teach to students about the environment and the relationships between personal consumptions and environmental effects. Social awareness of the role of consumers can be enhanced if education can play a more important role in raising awareness of daily issues, such as buying packaged food products. Governments and policy makers can adjust education plans, focusing more on professional development for teachers, and curriculum development with sustainability focus. This strategy could help address the issues of conventional consumption as well as packaging waste in Vietnam for the long run, starting from awareness and knowledge provided from formal education or equivalent form of education.

6.2 Willingness to Pay for Eco-Friendly Packaging

Existing studies have investigated consumers' sensitiveness to price with regard to eco-friendly products/packages in different countries [17, 47]. In the context of Vietnam, this study examined how consumers react to different price factors set for different types of packaging (RQ2). We adopted and embedded a choice-based experiment in our survey, which differentiated three different price levels (above market price, at market price and below market price). The results show that at the market price and below market price, consumers' willingness to pay is similar across different types of packages. The results further show that at a premium price, consumers are more willing to pay for biodegradable packaging (73.5 per cent), than for paper and plastic packaging. This indicates that consumers place more value on biodegradable packaging, which they may assume, bear more eco-friendliness than other types of packaging in the market. More in-depth consumer insights should be sought via a qualitative approach to elaborate, confirm or disconfirm this finding of the current study.

Our findings strengthen recent studies that have showed that consumers' willingness to pay premium prices for eco-friendly products [26, 96]. In this vein, [26], p. 3447) further suggest that consumers 'expressed their willingness to pay the additional price of up to 30 per cent for green products compared with conventional products'. Given this insight, manufacturers should go with eco-friendly packaging even when it costs more, since customers are willing to pay extra to minimise negative impacts on the environment. We further note that our choice-based experiment isolated the effects of both brand and product quality, which were controlled in the questions. This enables us to focus on consumer choices for price and different packaging types. In addition, we used a leading brand of packaged instant noodles in Vietnam, Hao Hao, in our investigation. Therefore, this knowledge is more useful for leading brands in the market rather than other smaller brands. The implications for packaged food manufacturers are that leading brands will enjoy a more competitive advantage if they implement eco-friendly packaging strategy, while smaller brands cannot enjoy the same benefit even though they apply eco-friendly packaging.

6.3 Consumer Perceptions of Eco-Friendly Packaging

We believe our work has also contributed to understanding consumer perceptions of eco-friendly packaging (RQ1). In particular, we find that biodegradable packaging has been perceived to be eco-friendly in the context of Vietnam. It further identifies packaging materials as a key dimension in the definition of eco-friendly packaging.

In consideration of prior work on a consumer definition of eco-friendly packaging [21], our findings support the inclusion of packaging materials as one of the most prominent packaging attributes to consumers in the packaged food market.

The results of the survey also show that there are segments of consumers who are still confused as to the environment-related characteristics of packaging used for food products. For instance, 14.3 per cent of surveyed consumers considered plastic to be eco-friendly. While there are increasing social awareness and concerns about plastic waste and plastic pollution [3], this finding indicates that there should be more public education to enhance public knowledge of plastic issues in Vietnam.

Finally, the findings also report that in Vietnam, a majority of consumers perceive paper packaging to be eco-friendly, while just a few do not perceive papers as more eco-friendly than other packaging types. In practice, this consumer perception may be confused by a paradox of paper packaging. While paper packaging is perceived eco-friendly in food packaging, it may produce more negative impacts on the environment if we take into account the whole packaging life cycle, from paper manufacturing, food packaging, transportation and the end-of-life management [21, 22]. To help clear consumer confusion as to packaging and environmental effects, manufacturers need to educate consumers about the environment-related characteristics of paper packaging, especially in relation to its whole packaging life cycle.

7 Conclusion

We began our research with the observation that there are pressures to phase out single-use plastic and further use eco-friendly packaging. Thus, there is a strong need for examining how demographics are associated with consumers' purchase intention and willingness to pay for eco-friendly packaging. Addressing this gap, this research used an online consumer survey, including a choice-based experiment, with a focus on packaged food products in Vietnam. The results show that education levels are moderately associated with green purchase intention, whereas gender and age are not significantly associated with green purchase intention. Furthermore, our findings are consistent with recent studies [26, 96] that show that consumers are willing to pay premium prices for green products. What is new in our research is the confirmation of the role education could play in consumer purchase behaviour with regard to eco-friendly packaging, e.g. packaging for instant noodles.

From practical perspectives, the research finds that education levels could exert some influences on green purchase intention related to eco-friendly packaging. The findings provide practical implications for both business and governance. For the business sector, it is not only about how good their eco-friendly packages are but also how to educate and engage with customers regarding eco-friendly packaging awareness. For the government, environmental knowledge should be integrated into the formal education systems. By increasing the education levels on environmental knowledge, it is expected to increase consumers' green purchase intention.

Furthermore, the current study provides a practical knowledge of consumer awareness and understanding of eco-friendly packaging. A large number of respondents in this study expressed their preferences for biodegradable packaging over other packaging types, such as paper and plastic. This shows that consumers do consider the differences of materials used to produce packaging, even when they buy a daily product such as packaged instant noodles. As consumers will buy this convenience product on a high frequency, the potential for manufacturers to reduce the impacts of packaging on the environment is likely if they adopt eco-friendly packaging strategy and educate consumers about the eco-friendly benefits of the packaging used. However, whilst the majority of consumers in this study placed value on biodegradable packaging, it is still not known whether or not they actually understand the distinctive characteristics of biodegradable packaging, in comparison with paper and plastic packaging. As indicated by prior research, the whole product life cycle in association with environmental impacts [22] should be considered to conclude if one type of packaging is more eco-friendly than others.

From the managerial perspective, the findings from this study are encouraging for green manufacturers and marketers because willingness to pay for perceived eco-friendly packaging is seen among all age and gender groups of consumers. Marketers have the choice to target a wide range of customers with varying ages and genders in green consumption. If brand owners set a reasonable price and clearly communicate their green packaging differentiation and benefits, they have the opportunity to persuade open-minded consumers to pay reasonably extra. Certainly, relevant marketing mix, such as reasonable pricing, green product availability and promotions must be made available to facilitate willingness for green consumption [18, 49]. In the particular instant noodles market, when the play field is currently dominated by plastic packaging, a brand with a biodegradable, appealing packaging and reasonable price will be able to differentiate itself.

8 Limitations and Future Directions

First, while our dataset well represents the Vietnamese consumer market in term of gender and age, it might limit in terms of education level (i.e. half of participants hold a university degree). It might further limit in terms of locations (i.e. the majority of surveyed consumers are from three major cities in Vietnam (Hanoi capital, Danang City and Ho Chi Minh City). Future research should address these limitations by recruiting a more comprehensive sample, covering balanced levels of education and a wide range of cities as well as suburban areas in Vietnam.

Second, the study examined packaged instant noodles, which is a low-involvement product. In practice, consumers may respond differently when evaluating packaging in relation to a high-involvement product. Future studies could supplement the findings by examining packaging in high-involvement product categories [97].

Third, this study was focused on demographic factors and not on other psychological factors inherent in consumers, such as attitude, environmental concern and environmental self-identity. Future research can help shine light on how psychological factors interact with demographics to gain more useful insights for segmenting the eco-friendly and non-eco-friendly consumers.

Finally, as our study focused only on the instant noodle product category, further syndicate research that can combine several packaged food products for large-scale investigations can provide food manufacturers and marketers deeper perspectives on how different demographic segments react to different packaged foods in the market. This knowledge will help food businesses as well as food authorities to develop long-term strategies to serve consumer needs in different demographic segments.

Acknowledgements We thank Dr. Subramanian Senthilkannan Muthu and the anonymous reviewers for their comments on our manuscript.

This work was supported by RMIT University Vietnam under Grant 52015.

Appendix 1: Survey Questionnaire (in English)

You are invited to participate in a web-based online survey on consumer purchase intention for packaged food products with regard to eco-friendly packaging. **You do not have to respond to this survey**. If you choose to participate, you may withdraw from the study at any time.

Your participation in this survey is voluntary. Participants who consent to participate will be asked to respond to ten online questionnaires, which will take around 10 minutes. These questionnaires include questions about your experiences buying packaged instant noodles. All information will be recorded anonymously. You will not be asked to provide your name, identification number or any other type of information that might personally identify you.

Completion of this survey implies your consent to serve as a participant in this research.

(1) What is your gender?

○ (1) Male○ (2) Female○ (3) Prefer not to say

(2) What is your age? Tick the number that best applies to you

○ (1) 20–25 years old

○ (2) 26–30 years old

○ (3) 31–36 years old

○ (4) 36–40 years old

○ (5) 41–45 years old

○ (6) 46–50 years old

○ (7) 51 years and above

(3) What is the highest level of school you have completed? Tick the number that best applies to you

○ (1) No school

○ (2) Completed primary school

○ (3) Completed junior high school

○ (4) Completed senior high school

○ (5) Associate or vocational degrees

○ (6) University degree

○ (7) Post-graduate degree

(4) How often do you buy packaged instant noodles? Tick the number that best applies to you

○ (1) Very rarely

○ (2) Less than once a month

○ (3) Once a month

○ (4) Two or three times a month

○ (5) Once a week

○ (6) Two or three times a week

○ (7) Almost daily

(5) How many packets of instant noodles do you often buy at a time? Tick the number that best applies to you

○ (1) 1–2 packets

○ (2) 2–5 packets

○ (3) 5–10 packets

○ (4) 10–20 packets

○ (5) half box (24 packets)

○ (6) full box (48 packets)

○ (7) more than one box

(6) Please indicate your intention to buy instant noodles in environmentally friendly packaging. Tick the number that best applies

Unlikely : ___:___:___:___:___:___:___: Likely

1 2 3 4 5 6 7

Improbable : ___:___:___:___:___:___:___: Probably

1 2 3 4 5 6 7

Impossible : ___:___:___:___:___:___:___: Possible

1 2 3 4 5 6 7

(7) Please indicate your willingness to pay more for environmentally friendly packaging. Choose the number that best applies

	Strongly disagree	2	3	4	5	6	Strongly agree
It is acceptable to pay some extra money for instant noodles that are packaged in an environmentally friendly way.	○	○	○	○	○	○	○
I would accept paying more taxes as a consumer to pay for environmentally friendly packaging.	○	○	○	○	○	○	○
I would be willing to spend extra money for instant noodles that are packaged in a less environmentally harmful way.	○	○	○	○	○	○	○

In the three following questions 8, 9 and 10, you will view three instant noodle product scenarios, using the Hao Hao brand. The colours and visual images on packaging are the same but the materials used for packaging are different (plastic, paper and biodegradable). Please read the product descriptions carefully and indicate your responses to the following statements.

(8)

This is Hao Hao instant noodles in a **plastic bag**. Assume the market price is 3,000 Vietnam dongs (12 cents US). Please indicate your responses (Yes/No) to the following statements.

	Yes/No
I would be prepared to pay 2,000 Vietnam dongs (9 cents US) for this	
I would be prepared to pay 3,000 Vietnam dongs (12 cents US) for this	
I would be prepared to pay more than 3,000 Vienam dongs (12 cents US) for this	
This is an environmentally friendly package	

(Q9)

This is Hao Hao instant noodles in a **paper** bag. Assume the market price is 3,000 Vietnam dongs (12 cents US). Please indicate your responses to the following statements.

	Yes/No
I would be prepared to pay less than 3,000 Vietnam dongs (12 cents US) for this	
I would be prepared to pay 3,000 Vietnam dongs (12 cents US) for this	
I would be prepared to pay more than 3,000 Vietnam dongs (12 cents US) for this	
This is an environmentally friendly package	

(Q10)

This is Hao Hao instant noodles in a **biodegradable** bag. Assume the average market price is 3,000 Vietnam dongs (12 cents US). Please indicate your responses to the following statements.

	Yes/No
I would be prepared to pay less than 3,000 Vietnam dongs (12 cents US) for this	
I would be prepared to pay 3,000 Vietnam dongs (12 cents US) for this	
I would be prepared to pay more than 3,000 Vietnam dongs (12 cents US) for this	
This is an environmentally friendly packagc	

Thank you very much for your time and your participation in the survey.

References

1. Peattie K (2010) Green consumptions: behaviour and norms. Annu Rev Environ Resour 35:195–228
2. World Population Review (2021) Vietnam population 2021. <https://worldpopulationreview.com/countries/vietnam-population>.
3. Ipsos Business Consulting (2019) Plastics: a growing concern—a Vietnam perspective. https://www.ipsos.com/sites/default/files/2019-09/vn_plastic_waste_deck_-_final_-_eurocham_-_en.pdf
4. Euromonitor (2018), Ethical living: plastic - lose it or re-use it?

5. Chau MQ, Hoang AT, Truong TT, Nguyen XP (2020) Endless story about the alarming reality of plastic waste in Vietnam. Energy Sources Part A Recover Util Environ Eff 1–9
6. Nguyen XC, Tran TPQ, Nguyen TTH, La DD, Nguyen VK, Nguyen TP, Nguyen X, Chang S, Balasubramani R, Chung WJ (2020) Call for planning policy and biotechnology solutions for food waste management and valorization in Vietnam. Biotechnol Rep 28:e00529
7. De Koning JIJC, Crul MRM, Wever R, Brezet JC (2015) Sustainable consumption in Vietnam: an explorative study among the urban middle class. Int J Consum Stud 39(6):608–618
8. Tien NH, Phuc NT, Thoi B, Duc L, Thuc T (2020) Green economy as an opportunity for Vietnamese business in renewable energy sector. Int J Res Financ Manag 3(1):26–32
9. Le Van Q, Viet Nguyen T, Nguyen MH (2019) Sustainable development and environmental policy: the engagement of stakeholders in green products in Vietnam. Bus Strateg Environ 28(5):675–687
10. Nguyen TLH, Nguyen TTH, Nguyen TTH, Le THA, Nguyen VC (2020) The determinants of environmental information disclosure in Vietnam listed companies. J Asian Financ Econ Bus 7(2):21–31
11. Tien NH, Hiep PM, Dai NQ, Duc NM, Hong TTK (2020) Green entrepreneurship understanding in Vietnam. Int J Entrep 24(2):1–14
12. Gilg A, Barr S, Ford N (2005) Green consumption or sustainable lifestyles? Identifying the sustainable consumer. Futures 37(6):481–504
13. Nguyen NT, Nguyen LHA, Tran TT (2021) Purchase behavior of young consumers toward green packaged products in Vietnam. J Asian Financ Econ Bus 8(1):985–996
14. Su DN, Duong TH, Dinh MTT, Nguyen-Phuoc DQ, Johnson LW (2021) Behavior towards shopping at retailers practicing sustainable grocery packaging: the influences of intra-personal and retailer-based contextual factors. J Clean Prod 279:123683
15. Sustainable Packaging Coalition (2011) Definition of sustainable packaging
16. Magnier L, Crié D (2015) Communicating packaging eco-friendliness: an exploration of consumers' perceptions of eco-designed packaging. Int J Retail & Distrib Manag 43(4/5):350–366
17. Hao Y, Liu H, Chen H, Sha Y, Ji H, Fan J (2019) What affect consumers' willingness to pay for green packaging? Evidence from China. Resour Conserv Recycl 141:21–29
18. Luu DT (2019) Willingness to pay and actual purchase decision for organic agriculture products in Vietnam. Econ J Emerg Mark 11(2):123–134
19. Lewis H, Stanley H (2012) Marketing and communicating sustainability. In: Packaging for sustainability. Springer, Berlin pp 107–153
20. Rokka J, Uusitalo L (2008) Preference for green packaging in consumer product choices—do consumers care? Int J Consum Stud 32(5):516–525
21. Nguyen AT, Parker L, Brennan L, Lockrey S (2020) A consumer definition of eco-friendly packaging. J Clean Prod 252:119792
22. Bertolini M, Bottani E, Vignali G, Volpi A (2016) Comparative life cycle assessment of packaging systems for extended shelf life milk. Packag Technol Sci 29(10):525–546
23. Fernqvist F, Olsson A, Spendrup S (2015) What's in it for me? Food packaging and consumer responses, a focus group study. Br Food J 117(3):1122–1135
24. Seo S, Ahn H, Jeong J, Moon J (2016) Consumers' attitude toward sustainable food products: ingredients versus packaging. Sustainability 8(10):1073
25. Wiederhold M, Martinez LF (2018) Ethical consumer behaviour in Germany: the attitude-behaviour gap in the green apparel industry. Int J Consum Stud 42(4):419–429
26. Chekima B, Wafa S, Igau O, Chekima S, Sondoh S Jr (2016) Examining green consumerism motivational drivers: does premium price and demographics matter to green purchasing? J Clean Prod 112:3436–3450
27. Liobikienė G, Mandravickaitė J, Bernatonienė J (2016) Theory of planned behavior approach to understand the green purchasing behavior in the EU: a cross-cultural study. Ecol Econ 125:38–46
28. ElHaffar G, Durif F, Dubé L (2020) Towards closing the attitude-intention-behavior gap in green consumption: a narrative review of the literature and an overview of future research directions. J Clean Prod 275:122556

29. Bennett G, Williams F (2011) Mainstream green: moving sustainability from niche to normal. https://www.ogilvy.com/ideas/mainstream-green
30. De Silva DG, Pownall RA (2014) Going green: does it depend on education, gender or income? Appl Econ 46(5):573–586
31. Wang L, Wong PP, Narayanan EA (2020) The demographic impact of consumer green purchase intention toward green hotel selection in China. Tour Hosp Res 20(2):210–222
32. Durif F, Roy J, Boivin C (2012) Could perceived risks explain the 'green gap'in green product consumption? Electron Green J 1(33)
33. Young W, Hwang K, McDonald S, Oates CJ (2010) Sustainable consumption: green consumer behaviour when purchasing products. Sustain Dev 18:20–31
34. Kim Y, Oh S, Yoon S, Shin HH (2016) Closing the green gap: the impact of environmental commitment and advertising believability. Soc Behav Personal Int J 44(2):339–351
35. Groening C, Sarkis J, Zhu Q (2018) Green marketing consumer-level theory review: a compendium of applied theories and further research directions. J Clean Prod 172:1848–1866
36. Joshi Y, Rahman Z (2015) Factors affecting green purchase behaviour and future research directions. Int Strat Manag Rev 3(1/2):128–143
37. Martenson R (2018) When is green a purchase motive? Different answers from different selves. Int J Retail & Distrib Manag 46(1):21–33
38. Frank P, Brock C (2018) Bridging the intention–behavior gap among organic grocery customers: the crucial role of point-of-sale information. Psychol Mark 35(8):586–602
39. Kulshreshtha K, Tripathi V, Bajpai N, Dubey P (2017) Discriminating market segments using preferential green shift: a conjoint approach. Foresight
40. Gruber V, Schlegelmilch BB (2014) How techniques of neutralization legitimize norm- and attitude-inconsistent consumer behavior. J Bus Ethics 121(1):29–45
41. Jacobs K, Petersen L, Hörisch J, Battenfeld D (2018) Green thinking but thoughtless buying? An empirical extension of the value-attitude-behaviour hierarchy in sustainable clothing. J Clean Prod 203:1155–1169
42. Reimers V, Magnuson B, Chao F (2017) Happiness, altruism and the Prius effect: how do they influence consumer attitudes towards environmentally responsible clothing? J Fash Mark Manag 21(1):115–132
43. Olson EL (2013) It's not easy being green: the effects of attribute tradeoffs on green product preference and choice. J Acad Mark Sci 41(2):171–184
44. Barber Nelson A, Taylor DC, Remar D (2016) Desirability bias and perceived effectiveness influence on willingness-to-pay for pro-environmental wine products. Int J Wine Bus Res 28(3):206–227
45. Schäufele I, Hamm U (2018) Organic wine purchase behaviour in Germany: exploring the attitude-behaviour-gap with data from a household panel. Food Qual Prefer 63:1–11
46. Konuk F, Rahman S, Salo J (2015) Antecedents of green behavioral intentions: a cross-country study of Turkey, Finland and Pakistan. Int J Consum Stud 39(6):586–596
47. Prakash G, Pathak P (2017) Intention to buy eco-friendly packaged products among young consumers of India: a study on developing nation. J Clean Prod 141:385–393
48. World Bank (2020). Vietnam overview. <https://www.worldbank.org/en/country/vietnam/overview>.
49. Nguyen N, Greenland S, Lobo A, Nguyen HV (2019) Demographics of sustainable technology consumption in an emerging market: the significance of education to energy efficient appliance adoption. Soc Responsib J
50. Nguyen TN, Lobo A, Nguyen BK (2017) Young consumers' green purchase behaviour in an emerging market. J Strat Mark 26:1–18
51. Euromonitor (2020a) Lifestyle consumer types in Vietnam
52. Lange H, Meier L (2009) The new middle classes: globalizing lifestyles, consumerism and environmental concern. Springer Science and Business Media
53. Vergragt P, Akenji L, Dewick P (2014) Sustainable production, consumption, and livelihoods: global and regional research perspectives. J Clean Prod 63:1–12
54. Euromonitor (2020b) New frontiers in packaging technology

55. Vietnam Biz (2020) 'Đại chiến mì gói thời 'bình thường mới' [Instant noodle battle field in the "new normal" age', Vietnam Biz
56. Euromonitor (2020c) Rice, pasta and noodles in Vietnam
57. World Instant Noodle Association (2020), Global demand for instant noodles. <https://instantnoodles.org/en/noodles/market.html>.
58. Nguyen TN, Lobo A, Nguyen BK (2018) Young consumers' green purchase behaviour in an emerging market. J Strateg Mark 26(7):583–600
59. Nguyen AT (2018) Exploring consumers' green purchase intention for a packaged food product with regard to eco-friendly packaging: the case of packaged instant noodles in Vietnam. Doctoral dissertation, RMIT University Melbourne
60. Rydström, C (2020) Sustainable consumption in Vietnam: an examination of the behavior of young consumers through the motivation, opportunity, and ability model
61. Nguyen HV, Nguyen CH, Hoang TTB (2019) Green consumption: closing the intention-behavior gap. Sustain Dev 27(1):118–129
62. Parker L, Watne TA, Brennan L, Duong HT, Nguyen D (2014) Self expression versus the environment: attitudes in conflict. Young Consumers
63. Nguyen T, Lobo A, Greenland S (2017) The influence of cultural values on green purchase behaviour. Mark Intell Plan 35(3):377–396
64. Nguyen X, Tran H, Nguyen Q, Luu T, Dinh H, Vu H (2020) Factors influencing the consumer's intention to buy fashion products made by recycled plastic waste. Manag Sci Lett 10(15):3613–3622
65. Nguyen T, Nguyen V, Lobo A, Dao T (2017) Encouraging Vietnamese household recycling behaviour: insights and implications. Sustainability 9(2):179–194
66. Nguyen T, Lobo A, Nguyen H, Phan T, Cao T (2016) Antecedents influencing conservation behaviour: perceptions of Vietnamese consumers. J Consum Behav 15(6):560–570
67. De Koning JIJC, Ta TH, Crul MR, Wever R, Brezet JC (2016) GetGreen Vietnam: towards more sustainable behaviour among the urban middle class. J Clean Prod 134:178–190
68. Tien NH, Ngoc NM, Anh DBH, Huong ND, Huong NTT, Phuong TNM (2020) Sustainable development of tourism industry in post Covid-19 period in Vietnam. Sustain Dev 1:5
69. Rea LM, Parker RA (2014) Designing and conducting survey research: a comprehensive guide. John Wiley & Sons
70. Nardi PM (2018) Doing survey research: a guide to quantitative methods. Routledge
71. Yeh C-H, Hartmann M, Hirsch S (2018) Does information on equivalence of standards direct choice? Evid Org Labels Differ Ctries-Of-Orig Food Qual Prefer 65:28–39
72. Chen Q, Anders S, An H (2013) Measuring consumer resistance to a new food technology: a choice experiment in meat packaging. Food Qual Prefer 28(2):419–428
73. MacKenzie SB, Lutz RJ, Belch GE (1986) The role of attitude toward the ad as a mediator of advertising effectiveness: a test of competing explanations. J Mark Res 23:30–143
74. Kulviwat S, Bruner G II, Al-Shuridah O (2009) The role of social influence on adoption of high tech innovations: the moderating effect of public/private consumption. J Bus Res 62(7):706–712
75. Sichtmann CaS, S. (2007) Limit conjoint analysis and Vickrey auction as methods to elicit consumers' willingness-to-pay. Eur J Mark 41(11):1359–1374
76. Wertenbroch K, Skiera B (2002) Measuring consumers' willingness to pay at the point of purchase. J Mark Res 39(2):228–241
77. Biswas A, Roy M (2015) Leveraging factors for sustained green consumption behaviour based on consumption value perceptions: testing the structural model. J Clean Prod 95:332–340
78. Chen KK (2014) Assessing the effects of customer innovativeness, environmental value and ecological lifestyles on residential solar power systems install intention. Energy Policy 67:951–961
79. Wang P, Liu Q, Qi Y (2014) Factors influencing sustainable consumption behaviours: a survey of the rural residents in China. J Clean Prod 63:152–165
80. Laroche M, Bergeron J, Barbaro-Forleo G (2001) Targeting consumers who are willing to pay more for environmentally friendly products. J Consum Mark 18(6):503–520

81. Krystallis A, Chryssohoidis G (2005) Consumers' willingness to pay for organic food: factors that affect it and variation per organic product type. Br Food J 107(5):320–343
82. Brennan L, Binney W, Aleti T, Parker L (2014) Why validation is important: an example using the NEP scales. Mark Soc Res 22(2):15–31
83. Saunders M, Lewis P, Thornhill A (2012) Research methods for business students, 6th edn. Pearson Education Limited
84. Central Intelligence Agency (2021) The world factbook: Vietnam. <https://www.cia.gov/the-world-factbook/countries/vietnam/>.
85. Tabachnick BG, Fidell LS (2007) Using multivariate statistics, 5th edn. Allyn & Bacon/Pearson Education, Boston, MA
86. Sang Y-N, Bekhet HA (2015) Modelling electric vehicle usage intentions: an empirical study in Malaysia. J Clean Prod 92:75–83
87. Zhao H-h, Gao Q, Wu Y-p, Wang Y, Zhu X-d (2014) What affects green consumer behavior in China? A Case Study Qingdao J Clean Prod 63:143–151
88. Patel J, Modi A, Paul J (2017) Pro-environmental behavior and socio-demographic factors in an emerging market. Asian J Bus Ethics 6(2):189–214
89. Shiel C, do Paco A, Alves H (2020) Generativity, sustainable development and green consumer behaviour. J Clean Prod 245:118865
90. Otto S, Kaiser FG (2014) Ecological behavior across the lifespan: why environmentalism increases as people grow older. J Environ Psychol 40:331–338
91. Straughan R, Roberts J (1999) Environmental segmentation alternatives: a look at green consumer behavior in the new millennium. J Consum Mark 16(6):558–575
92. Wong KW-K (2013) Partial least squares structural equation modelling (PLS-SEM) techniques using smart-PLS. Mark Bull 24:1–32
93. Diamantopoulos A, Schlegelmilch B, Sinkovics R, Bohlen G (2003) Can socio-demographics still play a role in profiling green consumers? A review of the evidence and an empirical investigation. J Bus Res 56(6):465–480
94. Paul J, Rana J (2012) Consumer behaviour and purchase intention for organic food. J Consum Mark 29(6):412–422
95. Yuan X, Zuo J (2013) A critical assessment of the higher education for sustainable development from students' perspectives–a Chinese study. J Clean Prod 48:108–115
96. Dekhili S, Achabou MA (2014) Eco-labelling brand strategy. Eur Bus Rev
97. Daae J, Boks C (2015) A classification of user research methods for design for sustainable behaviour. J Clean Prod 106:680–689

Printed in the United States
by Baker & Taylor Publisher Services